# Study Guide

# Introductory Chemistry and Fundamentals of Introductory Chemistry

Second Edition

## Darrell D. Ebbing

## R. A. D. Wentworth

### Susan M. Schelble
University of Southern Colorado

Houghton Mifflin Company    Boston    New York

Senior Sponsoring Editor: Richard Stratton
Assistant Editor: Marianne Stepanian
Senior Manufacturing Coordinator: Priscilla J. Abreu
Marketing Manager: Penny Hoblyn

Copyright©1998 Houghton Mifflin Company. All rights reserved.

No part of this work may be reproduced or transmitted in any form or by any means, electronic or mechanical, including photocopying and recording, or by any information storage or retrieval system without the prior written permission of Houghton Mifflin Company unless such copying is expressly permitted by federal copyright law. Address inquiries to College Permission, Houghton Mifflin Company, 222 Berkeley Street, Boston, MA 02116.

Printed in the U.S.A.

ISBN: 0-395-87120-4

23456789-CS-01 00 99

# Contents

**Preface** iv

**Chapter 1** Introduction to Chemistry 1

**Chapter 2** Measurement in Chemistry 9

**Chapter 3** Matter and Energy 32

**Chapter 4** Atoms, Molecules, and Ions 52

**Chapter 5** Chemical Formulas and Names 69

**Chapter 6** Chemical Reactions and Equations 83

**Chapter 7** Chemical Composition 102

**Chapter 8** Quantities in Chemical Reactions 118

**Chapter 9** Electron Structure of Atoms 130

**Chapter 10** Chemical Bonding 150

**Chapter 11** The Gaseous State 169

**Chapter 12** Liquids, Solids, and Attractions Between Molecules 189

**Chapter 13** Solutions 207

**Chapter 14** Reaction Rates and Chemical Equilibrium 230

**Chapter 15** Acids and Bases 251

**Chapter 16** Oxidation-Reduction Reactions 269

**Chapter 17** Nuclear Chemistry 285

**Chapter 18** Organic Chemistry 303

**Chapter 19** Biochemistry 332

# *Preface*

This ancillary book is designed to reinforce the material in *Introductory Chemistry*, second edition, by Darrell D. Ebbing and R. A. D. Wentworth. It will also provide you with tips for success in mastering the material for your introductory chemistry course. This *Study Guide* will follow the chapter names (and numbers) as well as the sections in the text. Salient points will be highlighted from each section.

At the beginning of each chapter, you will be given a list of key skills typically needed for mastering the material in the chapter. The time needed for each chapter will vary.

The authors of *Introductory Chemistry* have some very nice features in each chapter called "Chemical Perspectives." These short essays are intended to be enjoyed by you without fear of learning the material for testing purposes. This *Study Guide* will take a similar approach. Students are encouraged to read the "Chemical Perspectives," but the material contained within these subunits of the text will not be expanded upon in this supplement.

This *Study Guide* will highlight several aspects of learning that should be mastered as you proceed through this course. Each chapter contains:

**Summary of Verbal Knowledge**: This consists of a definition of terms found at the end of each text chapter. A glossary at the end of *Introductory Chemistry* contains an alphabetical summary of terms for the entire course.

**Review of Mathematical and Calculator Skills**: Algebraic skills are essential for success in this course. You may find that your skills are rusty. A review of some simple skills will be covered in Chapter 1 of the *Study Guide.* In subsequent chapters, skills needed to solve problems in each chapter will be reviewed. These should expand upon the concepts in Appendices A and B in the textbook. The art of using estimation and reasoning skills to assess the validity of calculations will be emphasized.

**Application of Skills and Concepts**: This will be done on a sectional basis for each chapter. Problem solutions are highlighted in "yellow" in the textbook, and this *Study Guide* will expand on solving problems similar to those at the end of the textbook. Stepwise solutions will be provided and similar sample problems should provide you with practice.

**Integration of Multiple Skills**: This involves the pursuit of "higher level thinking" skills. At the end of each chapter you will be introduced to problems that can be solved by making use of language and mathematical skills learned during the course of the chapter and cumulatively over the course of the book.

**Self-Exam of Mastery**: Finally, you will find sample exams at the end of each chapter that you can use to test your learning as you proceed through the course of *Introductory Chemistry*. The exam at the end of each chapter will be a typical 100 point exercise for students taking this course. The questions will vary in difficulty and will include memorization (easiest), application (moderately difficult), and multistep reasoning (most difficult).

The material in each chapter will build on that material which has already been covered. You will find success if you practice skills and review concepts frequently (at least four times a week). The practice problems at the end of the textbook and this *Study Guide* are designed to help you achieve mastery of the principles of chemistry. This course should be approached in the same manner you would use to learn a skill in sports or mastering a musical instrument. Try to be an *active learner* by learning the vocabulary and practicing the problems presented in the textbook and this *Study Guide*.

*Susan M. Schelble*

# 1 Introduction to Chemistry

The information contained within Chapter 1 is designed to introduce you to the subsequent course work and to provide historical and experiential connections to the study of chemistry. Some key ideas about how the study of any science, including chemistry, are described. Terms both new and familiar are defined. The keys for success in this chapter follow:

- Read chapter for an understanding of how the science of chemistry interfaces with other sciences and history.
- Even though no mathematical skills are required to master any problems in this chapter, now is a good time to review some basic algebraic manipulations. Rearrangement of mathematical equations will be covered in the "Review of Mathematical and Calculator Skills" section below.

## Summary of Verbal Knowledge

**matter:** This refers to material things around you. It can be a single chemical substance or a complex mixture of substances.
**chemistry:** A branch of science that undertakes the systematic study and explanation of matter. It also examines different forms of matter, reactions of matter, and energy changes accompanying these reactions.
**experiment:** The observation of natural phenomena under controlled conditions.
**law:** A simple generalization from many experiments.
**hypothesis:** A tentative explanation of a law or observed regularity in nature.
**theory:** A tested explanation of some body of natural phenomena.
**scientific method:** The general process involving experimentation (controlled observation) and explanation (hypothesis and theory) whereby scientific knowledge grows.

## Review of Mathematical and Calculator Skills

Algebraic equations often involve "solving" for some unknown. In the typical high school or college algebra course this unknown would often be an "x" or a "y." In equations used for the typical chemistry course the unknown is usually some letter from a symbol assigned to some measurable quantity. For example, "K" stands for kelvins (units on the Kelvin temperature scale); "g" often stands for grams (a unit of mass in the metric system); "J" stands for joules (a unit of heat in the metric system), etc. Very shortly you will encounter these and other symbols used in the field of chemistry.

## Chapter 1

In the application of algebra there are two main types of equation manipulations that you will need to recall. They are

- Isolation of an unknown using addition (or subtraction).
- Isolation of an unknown using multiplication (or division).

Let's look at each individually. Below is an equation (EQ 1.1) which you will use in Chapter 2 to convert from units of degrees Celsius (°C) to kelvins (K) and vice versa.

$$K = °C + 273 \qquad (EQ\ 1.1)$$

If you are presented with information about the quantity of °C, then that number can be substituted into the above equation as is, and a solution can be obtained using arithmetic (addition).

**Example 1.1:** If °C is 25°C, what is the value of K?

**Solution:** Substitute 25 for °C in EQ 1.1.

$$K = 25 + 273$$
$$K = 298$$

What if °C is negative? This still allows use of EQ 1.1

**Example 1.2:** If °C is -78°C, what is the value of K?

**Solution:** Substitute -78 for °C in EQ 1.1. Remember that addition of a negative number to a positive number is the same as subtraction.

$$K = -78 + 273$$

Remember the *commutative* property of algebra. It says the order of the factors for addition/subtraction or multiplication/division does not change the result. Therefore, the above equation can be rewritten.

$$K = 273 - 78$$
$$K = 195$$

This is a good time to take out the calculator you will be using for this course and try some easy functions. If your calculator came with an instruction book, this will be a handy reference, since each manufacturer has some slight variations. Try to determine the solution for K when °C = -78. Some calculators will allow you to enter the mathematical steps as shown above in the following illustration.

# Introduction to Chemistry

| Operation | | Display |
|---|---|---|
| 1. Press | [78] | 78 |
| 2. Press | [+/-] | -78 |
| 3. Press | [+] , [273] | 273 |
| 4. Press | [=] | 195 |

Others will require that the first number you enter is positive, and then the sign is changed by hitting the +/- button on your calculator. This is illustrated below. Try this simple calculation on your calculator so that you are familiar with its function before moving on to more difficult calculations.

| Operation | | Display |
|---|---|---|
| 1. Press | [-] , [78] | -78 |
| 2. Press | [+] , [273] | -78 + 273 |
| 3. Press | [EXE] | 195 |

Your calculator may not follow either of these steps exactly, but it should be similar to one of the above forms. Try the first two problems presented here with your calculator and become comfortable with the operations that you will frequently use in this course.

A second type of problem using EQ 1.1 would be one where the value of K is known, and °C needs to be determined.

**Example 1.3:** If K = 398, what is the value of °C?

> **Solution:** Finding a solution now requires that the equation be rearranged so that the unknown (°C) can be isolated on one side of the equation. As noted in Appendix A of *Introductory Chemistry*, this can be done by subtracting 273 from both sides of the equation.

$$K = °C + 273$$
$$\underline{-273 \quad - 273}$$
$$K - 273 = °C$$

Substitute 398 for K.

$$398 - 273 = °C$$
$$125 = °C$$

**Example 1.4:** Find x, if 34 + x = 10.

**Solution:** Subtract 34 from both sides, to isolate x.

$$x = 10 - 34$$
$$x = -24$$

---

**Practice Problems**

**1.1** Using EQ 1.1, find K if °C = 100.

**1.2** Using EQ 1.1, find K if °C = -180.

**1.3** Solve for x if x + 6 = 2.

**1.4** Solve for y if y - 7 = 10.

**1.5** Solve for x if 29 = x - 16.

---

The second type of equation that sometimes requires rearrangement has the unknown as part of a multiplication or division operation as shown below (EQ 1.2) for determining amount of heat (q) from heat capacity (C) and change in temperature ($\Delta T$).

$$q = C \times \Delta T \qquad \text{(EQ 1.2)}$$

**Example 1.4:** Consider a situation where you have a substance with a heat capacity of 25.5 (at this introductory/review stage units will be omitted) and a change in T of 4.32. What is q?

**Solution:** The value of q can be calculated by straightforward multiplication (after substituting known values into the equation).

$$q = 25.5 \times 4.32$$
$$q = 110$$

**Example 1.5:** In the situation where q is known to be 343 and change in temperature is 22.4, what is the value of C?

**Solution:** The given values can be substituted for q and $\Delta T$.

$$343 = C \times 22.4$$

The above equation can be rearranged by dividing both sides by 22.4 (remember that a factor divided by itself is "1"). The solution of the operation below leads to C (22.4 ÷ 22.4 is 1 and 1 × C is C).

$$\frac{343}{22.4} = \frac{C \times 22.4}{22.4}$$

$$15.3 = C$$

---

**Practice Problems**

**1.6** Using EQ 1.2, find q if C = 5.26 and $\Delta T$ = 8.45.

**1.7** Using EQ 1.2, find $\Delta T$ if q = 207 and C = 10.7.

**1.8** Solve for x if $\frac{x}{5} = 2$.

**1.9** Solve for y if 5y = 2.

---

## Application of Skills and Concepts

### 1.1 The Science of Chemistry

This section includes important definitions. These should be understood in context if not memorized. One very important introduction is the illustration of how a chemical reaction is represented:

copper + nitric acid → blue-green solution + red-brown gas

The study of how reactions are written will be undertaken in depth in Chapter 6, but a clear idea on how chemical reactions are represented is important. In the above note the following:

- The substances on the left side of the arrow represent the "reactants."
- The arrow means "reacts chemically to give."
- The substances on the right side of the arrow represent the "products."

This section also introduces the idea that energy changes also accompany most chemical reactions.

### 1.2 A Short History of Chemistry

The topics in this section are designed to put modern knowledge of chemistry in an historical perspective. The text describes efforts by scientists at various stages of history. You should be able to comprehend this material by reading through it thoroughly.

### 1.3 The Scientific Method

This section contains some basic concepts about the modern approach to chemistry, an experimental science. It will be useful as you master this course and in your laboratory endeavors. Make sure you are able to comprehend the differences between scientific terms defined in this section.

# 6    Chapter 1

## Integration of Multiple Skills

Chapter 1 is an introduction to the subject matter that will be contained in the text and as such does not require a great deal of problem solving abilities. However, the integration of multiple types of algebraic maneuvers in one equation will be covered before proceeding to subsequent chapters. This should prepare you for upcoming material and make it easy for you to follow solutions as they are presented.

The first topic to be reviewed will be the addition (subtraction) of addends with the same coefficients.

**Example 1.6:** Solve for x if $5x + 2x = 20 - 10$

**Solution:** Both 2x and 5x have the coefficient "x," so they can be combined, and the "plain" numbers (those without coefficients) can be combined.

> You should notice that "5x + 2x" = 7x (**not 10x**).

$$5x + 2x = 7x$$
$$7x = 10$$
$$x = \frac{10}{7}$$

## Practice Problems

**1.10** Solve for x if $12x - 5 = 6x + 2$.

**1.11** Solve for x if $\frac{x}{2} - 5 = 13 - 7x$.

## Self-Exam

**I. (4 ea)** Write **T (true)** or **F (false)** in each corresponding blank.

_____ 1. The growth of a tree is an example of a chemical reaction.

_____ 2. Air, water, and the ground you walk on are all chemical substances.

_____ 3. The interplay of experiment and theory is generally known as a *reaction*.

_____ 4. When nitric acid is added to copper metal a chemical reaction occurs.

_____ 5. Chemical reactions **only** involve changes in substances, **never** changes in energy.

_____ 6. The superconducting magnet used in an MRI unit operates at room temperature.

_____ 7. Bronze is a mixture of copper metal and mercury.

_____ 8. A scientist should **never** be willing to revise his/her concept of nature.

# Introduction to Chemistry                                                                  7

_____ 9. When mercury(II) oxide is heated, it decomposes to liquid mercury and oxygen gas.

_____ 10. For the reaction in 9, the mass of mercury(II) oxide is greater than the sum of the mass of liquid mercury and the mass of oxygen gas.

**II. (4 ea) Multiple Choice. Place the letter of the BEST answer in the corresponding blank.**

_____ 1. Which of the following is **not** an example of matter?
   A. Heat     B. A book     C. Air     D. Gold     E. Water

_____ 2. What do chemists often use to mean "reacts chemically to give"?
   A. A colon     B. A semicolon     C. An equals sign
   D. A star      E. An arrow

_____ 3. Which of the following is **not** a form of energy?
   A. Heat     B. Light     C. Sound     D. Air

_____ 4. The **central science** is:
   A. Biology     B. Chemistry     C. Geology
   D. Physics     E. Astronomy

_____ 5. Which early scientist believed that all material things were made up of the elementary substances: fire, air, water, and earth?
   A. Aristotle     B. Lavoisier     C. Dalton
   D. Leucippus    E. Democritus

_____ 6. The application of knowledge to practical ends is:
   A. Alchemy           B. Technology         C. Theoretical chemistry
   D. Basic chemistry   E. Process chemistry

_____ 7. What scientist's work is generally cited as the beginning of "modern chemistry"?
   A. Aristotle     B. Lavoisier     C. Dalton
   D. Leucippus    E. Democritus

_____ 8. The 19th century revival of the ancient Greeks' notion of atoms was done by:
   A. Aristotle     B. Lavoisier     C. Dalton
   D. Leucippus    E. Democritus

_____ 9. The controlled observation of some natural phenomena is:
   A. Hypothesis     B. Theory         C. Experiment
   D. Law            E. Methodology

_____ 10. The value of "x" in $22 = 44x$ is:
   A. 0.5     B. 2     C. 22     D. -22     E. -2

**III. Provide Complete Answers**

1. (5 pts) Describe how Lavoisier discovered that a component of air was in the orange compound mercury(II) oxide.

2. (5 pts) If the value for kelvins (K) is 42, what is the temperature in °C?

3. (5 pts) Solve for "x": $2x - 5 = \dfrac{x}{10} + 4$

4. (5 pts) Solve for "y": $8 - \dfrac{10}{y} = 6$

## Answers to Practice Problems

| | | | | | | | |
|---|---|---|---|---|---|---|---|
| **1.1** | 373 | **1.2** | 93 | **1.3** | -4 | **1.4** | 17 |
| **1.5** | 45 | **1.6** | 44.4 | **1.7** | 19.3 | **1.8** | 10 |
| **1.9** | 0.4 | **1.10** | $\dfrac{7}{6}$ | **1.11** | $\dfrac{36}{15}$ | | |

## Answers to Self-Exam

**I. T/F**
1. T  2. T  3. F  4. T  5. F  6. F  7. F  8. F  9. T  10. F

**II. MC**
1. A  2. E  3. D  4. B  5. A  6. B  7. B  8. C  9. C  10. A

**III. Complete Answers**

1. He heated the liquid mercury in a measured volume of air until all the silvery mercury was converted to mercury(II) oxide. Because a portion of the air had disappeared, he knew something in the air (later shown to be oxygen) had combined with the mercury.

2. -231°C

3. x = 4.7

4. y = 5

# 2
# Measurement in Chemistry

This chapter dives into the mathematical skills you will need to succeed in chemistry. You may already be very well versed in using numbers and performing calculations more advanced than those you encountered in Chapter 1. However, if you feel you need to hone your number usage, now is the time. Keys for success in this chapter follow:

- Review all algebra and calculator techniques.
    The amount of time needed will depend on your knowledge in this area.
- Become familiar with metric units you will use in the study of chemistry.
- Start to memorize terms and definitions needed for this course.
- Develop your *number sense*.

Remember that you want to learn each skill highlighted in the textbook and in this *Study Guide*. By the time you have performed enough problems to review all necessary skill areas, you should be able to take the practice exam at the end of this chapter **without** use of hints or the answer key.

## Summary of Verbal Knowledge

**scientific notation:** The representation of a number in the form $A \times 10^n$, where A is a number with a single nonzero digit to the left of the decimal point, and *n* is a whole number (either positive or negative).
**metric system:** This is a system of measurement where larger and smaller units of a given quantity are related by multiples of 10. It is used in all scientific work throughout the world.
**SI base unit:** This stands for units in the International System, or SI (from the French le Système). The major base units needed for this course are noted below.
**SI prefix:** These note multiplication of base units by powers of 10.
**meter:** The SI unit for length. It is equal to approximately 39 inches and has the symbol **m**.
**volume:** This is a *derived* unit. Derived units are very common, and they are defined in terms of SI base units. In the case of volume it can be determined by cubing length. In the SI system it is meters cubed, $\mathbf{m^3}$.
**liter:** This is a derived metric unit for volume equivalent to 1 decimeter cubed, $\mathbf{dm^3}$.
**kilogram:** This is the SI unit for mass. It is equal to approximately 2.2 pounds and has the symbol **kg**.
**significant figures:** The digits in a measured number (or in the result of calculations with measured numbers) that include all *certain* digits plus a final one that is somewhat *uncertain*.
**rounding:** This is the process of dropping insignificant digits in a calculation.
**Celsius scale:** The temperature scale for science and common usage in much of the world.
**Fahrenheit scale:** The temperature scale for common usage in the United States.

**Kelvin scale:** This is an *absolute* temperature scale. It is the SI base unit of temperature, where the units have the same magnitude as units on the Celsius scale. On the Kelvin scale there are no negative numbers.
**kelvin:** The unit for measurement on the Kelvin scale. It has the symbol K.
**dimensional analysis:** The general problem solving method that can be used throughout this and other science courses. It makes use of units of quantities to solve problems.
**conversion factor:** A factor equal to 1 that converts a quantity in one unit to an equivalent quantity in another unit.
**density:** A derived unit of measurement that is mass of a substance per unit volume. It is often measured as grams per milliliter, g/mL (which = $g/cm^3$), or grams per liter, g/L.

## Review of Mathematical and Calculator Skills

Building on the algebra covered in Chapter 1, now is a good time to expand on the *distributive law*. Recall how it works in multiplication by looking at Example 2.1.

**Example 2.1:** The equation 5(x + 0.2) is equal to what factors?

**Solution:** The factor "5" when used in multiplication is distributed across all the terms in parentheses. It results in multiplying "x" times "5" and "0.2" times "5."

$$5(x + 0.2) = 5x + 1$$

**Example 2.2:** The equation $\dfrac{8x - 2.4}{0.4}$ is equal to what factors?

**Solution:** This same rule applies for division as shown. The factor "0.4" in the denominator is distributed across each term in the numerator. Notice that for the term "8x" only the coefficient "8" is divided by the divisor.

$$\frac{8x - 2.4}{0.4} = 20x - 6$$

Try these simple calculations on your calculator. By now you should find some of the basic operations of your calculator are quite routine.

It is time to introduce a new operation which will come in handy as you work with *scientific notation*. The new operation will involve working with exponents, specifically exponents of the base 10. Before moving to the specific case, let's review some basics of working with exponents. Do you remember what a notation like $x^4$ means? It is explained in EQ 2.1.

$$x^4 = x \text{ times } x \text{ times } x \text{ times } x \text{ (or } x^4 = x \cdot x \cdot x \cdot x) \qquad \text{(EQ 2.1)}$$

Contrast the above with the meaning of 4x which is demonstrated in EQ 2.2.

$$4x = x + x + x + x \qquad \text{(EQ 2.2)}$$

Now that you have recalled the specifics of working with positive exponents, let's review what negative exponents mean. Take, for example, $x^{-4}$. Look at the equalities below in EQ 2.3 to get an understanding of what a negative exponent implies.

$$x^{-4} = \frac{1}{x^4} = \left(\frac{1}{x}\right)\left(\frac{1}{x}\right)\left(\frac{1}{x}\right)\left(\frac{1}{x}\right) \qquad (EQ\ 2.3)$$

Notice that when a term containing an exponent is moved from the numerator to the denominator, it changes signs. This is also true for a term being moved from the denominator to the numerator.

Now it is time to apply the above exponent rules for the base 10. This is the base that will be used in scientific notation and later in the course when you encounter logarithms and the pH scale. If you understand the above examples using x, you will find it easy to apply exponent rules to base 10. Look at the following worked-out solutions:

**Example 2.3:** What is the value of $10^5$? What is the value of $10^{-5}$?

**Solution:** $10^5 = 10 \times 10 \times 10 \times 10 \times 10 = 10000$

$$10^{-5} = \frac{1}{10^5} = \left(\frac{1}{10}\right)\left(\frac{1}{10}\right)\left(\frac{1}{10}\right)\left(\frac{1}{10}\right)\left(\frac{1}{10}\right) = 0.00001$$

This is an important time to check your *number sense*. Notice in the two examples above that when the exponent is positive, $10^n$ is greater than 1. Conversely, when the exponent is negative, $10^n$ is less than one. This is a generalization that will hold true as you deal with exponents where the base is 10. A final definition for base 10 exponents is that $10^0$ (in fact, any base $x^0$) is defined as being equal to 1. The following table summarizing the above should prove useful to you.

**Table 2.1** Summary of values for $10^n$ with various sizes of n.

| n | $10^n$ |
|---|---|
| > 0 | > 1 |
| < 0 | < 1 |
| = 0 | 1 |

Now is a good time to review the use of exponents with your calculator. If you have a scientific calculator or a graphic calculator, it should have a $10^x$ button. For most brands this will be a secondary function of the button (i.e., you will need to use the shift key). Try this for a positive number, $10^7$, a negative number, $10^{-3}$, and for $10^0$. When you do these calculations, verify that the results are consistent with the rules in Table 2.1. If you have a regular scientific calculator, try the following process for $10^7$.

## 12   Chapter 2

|   | Operation | Display |
|---|---|---|
| 1. Press | SHIFT , log | 0 (but $10^x$ may appear on screen corner) |
| 2. Press | 7 | 7 |
| 3. Press | = | 10000000 |

If you have a graphic calculator, the order of operation may vary from that above. For $10^{-3}$ try to follow the procedure listed below.

|   | Operation | Display |
|---|---|---|
| 1. Press | SHIFT , log | 10_ (appears on screen corner) |
| 2. Press | - , 3 | 10-3_ |
| 3. Press | EXE | 0.001 |

After checking to see that you can find the value of $10^0$ (it should equal 1) on your calculator, try the following sample problems. Note from the examples given that n does not have to be an integer in all cases (n = integer in scientific notation) but can also be a decimal.

---

### Practice Problems

2.1 Simplify $0.82 (12x - 5.7) =$

2.2 Simplify $\dfrac{0.6x + 5.8}{2} =$

2.3 Solve for each if x = 3: $x^4$; $4x$.

2.4 Solve for each if x = 2: $x^{-5}$; $-5x$.

2.5 Determine the value of $10^{5.2}$.

2.6 Determine the value of $10^{-5.2}$.

2.7 Determine the value of $10^{0.7}$.

2.8 Determine the value of $10^{-0.26}$.

# Application of Skills and Concepts

## 2.1 Measured Numbers and Units

There are two major points that you should grasp from this section. They both relate to measurable quantities. The first is the *quantity* or *magnitude* of the measurement. The second is the *unit* of the measurement. Both the quantity and the unit should be noted when making and recording measurements. They will also be used when doing calculations. When doing problems in the *dimensional analysis* section it will be very important to keep accurate track of *units* as well as *magnitude*.

## 2.2 Writing Measurements in Scientific Notation

The basic form of numbers written in scientific notation is $A \times 10^n$, where A is a number with a single nonzero digit to the left of the decimal point, and n is a whole number (either positive or negative). There may be an unspecified number of digits to the right of the decimal. Three skills need to be mastered in this section of Chapter 2. They are:

- Writing a decimal number in scientific notation.
- Writing a scientific notation in decimal form.
- Using scientific notation with your calculator.

When addressing the first of these skills, it is necessary to determine what n is. In a number such as 1426, it is easy to determine A, by the rule which says there is one nonzero digit to the left of the decimal. Therefore A is 1.426. To determine n, you only need to look at how many places the decimal was moved from the original number.

$$1426. \qquad 1.426$$

It is easy to see that the decimal was moved three places to the *left*. Therefore, the magnitude of n is 3. You may decide to use the memory approach and *remember* that when going from decimal to scientific notation, the decimal moves to the *left* and the sign on the exponent is positive. However, it is better to *think* through this problem. Since 1426 is greater than one, the exponent must be greater than zero, or positive. The value of A is +3, and 1426 in scientific notation is **$1.426 \times 10^3$**.

Likewise, a number smaller than one, such as 0.0002436, would involve moving the decimal four places to the right (shown below) giving A equal to 2.436.

$$0.0002436 \qquad 2.436$$

Again, rather than memorize the direction of decimal movement, rationalize that n must be a negative number because 0.0002436 is smaller than one. Since the decimal moved 4 places, the magnitude of n is 4 and the sign on the exponent is negative. When expressed in scientific notation, 0.0002436 is **$2.436 \times 10^{-4}$**.

The opposite skill of converting numbers written in scientific notation to decimal form is also valuable.

**Example 2.4:** What is the decimal value of $5.1 \times 10^3$?

**Solution:** Since n is positive, then the number must be greater than one. The decimal must move three (magnitude of n) places to the right.

$$5.1 \times 10^3 = 5100$$

**Example 2.5:** What is the decimal value of $5.1 \times 10^{-6}$?

**Solution:** In this case the decimal is moved 6 places to the left to give a number less than one.

$$5.1 \times 10^{-6} = 0.0000051$$

Using your calculator with numbers written in scientific notation is one of the most important skills you need to master early in this course. Carefully follow the examples below. Also make sure you read the section on **what not to do.**

Your calculator should have a button for dealing with exponents labeled "EXP" or "EE." This key is used to enter the "× 10" portions of a number in the form $A \times 10^n$. For example, $2.5 \times 10^4$ is 25000, which you know from the problems you just did. Now you need to be able to enter this number into your calculator and have it recognize it as 25000. If you are using a scientific calculator, the procedure should look like that shown below. (Remember you may have a button EE instead of EXP).

| | Operation | Display |
|---|---|---|
| 1. Press | [ 2.5 ] [ EXP ] | $2.5^{00}$ |
| 2. Press | [ 4 ] | $2.5^{04}$ |
| 3. Press | [ = ] | 25000 |

If you are using a graphic calculator, the procedure for entering this number may look like the following sequence.

## Measurement in Chemistry

|  | Operation |  | Display |
|---|---|---|---|
| 1. Press | [2.5] | [EXP] | 2.5E |
| 2. Press | [4] |  | 2.5E4 |
| 3. Press | [EXE] |  | 25000 |

It is very likely that your calculator displays "$2.5 \times 10^4$" as "2.5E4." These numbers are equivalent, but you should make sure that you write the number in standard scientific notation not in the computer language format that your calculator displays. Be sure that you can convert between these two forms. The computer language form is not acceptable in written documents and very few instructors will accept this form on exams etc.

Your calculator enters a factor of 10 when you push the EXP or EE button. Therefore you do not want to enter $2.5 \times 10^4$ as shown in the sequence below. Notice that it gives you the **wrong** answer. If you are carrying out a calculation in this manner, it will lead to endless frustrations.

|  | Operation |  | Display |
|---|---|---|---|
| 1. Press | [2.5] | [x] | $2.5^x$ |
| 2. Press | [10] | [EXP] | $10^{00}$ |
| 3. Press | [4] |  | $10^{04}$ |
| 4. Press | [EXE] or [=] |  | 250000 |

The answer, "250000" is **incorrect** for $2.5 \times 10^4$. Make sure you don't make this error.

When dealing with negative exponents, you will encounter one of two possibilities shown below. Consider $2.5 \times 10^{-4}$, which you know is 0.00025 in decimal form. Most scientific calculators will need to be operated as follows to record the correct number.

|  | Operation |  | Display |
|---|---|---|---|
| 1. Press | [2.5] | [EXP] | $2.5^{00}$ |
| 2. Press | [4] | [+/-] | $2.5^{-04}$ |
| 3. Press | [=] |  | 0.00025 or $2.5^{-04}$ |

If you are using a graphic calculator, the entry will likely follow the sequence that follows.

16    Chapter 2

|  | Operation | Display |
|---|---|---|
| 1. Press | [2.5] , [EXP] | 2.5E |
| 2. Press | [-] , [4] | 2.5E-4 |
| 3. Press | [EXE] | 0.00025 or $2.5^{-04}$ |

**Example 2.6:** What is the decimal value of $\dfrac{1.946 \times 10^{23}}{1.234 \times 10^{21}}$?

**Solution:** Enter $1.946 \times 10^{23}$ in the correct manner for your calculator followed by a ÷ sign and then enter $1.234 \times 10^{21}$ followed by the = sign. The answer you should see displayed is **157.7**. This may be shown in computer form, **1.577E2**, or in scientific form, **$1.577 \times 10^2$**. These numbers are all equivalent. You may also find that you have more digits displayed after the decimal point than we have noted here. The reason for this will be discussed in Section 2.4 of this chapter.

Before proceeding to the next sample problem, take a few moments to check your *number sense*. When working with scientific notation, you can estimate the value of a calculation by quickly working with the exponents. For example, on the previous calculation, $10^{23} \div 10^{21}$, the answer can be approximated to equal 100 (i.e., $10^2$) because when dividing similar bases you need only subtract the exponents. In other words, 23 - 21 = 2, and the approximate answer is thus $10^2$, which means 157.7 is a *reasonable* answer.

**Example 2.7:** What is the decimal value of $(1.326 \times 10^{-34})(6.02 \times 10^{23})$?

**Solution:** Before you begin to put the answer in your calculator, try to estimate the answer using exponents. Remember that when *multiplying* numbers with the same base, you *add* the exponents. In the case below, -34 + 23 = -11. Therefore, you would expect an answer like $A \times 10^{-11}$, which is a very small number. That is indeed the case.

$$(1.326 \times 10^{-34})(6.02 \times 10^{23}) = 7.983 \times 10^{-11}$$

One final note about *number sense*. A *large positive* number in the exponent means a very *large* number. Conversely, a *large negative* number is a very *small* number. For example, $1.25 \times 10^{-6}$ is smaller than $8.67 \times 10^{-2}$. Write these numbers in decimal form to prove this to yourself.

## Practice Problems

**2.9** Write each in scientific notation: 1001; 0.000000000154

**2.10** Write each in decimal form: $7.21 \times 10^2$; $1.36 \times 10^{-9}$

**2.11** Find the product: $(7.24 \times 10^6)(1.52 \times 10^{-3}) =$

**2.12** Find the quotient: $\dfrac{5.4 \times 10^3}{6.8 \times 10^6} =$

**2.13** Determine the solution: $\dfrac{(6.24 \times 10^8)(1.21 \times 10^{-2})}{(5.24 \times 10^{-26})(8.92 \times 10^{31})} =$

**2.14** Rank the numbers from smallest (1) to largest (3):
$1.626 \times 10^{-34}$, $6.626 \times 10^{-34}$, $1.626 \times 10^{-24}$

## 2.3 Units of Length, Volume, and Mass

Some of the basic units for the metric system are summarized in Table 2.2. Refer to this and begin to commit the symbols to memory. The applied units are listed in Table 2.3.

Table 2.2 Some Units Used in the SI System of Measurement

|        | Mass     | Length | Volume | Time   |
|--------|----------|--------|--------|--------|
| Name   | kilogram | meter  | liter  | second |
| Symbol | kg       | m      | L      | s      |

Table 2.3 Prefixes Commonly Used in the SI System

| Prefix | Symbol | Size | Example |
|--------|--------|------|---------|
| giga-  | G | $10^9$ | gigabyte (gB) $1 \times 10^9$ Bytes<br>gigawatt (gW) $1 \times 10^9$ Watts |
| mega-  | M | $10^6$ | megahertz (MHz) $1 \times 10^6$ cycles/s<br>megaton (Mton) $1 \times 10^6$ tons |
| kilo-  | k | $10^3$ | kilogram (kg) $1 \times 10^3$ grams<br>kilocalorie (kcal) $1 \times 10^3$ calories |
| deci-  | d | $10^{-1}$ | decimeter (dm) $1 \times 10^{-1}$ meters<br>decidollar (dime) $1 \times 10^{-1}$ dollars |
| centi- | c | $10^{-2}$ | centimeter (cm) $1 \times 10^{-2}$ meters<br>centidollar (cent) $1 \times 10^{-2}$ dollars |
| milli- | m | $10^{-3}$ | milliliter (mL) $1 \times 10^{-3}$ liters<br>millimole (mmol) $1 \times 10^{-3}$ moles |
| micro- | μ | $10^{-6}$ | microgram (μg) $1 \times 10^{-6}$ grams<br>microsecond (μs) $1 \times 10^{-6}$ seconds |
| nano-  | n | $10^{-9}$ | nanometer (nm) $1 \times 10^{-9}$ meters<br>nanosecond (ns) $1 \times 10^{-9}$ seconds |
| pico-  | p | $10^{-12}$ | picometer (pm) $1 \times 10^{-12}$ meters<br>piosecond (ps) $1 \times 10^{-12}$ seconds |

## 18   Chapter 2

The metric system, which uses the base 10 to modify base units, has an advantage over others. The prefix used to modify the base unit is the **same** no matter what is being modified. For example, the prefix *kilo-* (see Table 2.3 for more details) always means $10^3$ or 1000 times the unit it modifies. The kilogram (kg) means 1000 g, and the kilometer (km) means 1000 meters. The prefixes in Table 2.3 are ordered from large to small. Take a look at these and try to become familiar with them while developing an intuition for their relative sizes.

From Table 2.3 it is easy to see that 1 kg = $1 \times 10^3$ or 1000 g.

**Example 2.8:** How many kilograms are equal to 1 g?

> **Solution:** Hopefully you are developing your *number sense* and you are able to deduce that 1 g is but a fraction of a kilogram. In fact 1 g = $1 \times 10^{-3}$ kilogram. If it takes 1000 g to make up a kilogram, then 1g is *1 thousandth of* a kilogram or $1 \times 10^{-3}$ g. Notice that the magnitude on the exponent stayed the same, but the sign changed (from negative to positive). This is a generalization that you can apply.

From a practical standpoint it is good to take a quick look at some of the units you will encounter in the laboratory portions of this course.
- You will measure length in centimeters ($1 \times 10^{-2}$ m)
  1 cm is equal to a little more than a third of an inch.
- You will measure mass in grams rather than the SI standard of kilogram.
  It takes about 2200 g to equal a pound, so this is a fairly small unit.
- Volume will be measured in milliliters (mL) rather than the larger SI standard liters.
  It takes about 473 mL to equal a pint.
  You will also use the unit $cm^3$ (often denoted as "cc," cubic centimeters.)
  Remember that by definition **1 mL = 1 $cm^3$**.

A final word on *mass* versus *weight*. This is addressed in your textbook, *Introductory Chemistry*. Some instructors may want to make a strict distinction between the two terms. However, this *Study Guide* and your textbook will generally use the terms *weight* and *mass* interchangeably, since most references will be to measurements on earth, where these terms are interconvertable.

---

**Practice Problems**

**2.15** How many microseconds are in a second?

**2.16** How many nanometers are in a meter?

---

## 2.4 Significant Figures and Uncertainty in Measurement

This topic often lends itself to confusion. The logic behind rules for uncertainty in measurements and applied use of significant figures are covered very nicely in the text. After reading them carefully, use this section of the *Study Guide* to apply the rules. Practice will be the quickest way to learn and understand the rules.

Suppose you are measuring an object so that you might determine its area. You measure the length and find it to be 2.4 cm and the width to be 1.7 cm. You can enter these numbers in your calculator, and it will display 4.08, as shown in EQ 2.4.

$$1.7 \text{ cm} \times 2.4 \text{ cm} = 4.08 \text{ cm}^2 \qquad \text{(EQ 2.4)}$$

When dealing with the above equation, several aspects of the numbers need to be addressed.

1. How *certain* is each number in the measurement?
2. How many *significant figures* are in each number of the measurement?
3. How many of the digits in the product 4.08 are *significant*?

Addressing the first question, only the "1" is *certain* in 1.7, and the "2" is *certain* in 2.4.

In answering the second question, both the "1" and "7" are *significant* in 1.7, and the "2" and "4" are *significant* in 2.4. The number 1.7 cm implies that the device used for making the measurement was accurate to the nearest centimeter and that the "7" was estimated. This is how you should perform measurements in the laboratory. Always estimate one decimal place beyond that which you can accurately read off the gradings on the measuring device.

Addressing the third question, it is important to remember that the solution for any product and/or quotient can only have as many *significant figures* as the number of *significant figures* in the *least* significant factor. For EQ 2.4 each measurement has "2" significant figures; therefore, the product can **only** have 2. The product, 4.08, must be rounded to 4.1, and the equation should look like EQ 2.5.

$$1.7 \text{ cm} \times 2.4 \text{ cm} = 4.1 \text{ cm}^2 \qquad \text{(EQ 2.5)}$$

There are two types of numbers you need to be able to handle.

- Decimal form numbers
- Scientific notation numbers

The following examples demonstrate the way to handle zeros in numbers greater than 1.

- 124      (3 significant figures)
- 1240     (3 significant figures)
- 1240.    (4 significant figures)

Notice that **all the nonzero digits are significant**, but the zero in the "ones" column is only significant if there is a decimal after it. Another way this might be indicated is to use some word like exactly. For example, "100" has **1** significant figure, but "exactly 100" has **3** significant figures.

A zero between two nonzero digits is always significant, as the following examples show.

- 1204     (4 significant figures)
- 1013     (4 significant figures)
- 1240.5   (5 significant figures)

For numbers in the decimal form that are smaller than 1, zeros are **not** significant until you reach the first nonzero digit. After that they are **always** significant. Some examples follow:

- 0.00124        (3 significant figures)
- 0.240          (3 significant figures)
- 0.00012030     (5 significant figures)

# 20  Chapter 2

Probably the easiest numbers to deal with for identification of significant figures are those written in scientific notation. In this case, all the numbers in A of $A \times 10^n$ are significant.

- $1.2 \times 10^4$ (2 significant figures)
- $1.200 \times 10^4$ (4 significant figures)

Remember that when changing numbers from the decimal system to the scientific system, **only** the significant digits should be included, which the following examples indicate.

- $100 = 1 \times 10^2$
- $100. = 1.00 \times 10^2$

In order to report numbers from a calculation with the correct number of significant figures it is necessary to make a quick review of rounding. The rules are covered in *Introductory Chemistry* and should be reviewed, if needed.

- When rounding 2.82 to 2 significant figures, drop the "2" to yield 2.8.
- When rounding 2.82 to 1 significant figure, drop the "82" but round up to 3.

Exact numbers are the last category that you must consider. They have an *infinite* number of significant figures. They are easy to recognize, because they are counting numbers, or parts of exact definitions. In the first case, consider "40" people. If this refers to counting, it is exact, and when used in a calculation, it is considered to have an infinite number of significant figures. In the second case a definition such as 1 calorie = 4.184 joules exactly means 4.184 has infinite significant figures not 4.

---

## Practice Problems

**2.17** A measurement is made with a buret (a device for measuring volume) which is graded to the nearest tenth of a milliliter. The reading is 5.23 mL. Which digits are *certain*? Which are *significant*?

**2.18** How many significant figures are in 1000; 1001; 0.000102; 1.0000?

**2.19** How many significant figures are in $1.242 \times 10^4$, $2.4 \times 10^{-5}$?

**2.20** Write the numbers in problem 2.18 as correct scientific notation.

**2.21** Round 5.899 to 2 significant figures; to 3; to 1.

**2.22** Round 2290 to 2 significant figures; to 3; to 1.

**2.23** Which of the following highlighted numbers have infinite significant figures?
   (a) **1000 g** = 1 kg
   (b) **1000 mi** = distance from San Diego to Oregon
   (c) **60 s** = 1 min.
   (d) **60 lb** = weight of a small motor scooter

---

## 2.5 Significant Figures in Arithmetic Results

The rules for finding the number of significant figures in multiplication (division) are summarized in Rule 1 in your textbook. The calculation in EQ 2.5 is one example.

**Example 2.9:** Find the solutions for $\dfrac{(2.426 \times 10^6)(4.7 \times 10^3)}{143}$

**Solution:** To determine the number of significant figures for the solution, consider each factor.

| 2.426 | (4 significant figures) |
| 4.7 | (2 significant figures) |
| 143 | (3 significant figures) |

The smallest number is 2; thus the solution can only have 2 significant figures. When you solve the above with your calculator, it will likely report "79735664.34." Only two of these numbers are significant. The rounded answer you should record is

$$8.0 \times 10^7$$

**Example 2.10:** How many days are there in 2.42 weeks?

**Solution:** In this case the solution involves multiplying 2.42 weeks (3 significant figures) by 7 days/week (the 7 is an "exact" number, with infinite significant figures). The solution must contain 3 significant figures, since this factor has the least. When you do this calculation with your calculator, it will likely record "16.94." Round to

**16.9 days**

Rules for addition and subtraction are slightly different. The important thing is looking at the number of digits past the decimal point (see Rule 2 in your textbook). The sum (or difference) can only have as many digits past the decimal point as the factor in the calculation with the least number of digits past the decimal point.

When adding 23.5 + 3.543 + 6.29, the solution can only have one digit to the left of the decimal point. When you add these numbers with your calculator, it will read "33.333." This should be rounded and reported as "33.3."

---

## Practice Problems

**2.24** $\dfrac{(12.34)(6.345)}{2.3456} =$

**2.25** $(6.3)(4.53 \times 10^3)(0.01) =$

**2.26** $\dfrac{(421)(60)}{22.32} =$

**2.27** Find the solution for: 2.345 + 23.4 - 16.52 =

**2.28** Find the solution for: 29.454 - 32.427 + 0.56 =

## 2.6 Temperature and Changing Temperature Scales

There are three scales which you will encounter when measuring temperature. You are probably most familiar with the Fahrenheit scale, which has common usage in the United States. The Celsius scale has common usage in most of the world. It is also used in the world of science. Intuitively, you should know the following about the three scales:

- The size of the degrees on the Celsius scale equals the size of degrees on the Kelvin scale.
- The size of the degrees on the Celsius scale is larger than that on the Fahrenheit scale.
- The Fahrenheit and Celsius scales have negative as well as positive numbers.
- The Kelvin scale has **only** positive numbers.

**Table 2.4** Some Common Temperature Readings on Various Scales

|  | °F | °C | K |
|---|---|---|---|
| **Body temperature** | 98.6 | 37.0 | 310 |
| **Room temperature** | 72.5 | 22.5 | 297 |
| **Boiling point water** | 212 | 100 | 373 |
| **Freezing point water** | 32 | 0 | 273 |
| **Freezing point nitrogen** | -292 | -180 | 93 |
| **Cold day in Alaska** | -40 | -40 | 233 |

There are two types of conversions you need to be able to make. The most important is the conversion of °C to K and vice versa. These are summarized in EQS 2.6 and 2.7. Memorize only one of these. Use algebra skills to convert to the other (see Chapter 1 of this book).

$$K = °C + 273 \qquad \text{(EQ 2.6)}$$

$$°C = K - 273 \qquad \text{(EQ 2.7)}$$

Conversion of °F to °C can be accomplished by using EQ 2.8. Use your algebra skills to convert EQ 2.8 to EQ 2.9. Only memorize one of these equations, and find the other by algebraic manipulations. If you forgot how to rearrange equations, review the mathematical skills sections at the beginning of Chapter 1 and Chapter 2.

$$°F = 1.8°C + 32 \qquad \text{(EQ 2.8)}$$

$$°C = \frac{°F - 32}{1.8} \qquad \text{(EQ 2.9)}$$

## Practice Problems

**2.29** Convert 10.5°C to kelvins.

**2.30** Convert 343K to °C.

**2.31** Convert -7.0°F to °C.

**2.32** Convert -4.0°C to °F.

## 2.7 Problem Solving and Dimensional Analysis

Mastery of this topic is probably the **most important** task of this chapter. It involves the conversion of one quantity to another by a series of multiplications.

An example of *dimensional analysis* is the problem of converting 2.42 weeks to days that you did earlier. Since this problem is so simple and involves units you are familiar with, it is very likely that you were able to solve it without setting it up by the dimensional analysis method. Below you will see it set up formally.

$$2.42 \text{ weeks} \times \frac{7 \text{ days}}{1 \text{ week}} = 16.9 \text{ days}$$

It is important to note that
- 7days/1week = 1, if 7 days = 1 week
- If 7days = 1 week, then 1 week = 7 days (*commutative law*)
- If 7days/1week = 1, then 1week/7days = 1
- Any number multiplied by 1 is unchanged.
- Anything divided by itself is one; therefore, week ÷ week in the above equation cancels.

When practicing problems in this section **always** write down the units, because these need to be canceled in each step. Before trying some problems, look at some of the useful conversions in Table 2.5. Some other equalities are to be found in Table 2.3. Your textbook also has some equalities listed on the back cover and throughout the text.

**Table 2.5** Some Equalities That May be Used in Dimensional Analysis Problems

| Length | Mass/Weight | Volume | Other |
|---|---|---|---|
| 1 ft = 12 in | 16 oz = 1 lb | 1 gal = 4 qts | 1 h = 60 min |
| 1 yd = 3 ft | 1 kg = 2.205 lb | 1 qt = 2 pts | 1 min = 60 s |
| 1 in = 2.54 cm | 1 lb = 453.6 g | 1 gal = 3.784 L | 1 mol = $6.022 \times 10^{23}$ |
| 1 mi = 1.609 km | 1 ton = 2000 lb | 1 pt = 2 cups | 1 atm = 760 torr |
| 1 Å = $1 \times 10^{-10}$ m | 1 amu = $1.661 \times 10^{-24}$ g | 1 cm$^3$ = 1 mL | 1 calorie = 4.184 joules |
| 1 mi = 5280 ft | | 1 dm$^3$ = 1 L | |
| 1 m = 3.281 ft | | | |

You may have noticed that the "Other" column in Table 2.5 is quite small. As this course proceeds, you will find many equalities that you want to add to it. You will likely find it useful to start to compile a notebook of equalities so they will be handy.

Look at some simple steps for approaching dimensional analysis problems.

- Make a mental map of where you are and where you want to go.
- When you have a story problem, list all the given information and any equalities that might be useful.
- Always write down all the *quantities* and all the *units*.
- Estimate a likely answer so that you can evaluate your calculation.

**Example 2.11:** Convert 7.21 inches to millimeters.

**Solution:** Below is a map of how to get from inches to millimeters.

$$\text{in} \rightarrow \text{cm} \rightarrow \text{mm}$$

Once the mental map is drawn, you can insert appropriate equalities to get from inches to millimeters. Notice that the equalities are used in such a way as to ensure that the *unit* in the numerator is in the denominator in the next step. This will always be true with dimensional analysis problems that are set up correctly. Because each step is designed to have a *unit* in the denominator that was in the previous numerator, the units cancel out until you reach to *unit(s)* desired.

$$7.21 \text{ in} \times \frac{2.54 \text{ cm}}{1 \text{ in}} \times \frac{10 \text{ mm}}{1 \text{ cm}} = 183 \text{ mm} = 1.83 \times 10^2 \text{ mm}$$

You may be wondering where the equality 10 mm = 1 cm came from. This equality was derived from information in Table 2.3. Now is a good time to make sure you can make this derivation. Since $1 \text{ m} = 1 \times 10^{-2}$ cm and $1 \text{ m} = 1 \times 10^{-3}$ mm, it is easy to see that a centimeter is 10 times as large as a millimeter ($10^{-2}$ is 10 times larger than $10^{-3}$). Therefore, 10 mm = 1 cm.

Finally, notice that you can get from where you are to where you want to go by more than one route. This is similar to highway travel. There is often more than one way to get from point A to point B. The route shown below is not as direct as the first example, but it eventually gets the job done, and you *arrive* at the same answer. The solution below also leads to a correct answer for Example 2.11.

$$\text{in} \rightarrow \text{ft} \rightarrow \text{m} \rightarrow \text{mm}$$

$$7.21 \text{ in} \times \frac{1 \text{ ft}}{12 \text{ in}} \times \frac{1 \text{ m}}{3.281 \text{ ft}} \times \frac{1 \text{ mm}}{10^{-3} \text{ m}} = 183 \text{ mm} = 1.83 \times 10^2 \text{ mm}$$

You may have solved the above problem by entering the factors into your calculator in the following order:

$$7.21, \div, 12, \div, 3.281, \div, \text{shift}, \log, 3, +/-, = 183$$

You might also have arrived at the correct answer by the following series of steps:
- Finding the product of all the factors in the numerator:

$$7.21, \times, 1, \times, 1, \times, 1 = 7.21$$

- Finding the product of all the factors in the denominator:

$$12, \times, 3.281, \times, \text{shift}, \log, 3, +/-, = 0.039372$$

- Dividing the product of factors in the numerator by the product of factors in the denominator.

$$7.21 \div 0.039372 = 183$$

Now you can ask yourself, " Is this a *reasonable answer*?" Since millimeters (look on your ruler) are much smaller than inches, the answer should be larger than "7.21," which it is.

**Example 2.12:** Convert 3.54 days to seconds.

**Solution:** One mental map for this conversion is shown below.

$$\text{days} \rightarrow \text{hours} \rightarrow \text{minutes} \rightarrow \text{seconds}$$

Substituting equalities for each conversion, the answer, using dimensional analysis can be found easily.

$$3.54 \text{ days} \times \frac{24 \text{ h}}{1 \text{ day}} \times \frac{60 \text{ min}}{1 \text{ h}} \times \frac{60 \text{ s}}{1 \text{ min}} = 306000 \text{ s} = 3.06 \times 10^5 \text{ s}$$

Assessing for the *reasonableness* of this answer, you can conclude that it makes sense since the number of seconds in 3.54 days is expected to be quite large.

---

**Practice Problems**

**2.33** Convert 470 nm to inches.

**2.34** Convert 1.19 tons to kilograms.

**2.35** Convert 22.7 pm to angstroms (Å).

**2.36** Convert 2.34 cm³ to cups.

**2.37** Convert 5 weeks to seconds.

**2.38** Convert $2.05 \times 10^{23}$ amu to ounces.

---

## 2.8 Density

Simply defined, this means mass/volume. All the following examples are correct expressions of the density (d) of gold.

$$\text{gold}: d = \frac{19.3 \text{ g}}{\text{cm}^3} = \frac{19.3 \text{ g}}{\text{mL}} = \frac{19.3 \text{ kg}}{\text{dm}^3} = \frac{19.3 \text{ kg}}{\text{L}} = \frac{161 \text{ lb}}{\text{gal}}$$

The units can vary, but in chemistry it generally is used as g/mL or g/cm³ when measuring liquids and solids and g/L when measuring gases.

In practice, you need to be able to use the concept of *density* to do the following:

- Determine the density, given the volume and the mass.
- Determine the mass, given the volume and the density.
- Determine the volume, given the mass and the density.

The following examples illustrate these three types of problems.

# 26   Chapter 2

**Example 2.13:** If you know that the mass of a nugget of gold is 17.8 g and it occupies a volume of 0.92 mL, what is its density?

**Solution:** You can determine its density by dividing the mass by the volume.

$$\frac{17.8\,g}{0.92\,mL} = 19\,g/mL$$

**Example 2.14:** Given that the density of sand = 2.3 g/mL, what is the weight of 850 mL of sand?

**Solution:** Use the given information in a dimensional analysis solution.

$$850\,mL \times \frac{2.3\,g}{mL} = 2.0 \times 10^3\,g$$

**Example 2.15:** Given the density of sand (Example 2.14), what volume would 970 g of sand occupy?

**Solution:** This third skill also involves a one step dimensional analysis problem. However, you must take care to invert the density from its traditional form (remember if 1 mL = 2.3 g, 2.3 g = 1 mL).

$$970\,g \times \frac{1\,mL}{2.3\,g} = 420\,mL$$

---

### Practice Problems

**2.39** A chunk of lead has a mass of $1.20 \times 10^4$ g and a volume of $1.05 \times 10^3$ cm$^3$. What is the density of lead?

**2.40** Liquid mercury has a density of 13.6 g/cm$^3$. What is the mass of 256 mL of mercury?

**2.41** What volume would $1.42 \times 10^2$ g of mercury occupy?

**2.42** A cube of a metal is 1.2 cm × 2.4 cm × 1.1 cm. It has a mass of 1.69 g. What is its density?

**2.43** How much would 95 mL of alcohol (d = 0.78 g/cm$^3$) weigh?

**2.44** How many milliliters of alcohol would 1.2 kg occupy?

---

## Integration of Multiple Skills

The solution of problems using density within dimensional analysis involves more than one approach, especially when it is part of a story problem. The example below illustrates this.

**Example 2.16:** How many gallons of chloroform (d = 1.48 g/mL) would 25 lbs of chloroform occupy?

**Solution:** To solve this problem, at some point the density will be needed to convert mass to volume. However, this is also a dimensional analysis problem going from

$$\text{pounds} \rightarrow \text{gallons}$$

Since density is in units of g/mL, one step of the conversion will be

$$\text{g} \rightarrow \text{mL}$$

Using the rest of the information given and information from tables, the overall plan is

$$\text{lb} \rightarrow \text{kg} \rightarrow \text{g} \rightarrow \text{mL} \rightarrow \text{L} \rightarrow \text{gal}$$

You can estimate the solution to this problem after setting it up as below. The product of the denominator is approximately 12. The number of pounds, 25 ÷ 12, is about 2.

$$25 \text{ lb} \times \frac{1 \text{ kg}}{2.205 \text{ lb}} \times \frac{1000 \text{ g}}{1 \text{ kg}} \times \boxed{\frac{1 \text{ mL}}{1.48 \text{ g}}} \times \frac{1 \text{ L}}{1000 \text{ mL}} \times \frac{1 \text{ gal}}{3.784 \text{ L}} = 2.0 \text{ gal}$$

Density, inverted

The solution of problems with exponents in dimensional analysis also requires the use of several skills. An example illustrates this.

**Example 2.17:** If the density of zinc is 7.14 g/cm$^3$, what mass would a cubic meter occupy?

**Solution:** To solve this problem, you need to convert the units in the denominator from cubic centimeters to cubic meters. The equality $1 \times 10^2$ cm = 1 m can be used to do this. However, in order to cube the *units*, the magnitude must also be cubed, as shown below. Remember that when raising an entire quantity within parenthesis to a power, the exponent *multiplies* each exponent inside the parenthesis.

Multiplies all exponents inside

$$\frac{7.14 \text{ g}}{\text{cm}^3} \times \left(\frac{10^2 \text{ cm}}{\text{m}}\right)^3 = \frac{7.14 \text{ g}}{\text{cm}^3} \times \frac{10^6 \text{ cm}^3}{\text{m}^3} = \frac{7.14 \times 10^6 \text{ g}}{\text{m}^3}$$

The solution of problems with conversions involving both the numerator and the denominator are the last type that we will look at in this category. For example:

**Example 2.18:** Convert $1.7 \times 10^5$ m/h to feet per second.

**Solution:** To solve this problem, set up the known rate in the traditional dimensional analysis form, and use a series of equalities to convert the numerator and the denominator.

numerator: m → km → mi → ft
denominator: hr → min → s

The entire solution follows:

$$\frac{1.7 \times 10^5 \cancel{m}}{1 \cancel{hr}} \times \frac{1 \cancel{km}}{1000 \cancel{m}} \times \frac{1 \cancel{mi}}{1.609 \cancel{km}} \times \frac{5280 \text{ ft}}{1 \cancel{mi}} \times \frac{1 \cancel{h}}{60 \cancel{min}} \times \frac{1 \cancel{min}}{60 \text{ s}} = \frac{155 \text{ ft}}{\text{s}}$$

The solution just completed offers a great chance to review the *commutative law* for multiplication/division. When entering the factors into your calculator you can do the multiplication and division in any order as long as each step is preceded by the correct operation.

- 1.7, EXP, 5, ×, 5280, ÷, 1000, ÷, 1.609, ÷, 60, ÷, 60, = 155
- 1.7, EXP, 5, ÷, 1000, ÷, 1.609, ×, 5280, ÷, 60, ÷, 60, = 155

## Practice Problems

**2.45** What is the density of gold in kg/m$^3$?

**2.46** How many gallons of sand (d = 2.3g/mL) have the same mass as 2.5 L of gold (d = 19.3 g/mL)?

**2.47** A chunk of metal with a mass of 14.17 g is dropped into a graduated cylinder with water at the 19.6-mL mark. The new reading is 21.4 mL. What is the density of the metal?

**2.48** Convert 2.5 μm$^3$ to mm$^3$.

**2.49** What volume of water (d = 1.0 g/mL) would be needed to equal the mass of 642 cm$^3$ of mercury (d = 13.6 g/cm$^3$)?

**2.50** Convert 62 mi/h to m/s.

## Self-Exam

**I. (4 ea) Write T (true) or F (false) in each corresponding blank.**

_____1. The metric system is based on units of 10.

_____2. One liter is larger than 2000 cm$^3$.

_____3. Mass, time, length, and temperature all have base units in the SI system.

_____4. A substance at 80 K is colder than one at -78°C.

_____5. The prefix *micro*-means one thousandth or 1 × 10$^6$.

_____6. 6x = x$^6$.

_____7. 1 ÷ x$^6$ = x$^{-6}$.

_____8. A scientist should **never** be willing to revise his/her concept of nature.

_____9. Given that 1 calorie = 4.184 joules; a joule is a larger quantity than a calorie.

_____10. 1.2 L of sand (d = 2.3 g/mL) has a smaller mass than 1.2 L of gold (d = 19.3 g/mL).

**II. (4 ea) Multiple Choice.** Place the letter of the BEST answer in the corresponding blank.

_____1. How many significant figures are in 103.6?
    A. 1     B. 2     C. 3     D. 4     E. 0

_____2. How many significant figures are in 800.00?
    A. 0     B. 1     C. 3     D. 4     E. 5

_____3. How many significant figures are in 0.00420?
    A. 1     B. 2     C. 3     D. 4     E. 0

_____4. The solution for $x^{-2}$ if $x = 3$ is:
    A. 9     B. 6     C. -9     D. -1/3     E. 1/9

_____5. What is $4.26 \times 10^{-3}$ in decimal form?
    A. 426     B. 4260     C. 0.0426     D. 0.00426     E. 0.000426

_____6. What is 67240 in scientific notation?
    A. $6.7240 \times 10^4$     B. $6.724 \times 10^4$     C. $6.7240 \times 10^3$
    D. $7.7240 \times 10^{-4}$     E. $6.724 \times 10^{-3}$

_____7. If gold sells for $378/ oz, how much is 98.3 g worth?
    A. $1310     B. $3720     C. $23.60     D. $3.84     E. $2790

_____8. How many days are in $7.25 \times 10^5$ seconds?
    A. 1.20     B. 120     C. 8.39     D. 0.00000839     E. 2.89

_____9. A bike traveling at 19.4 mi/h is equivalent to:
    A. 1260 mi/hr     B. 1720 ft/min     C. 19400 km/h
    D. 8.67 m/s     E. Both B & D

_____10. A silvery liquid has a mass of 17.0 g and a volume of 1.25 mL. What is its density?
    A. 13.6 g/mL     B. 7.3 g/mL     C. 15.8 g/mL
    D. 0.073 g/mL     E. 18.3 g/mL

**III. Provide Complete Answers (show all work when necessary)**

1. (5 pts) The density of chloroform is 1.48 g/cm³. If you need 68.7 g of chloroform, what is the volume that you should measure?

2. (5 pts) How many liters are in $2.43 \times 10^2$ g of water (density of water = 1.00 g/cm³)?

3. (5 pts) A student wanted to identify an unknown chunk of metal. He has a graduated cylinder available and a balance. On the balance he found the mass of the metal was 11.02 g. He placed it in the graduated cylinder containing 12.4 mL of water. The new water level rose to 13.8 mL. Which of the metals (iron: d = 7.87g/cm³, lead: d = 11.4 g/cm³, platinum: d = 21.5 g/cm³) is the most likely identity of the unknown?

4. (5 pts) A student had a stack of three pennies which were measured with a ruler graded in millimeters. The student found that the stack height was 4.6 mm. The student then measured the height of a stack of 10 pennies and found the height was 15.0 mm. Which digit(s) in each of

these measurements is(are) *certain* and which is(are) *uncertain*? Calculate the average thickness of a penny from each of these experiments. Comment on the accuracy of each calculation.

## Answers to Practice Problems

**2.1**  $9.84x - 4.674$     **2.2**  $0.3x + 2.9$     **2.3**  $x^4 = 81; 4x = 12$

**2.4**  $x^{-5} = \dfrac{1}{32}; -5x = -10$     **2.5**  158489     **2.6**  $6.3096 \times 10^{-6}$

**2.7**  5.0118     **2.8**  0.5495

**2.9**  $1001 = 1.001 \times 10^3$; $0.000000000154 = 1.54 \times 10^{-10}$

**2.10**  $7.21 \times 10^2 = 721$; $1.36 \times 10^{-9} = 0.00000000136$

**2.11**  $1.10 \times 10^4$     **2.12**  $7.9 \times 10^{-4}$     **2.13**  1.62

**2.14**  $1.626 \times 10^{-34}$ (1); $6.626 \times 10^{-34}$ (2); $1.626 \times 10^{-24}$ (3)

**2.15**  $10^6\ \mu s$     **2.16**  $10^9\ nm$

**2.17**  Digits *certain*: 5, 2; *significant*: 5, 2, 3

**2.18**  1000 (1); 1001 (4); 0.000102 (3); 1.0000 (5)

**2.19**  $1.242 \times 10^4$ (4), $2.4 \times 10^{-5}$ (2)

**2.20**  $1 \times 10^3$; $1.001 \times 10^3$; $1.02 \times 10^{-4}$; $1.0000 \times 10^0$

**2.21**  2 significant figures = 5.9; 2 significant figures = 5.90; 1 significant figure = 6

**2.22**  2 significant figures = 2300; 3 significant figures = 2290; 1 significant figure = 2000

**2.23**  (a) and (c)

**2.24**  33.38 or $(3.338 \times 10^1)$     **2.25**  300 or $(3 \times 10^2)$     **2.26**  1000 or $(1 \times 10^3)$

**2.27**  9.2     **2.28**  -2.41     **2.29**  284 K

**2.30**  70°C     **2.31**  -22°C     **2.32**  25°F

**2.33**  $1.9 \times 10^{-5}$ in     **2.34**  $1.08 \times 10^3$ kg     **2.35**  $2.27 \times 10^{-1}$ Å

**2.36**  $9.89 \times 10^{-3}$ cups     **2.37**  $3 \times 10^6$ s     **2.38**  $1.20 \times 10^{-2}$ oz

**2.39**  11.4 g/mL     **2.40**  $3.48 \times 10^3$ g     **2.41**  10.4 mL

**2.42**  0.53 g/cm³     **2.43**  74 g     **2.44**  $1.5 \times 10^3$ mL

**2.45**  $1.93 \times 10^4$ kg/m³     **2.46**  5.5 gal     **2.47**  7.87 g/mL

**2.48**  $2.5 \times 10^{-9}$ mm³     **2.49**  8.73 L     **2.50**  28 m/s

## Answers to Self-Exam

### I. T/F
1. T  2. F  3. T  4. T  5. F  6. F  7. T  8. F  9. F  10. T

### II. MC
1. D  2. E  3. C  4. E  5. D  6. B  7. A  8. C  9. E  10. A

### III. Complete Answers
1. 46.4 mL

2. $2.43 \times 10^{-2}$ L

3. Volume: 13.8 mL - 12.4 mL = 1.4 mL
   density = 11.02 g/1.4 mL = 7.9 g/mL = 7.9 g/cm$^3$

   Therefore, the most likely metal is iron.

4. 4.6 mm/2 pennies = 1.5 mm/ penny
   4 is certain, both 4 and 6 are significant;
   Thus 2 significant figures in answer.

   15.0 mm/10 pennies = 1.50 mm/penny
   1 and 5 are certain, 1,5 and 0 are significant;
   Thus 3 significant figures in answer.
   This measurement is more accurate.

# 3
# Matter and Energy

This chapter has two major topics, as the title implies. Dealing with matter will be largely conceptual and will examine ways of classifying matter. The portion of the chapter dealing with energy is both conceptual and computational. Keys to success:

- You will need to recognize matter when grouped by *states* or by *type of substance*.
- Start to apply the mathematical skills you mastered in Chapters 1 and 2.
- Incorporate some of the chemical language into your memory bank.

## Summary of Verbal Knowledge

**states of matter:** Matter can exist in three physical states. These include solid, liquid, and gas.
**solid:** This is the form of matter having definite shape and volume.
**liquid:** This is the form of matter having definite volume but indefinite shape.
**gas:** This is the form of matter having indefinite shape and volume.
**material:** This is a particular type of matter.
**physical change:** A change in the form of matter but not in its chemical identity.
**chemical change:** A change in which one or more types of matter are changed into one or more new kinds of matter.
**physical property:** A characteristic that can be observed for a material but does not change its chemical identity.
**chemical property:** A characteristic of a property involving a chemical change.
**substance:** A type of material that *cannot* be separated into other types of materials by physical means.
**mixture:** A type of material that *can* be separated into other types of materials by physical means.
**homogeneous mixture:** This is also known as a solution. It is a mixture that is uniform in its properties throughout. An example would be salt water.
**heterogeneous mixture:** A mixture that consists of physically distinct parts with distinct physical properties. An example would be sand in water.
**element:** These are fundamental substances. They cannot be decomposed into simpler substances by *any* chemical means.
**symbol:** A one- or two-letter abbreviation for each element.
**compound:** These are substances made up of elements chemically combined. There are two or more elements chemically combined in a compound.
**law of definite proportions:** The fact that pure compounds always contain constant (definite) proportions of the elements by mass.
**law of conservation of mass:** Total amount of mass remains constant during a chemical reaction.
**energy:** The potential or capacity to move matter.

**kinetic energy:** The energy of matter in motion.
**potential energy:** The energy of matter with a potential for motion.
**joule:** A unit of energy used in the SI system.
**calorie:** A unit of energy *exactly* equal to 4.184 J. It is also the quantity of energy needed to raise the temperature of 1 g of water 1°C.
**specific heat:** Quantity of heat required to raise the temperature of 1 g of a substance 1°C.
**law of conservation of energy:** The fact that the total amount of energy remains constant even though it may be converted from one form to another.

## Review of Mathematical and Calculator Skills

The problems involving energy in the latter part of this chapter will require the use of skills you practiced in Chapters 1 and 2 of this book. You will need to be able to rearrange equations using the rules of algebra to isolate an *unknown* quantity. Look at the equation, EQ 3.1 below.

$$q = sm(T_2 - T_1) \qquad \text{(EQ 3.1)}$$

This equation is written with "q" isolated. The present equation can be solved for "q" if the other quantities, s, m, $T_2$, and $T_1$, are known.

**Example 3.1:** Solve for "m" (assuming q, s, $T_2$, and $T_1$ are known).

**Solution:** In this case you would rearrange the equation to isolate "m." This would mean dividing both sides of the equation by the factors "s" and "$(T_2 - T_1)$," or in essence moving these two factors from the numerator to the denominator as they are moved across the equal sign. This would lead to EQ 3.2.

$$\frac{q}{s(T_2 - T_1)} = m \qquad \text{(EQ 3.2)}$$

A second algebra skill you need to review is the handling of negative numbers. Remember that *minus* a *negative* number is a *positive* number. For example, -(-2) = +2.

Another skill you will revisit is the use of *dimensional analysis* covered in Chapter 2. Make sure to review this before trying heat problems found at the end of Chapter 3.

Finally, you need to check on your *number sense* regarding large and small numbers. Most of us have a very good intuition when it comes to positive numbers. For example, we know that 2 is **smaller** than 3. However, it is also true that -2 is **larger** than -3, which is not always intuitively obvious. If this seems hard to grasp, make a number line, as shown in Figure 3.1.

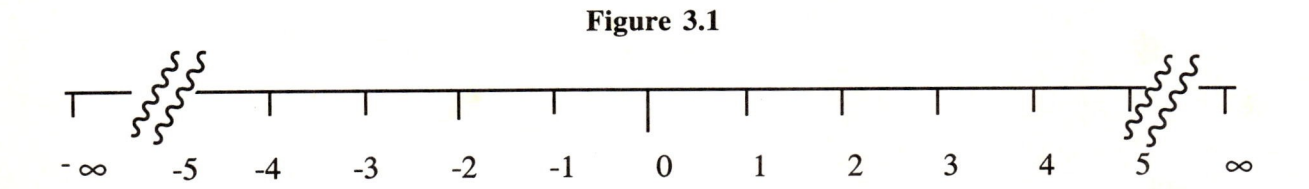

**Figure 3.1**

The numbers going from left to right increase in size. Hopefully you can relate this concept to a thermometer. You probably already are very aware that -25°F (a.k.a. "25 below zero") is colder than -8°F. The same applies for negative numbers. For example, -180 is smaller than -10.

## Practice Problems

**3.1** Isolate s from EQ 3.1.

**3.2** Isolate $T_2$ from EQ 3.1.

**3.3** Isolate $T_1$ from EQ 3.1.

**3.4** 20 - (-10) =

**3.5** 20 - (10) =

**3.6** 20 + (-10) =

**3.7** Which is larger, -180°C or -118°C?

## Application of Skills and Concepts

### 3.1 States of Matter

This concept will probably be easy for you to grasp because you can relate it to the three states of water, which are familiar to you.

**Solid**: known commonly as ice, which has a *definite* volume and shape.
**Liquid**: known commonly as water, which has a *definite* volume and *indefinite* shape.
**Gas**: known commonly as steam or water vapor, which has *indefinite* volume and shape.

Just like water, most materials can exist in all three states of matter (there are a few exceptions, such as hydrogen, which only exists as a gas). The **state** for each type of material will be dependent on conditions (temperature and pressure).

The important thing to grasp from this section is

*Changing state for a given material does not change the material itself, but only the arrangement of the individual particles (molecules/atom- see Chapter 4) that make up the material.*

Some temperatures are worth noting:

- Temperature where the solid form of a material becomes a liquid is known as the *melting point*.
- Temperature where the vapor pressure of a liquid is equal to the external pressure exerted on it is known as the *boiling point*.
    Practically, above this temperature, the material will exist as a gas (called *vapor* if it is normally found as a solid or liquid in nature).

Table 3.1 gives some examples of materials you may be familiar with, listing the melting points (m.p.) and boiling points (b.p.) at sea level atmospheric pressure (changing the pressure that a material exists at can change the m.p. and b.p. of a material). These parameters at sea level atmospheric pressure are known as *normal melting point* and *normal boiling point.*

Section 3.1 of *Introductory Chemistry* has a good explanation of how to determine the state of a substance, given the conditions (temperature and pressure).

**Temperature below m.p.**: material will be predominantly solid.
**Temperature above m.p., but below b.p.**: material will be predominantly liquid.
**Temperature above b.p.**: material will be vapor (gas).

**Table 3.1** The *Normal* Melting Points and Boiling Points of Some Common Materials

| Material | Melting Point (°C) | Boiling Point (°C) |
|---|---|---|
| Water | 0 | 100 |
| Butane | -138 | -0.5 |
| Iron | 1535 | 2750 |
| Rubbing alcohol | -89.5 | 82.4 |
| Mercury | -39 | 357 |
| Table salt | 801 | 1413 |
| Oxygen | -218 | -183 |

Use Table 3.1 to determine the state (solid, liquid, or gas) of compounds at given temperatures.

The second major concept you need to master when dealing with changes of state involves the transfer of *heat*. Again, this can be understood by drawing on your experiences with changes of state. You know that it takes *heat* to melt ice or make water boil (vaporize). You also know that if you decrease the temperature (remove heat) of water below its melting point, it will freeze. It is also true that when water vapor (steam) condenses (becomes liquid) that heat is removed from the steam. All these processes are summarized below:

| | |
|---|---|
| heat + solid (e.g., ice) → liquid (e.g., water) | (EQ 3.3) |
| heat + liquid (e.g., water) → vapor (e.g., steam) | (EQ 3.4) |
| vapor (e.g., steam) - heat → liquid (e.g., water) | (EQ 3.5) |
| liquid (e.g., water) - heat → solid (e.g., ice) | (EQ 3.6) |

## Practice Problems

**3.8** What is the state of water at -20°C?

**3.9** What is the state of oxygen at -198°C?

**3.10** What is the state of butane at 24°C?

**3.11** What would happen if enough heat were added to solid mercury to raise the temperature from -50°C to -20°C?

**3.12** What would happen if enough heat were removed from steam at 101°C to decrease its temperature to 90°C?

## 3.2 Physical and Chemical Changes and Properties

There are two important concepts you should get from this section.

- Distinguishing between a chemical and physical change.
- Distinguishing between a chemical and physical property.

Examples are probably the best way to make these distinctions. Section 3.1 examined one type of physical changes, which are changes of state. Remember that changing state does **not** alter the chemical composition of the matter.

The other broad category of physical changes involve making *mixtures* from substances or separating *mixtures* into substances. This topic will be covered in detail in Section 3.3.

Some examples of physical changes:
- Ice melting
- Liquification of nitrogen gas
- Mixing salt in water
- Evaporating water from mixture of salt water
- Distillation of a mixture of ether and carbon tetrachloride
- Filtration of sand in water

Chemical changes **do** involve changes in the identity of substances. For example, when hydrogen burns in oxygen, water is formed. Water is a different substance from either hydrogen or oxygen. The reverse of this reaction is also a chemical change. These processes and the others listed below are all examples of chemical changes.

- Water + electric current form hydrogen + oxygen
- Acid + base form water + salt and gives off heat
- Sodium + water form hydrogen gas + sodium hydroxide and gives off heat
- Sugar + oxygen form water + carbon dioxide and gives off heat
- Carbon dioxide + water with sunlight form glucose + oxygen (photosynthesis)
- Iron + oxygen form rust plus gives off heat
- Potassium chlorate with heat forms potassium chloride + oxygen

Physical *changes* and chemical *changes* are related to physical *properties* and chemical *properties*, respectively.

Some examples of physical properties follow:
- Melting point
- Boiling point
- Color
- State (solid, liquid, gas) at given pressure and temperature
- Odor
- Solubility of a given substance in various other substances

Some examples of chemical properties follow:
- Flammability
- Reactivity with various other chemicals
- Ability to decompose

> **Practice Problems**
>
> Try to determine if each of the following is a *chemical* or *physical* change.
>
> **3.13** Solid potassium chromate dissolves in water.
>
> **3.14** Water is evaporated from a mixture of silver nitrate in water to leave solid silver nitrate.
>
> **3.15** A mixture of potassium chromate in water is combined with a mixture of silver nitrate in water, and a red solid, silver chromate is formed.
>
> **3.16** The mixture of the red solid in water from problem 3.15 is poured through a filter paper, leaving the red solid behind.
>
> **3.17** Wood burns in the fireplace.
>
> Determine if each of the following properties are *chemical* or *physical*.
>
> **3.18** Ether is very flammable.
>
> **3.19** Ether will boil at 37°C.
>
> **3.20** Ammonia has a pungent smell.

## 3.3 Substances and Mixtures

The following outline summarizes Figure 3.12 in your textbook, *Introductory Chemistry*.

I. Matter (types of material)
   A. Substances
      1. Elements (examples in Section 3.4)
      2. Compounds (examples in Section 3.4)
   B. Mixtures
      1. Homogeneous
         a. Also known as solutions
         b. Examples
            i. Sugar water
            ii. Tincture of iodine (iodine in alcohol)
            iii. Oxygen in nitrogen
            iv. Acid in water
      2. Heterogeneous
         a. Mixture has physically distinct parts with different properties
         b. Examples
            i. Sand in water
            ii. Orange juice
            iii. Silver chromate in water (red insoluble solid)
            iv. Atmosphere
II. Conversion of Matter
   A. Substances to Mixtures
      1. Physical change
      2. Examples
         a. Dissolve sugar in water.
         b. Mix antifreeze with water.
         c. Dissolve iodine in carbon tetrachloride.

# 38  Chapter 3

  B. Mixtures to Substances
  1. Physical change
  2. Examples
     a. Filter water which contains solid lead iodide.
     b. Evaporate ocean water.
     c. Use a magnet to remove pieces of iron from sulfur.
     d. Separate components of crude oil into various compounds by distillation.
  C. Elements to Compounds
  1. Chemical change
  2. Examples
     a. Oxygen and iron form rust.
     b. Carbon and hydrogen form oil.
     d. Nitrogen, oxygen, and hydrogen form nitric acid.
     e. Sodium and chlorine form salt.
     f. Carbon, oxygen, nitrogen, and sulfur form proteins.
  D. Compounds to Elements
  1. Chemical change
  2. Examples
     a. Electrolysis of molten salt to sodium and chlorine.
     b. Shock of solid nitrogen triiodide to nitrogen and iodine.

## 3.4 Elements and Compounds

The inside covers of the textbook have summaries of *elements*. These are fundamental substances, of which there are only 112 known types. They are categorized in the periodic table in the front cover of your textbook, *Introductory Chemistry*, and in alphabetical order on the back cover. This is a good time to take a look at typical names of elements. Many will be familiar to you.

The elements are represented by *symbols* in the periodic table. These symbols are one-letter or two-letter representations. The first letter is always written in *upper case*, and if the symbol has a second letter, it is written in *lower case*.

Your instructor may require you to memorize some or all of the symbols for the elements. Whatever the requirements for your course, now is a good time to start to become familiar with the symbols of some of the common elements. Use the front and back covers of your textbook, and Table 3.1 in *Introductory Chemistry* to fill in the blanks in Table 3.2. Many symbols for elements such as "H" for "hydrogen" are easily deduced. Some such as "K" for "potassium" are derived from the Latin name, "kalium," for this compound. When possible, try to determine a practical symbol for the name or, in some cases, a practical name for a symbol.

While there are only 112 known *elements*, there are so many compounds known to chemists that it would take years to write them all down. The definition at the beginning of this chapter states that compounds contain two or more different elements chemically bound together. Table 3.3 lists a sampling of compounds and the elements in their makeup.

## Practice Problems
Try to determine the missing boxes in Table 3.2

**Matter and Energy** 39

Table 3.2 Some Elements or Symbols with Properties and Discoveries

| Problem | Element | Symbol | Properties | Discoverer | Date Discovered |
|---|---|---|---|---|---|
| 3.21 | Oxygen |  | Clear, colorless gas | Priestly, Scheele, Lavoisier | 1774 |
| 3.22 |  | N | Clear, colorless gas | Rutherford | 1772 |
| 3.23 | Gold |  | Shiny, yellow metal | Ancients | Unknown |
| 3.24 |  | Ag | Shiny, silver-white metal | Ancients | Unknown |
| 3.25 | Mercury |  | Liquid, silver metal | Ancients | Unknown |
| 3.26 |  | Fe | Shiny, silver-white metal | Ancients | Unknown |
| 3.27 | Iodine |  | Blue-black solid | Courtois | 1811 |
| 3.28 |  | F | Pale, yellow gas | Moissan | 1886 |
| 3.29 | Chromium |  | Shiny, silver-white metal | Vaquelin | 1797 |
| 3.30 |  | Ca | Shiny, silver-white metal | Sir H. Davy | 1808 |
| 3.31 | Uranium |  | Shiny, silver-white metal | Klaproth Peligot | 1789 1841 |
| 3.32 |  | Pu | Shiny, silver-white metal | Seaborg, McMillan Kennedy, Wahl | 1940 |
| 3.33 | Helium |  | Clear, colorless gas | Janssen, Ramsay | 1868 |
| 3.34 |  | Kr | Clear, colorless gas | Ramsay, Travers |  |
| 3.35 | Copper |  | Shiny, reddish metal | Ancients | Unknown |
| 3.36 |  | Zn | Shiny, silver-white metal | Marggraf | 1746 |

Table 3.3 Some Common Compounds and the Elements They Contain

| Compound | Elements |
|---|---|
| Water | H, O |
| Alcohol | C, H, O |
| Ether | C, H, O |
| Glucose | C, H, O |
| Table salt (sodium chloride) | Na, Cl |
| Sulfuric acid | S, H, O |
| Gasoline (octane) | C, H |
| Calcium chloride deicer | Ca, Cl |
| Rust | Fe, O |

# 40   Chapter 3

> **Practice Problems**
>
> Determine if the following are elements or compounds without looking at a periodic table.
>
> 3.37 Magnesium
>
> 3.38 Magnesium hydroxide
>
> 3.39 Bronze
>
> 3.40 Nickel

## 3.5 Law of Conservation of Mass

Simply, this states that *the total mass remains constant during a chemical reaction*. You were first introduced to this law in Chapter 1, with the work of Lavoisier. He had found that when the orange solid mercury(II) oxide was heated, it formed silvery liquid mercury and oxygen gas. With careful measurement, he determined that the total mass of mercury(II) oxide equaled the total mass of liquid mercury and oxygen gas. The following equations (EQ 3.7 & EQ 3.8) are examples.

$$\text{mercury(II) oxide} \rightarrow \text{mercury} + \text{oxygen} \qquad (EQ\ 3.7)$$
$$25.0\ g \qquad\qquad 23.2\ g \quad 1.8\ g$$

$$\text{mercury} + \text{oxygen} \rightarrow \text{mercury(II) oxide} \qquad (EQ\ 3.8)$$
$$10.5\ g \qquad 0.8\ g \qquad 11.3\ g$$

Notice that the mass sum on the left side of the arrow is equal to that on the right side.

> **Practice Problems**
>
> 3.41 How many grams of mercury(II) oxide are produced when 31.5 g of mercury combines with 2.4 g of oxygen?
>
> 3.42 When 5.0 g of mercury(II) oxide decomposes, 4.6 g of liquid mercury are obtained. How much oxygen was produced?

## 3.6 Types of Energy

Two major categories are important for you know:

- Potential (stored energy)
- Kinetic (motion energy)

Your book highlights some examples of both types of classification of energy.

- A rock on top of a cliff is obviously *potential* energy because it is not in motion.
- When it is falling over the cliff, it is obviously in motion, and the potential energy of the rock has been converted to *kinetic* energy.
- The rock at the bottom of the cliff has a lower *potential* energy than it did on the top.

A chemical reaction occurs when gasoline reacts with oxygen to form carbon dioxide and water (EQ 3.9):

octane + oxygen → carbon dioxide + water + heat     (EQ 3.9)

When dealing with chemical reactions, you should recognize two types.

- Heat (energy) is produced during the reaction.
- Heat (energy) is used as a reactant.

Examples of each type are shown in the following equations (EQ 3.10 and EQ 3.11):

hydrogen + oxygen → water + energy     **(EQ 3.10)**

energy + water → hydrogen + oxygen     **(EQ 3.11)**

You should have noticed that these two reactions are the reverse of each other. The reaction in EQ 3.10 has energy as a product; this means that some of the *potential* energy stored in the bonds of the reactants (hydrogen and oxygen) has been converted to *kinetic* energy.

The reaction in EQ 3.11 is comparable to pushing the rock back up the hill. Energy must be used to convert water back to hydrogen and oxygen.

In any chemical equation that provides you with energy information, you can determine if the substances that are reactants (left side of arrow) have *more* or *less* energy stored in their bonds than the products (right side of the arrow).

- If energy is a product (right side), then the *reactants* have *more* potential energy.
- If the energy is a reactant (left side), then the *products* have *more* potential energy.

---

**Practice Problems**

Label the substances (reactants or products) with the higher *potential* (stored chemical) energy.

**3.43** Nitroglycerin → carbon dioxide + nitrogen + water + oxygen + energy

**3.44** Energy + carbon dioxide + water → glucose

---

### 3.7 Heat and Heat Calculations

This section of Chapter 3 deals with understanding heat transfer between substances and making calculations dealing with the exchange of heat energy.

One of the topics regarding heat (this is a form of *kinetic* energy) is the understanding that each chemical substance is unique when it comes to the amount of heat energy needed to change its temperature. This uniqueness is known as *specific heat* of a substance. Table 3.4 summarizes the specific heats of a few selected substances.

You should understand the relationship of *heat* and *temperature* while noting that they are not identical. The *temperature* of a given substance is directly proportional to that substance's *kinetic* energy. This is the energy of motion, as shown in EQ 3.12.

$$K.E. = \tfrac{1}{2}mv^2 \qquad \text{(EQ 3.12)}$$

In this equation, K.E. stands for *kinetic energy*, m stands for *mass*, and v stands for *velocity*. You can see that increasing the *velocity* of particles making up a substance will increase its *kinetic energy*. It should be intuitively obvious that adding *heat* to a substance increases the velocity of the particles of a substance, thus increasing the temperature of the substance.

## 42   Chapter 3

How, then, are *temperature* and *heat* different? To understand this, consider an oven at 350°F. The temperature of all substances in the oven is the same; therefore, all substances have the same kinetic energy. Now think about which substance in this oven you would prefer to touch. The air molecules at 350°F would obviously be more touchable than the metal cake pan at 350°F. Even though these two substances have the same *temperature*, they contain different amounts of heat. The heat they contain and the way that the heat would be transferred to a colder object, namely, your fingers, has a lot to do with the substance you prefer to touch.

**Table 3.4** Some Substances and Their Specific Heats

| Substance | Specific Heat (J/g °C) | Specific Heat (cal/g °C) |
|---|---|---|
| Water | 4.18 | 1 |
| Aluminum | 0.901 | 0.215 |
| Copper | 0.384 | 0.0918 |
| Silver | 0.237 | 0.0567 |
| Lead | 0.129 | 0.0308 |
| Iron | 0.444 | 0.106 |
| Carbon (graphite) | 0.502 | 0.120 |
| Carbon (diamond) | 0.720 | 0.172 |
| Ether | 2.32 | 0.554 |
| Octane (gasoline) | 2.23 | 0.533 |
| Ethanol (ethyl alcohol) | 2.43 | 0.581 |
| Mercury | 0.139 | 0.0333 |

### Practice Problem

**3.45** Use the heat of kinetic energy and EQ 3.12 to determine the units of mass, length and time that are incorporated into the quantity joule.

Now that you have a conceptual understanding of *heat*, you are ready to try solving problems involving heat calculations. There are five main types of problems you need master.

- Conversion of one type of energy unit to another.
- Determination of the *amount of heat* needed to *raise* the temperature of a given quantity of a substance by a given amount of degrees. Conversely, the determination of the amount of heat *lost* by a substance when its temperature decreases.
- Determination of the *mass* of a known substance if the heat added to or removed from the substance and the temperature change are known.
- Determination of the *specific heat* of a substance if the mass of the substance, heat added to or removed from the substance, and change in temperature are known.
- Determination of the *temperature change* or *initial* or *final* temperature of a substance if the amount of heat added to or removed from the substance, the specific heat, and the mass are known.

The conversion of one type of energy unit to another is a *dimensional analysis* problem. There are two very common types you will encounter in this chapter.

A. Conversion of calories to joules.
B. Conversion of joules to calories.

The solutions for both these problems involve using the equality

$$1 \text{ cal} = 4.184 \text{ J}$$

Look at the solutions to the following problems to see how this equality can be applied.

**Example 3.2:** Convert 2.612 cal to joules.

    **Solution:** Use the equality with calorie in the denominator.

$$2.612 \text{ cal} \times \frac{4.184 \text{ J}}{1 \text{ cal}} = 10.93 \text{ J}$$

**Example 3.3:** Convert 5.146 J to calories.

    **Solution:** Use the equality with joules in the denominator.

$$5.146 \text{ J} \times \frac{1 \text{ cal}}{4.184 \text{ J}} = 1.230 \text{ cal}$$

The second type of problem requires the determination of the heat that needs to be added to increase the temperature of a substance or the heat that needs to be removed to decrease the temperature of a substance. This type is demonstrated by the following problem.

**Example 3.4:** Suppose you wanted to raise the temperature of 20.5 g of water from 24.5°C to 98.7°C. How much heat needs to be added to the water?

    **Solution:** This involves the use of EQ 3.13, which is written out below.

$$\text{heat} = \underline{\text{specific heat}} \times \underline{\text{mass}} \times \underline{\text{temperature change}} \qquad \text{(EQ 3.13)}$$

The temperature change can be calculated by subtracting the initial temperature ($T_1$) from the final temperature ($T_2$). Then substitute into EQ 3.13.

$$\text{change in } T = T_2 - T_1 = 98.7°C - 24.5°C = 74.2°C$$

$$q = \frac{4.184 \text{ J}}{\text{g } °C} \times 20.5 \text{ g} \times 74.2°C = 6.37 \times 10^3 \text{ J}$$

The third type of problem also begins with EQ 3.13, but it needs to be rearranged so that the *mass* term is isolated. See EQ 3.2, which you should be able to apply to solve the following:

**Example 3.5:** Suppose that adding $2.24 \times 10^4$ J of heat to a piece of iron raises its temperature from 24.2°C to 42.6°C. What is the mass (m) of the iron?

    **Solution:** Determine change in T. Substitute values to solve.

$$\frac{2.24 \times 10^4 \text{ J}}{\frac{0.444 \text{ J}}{\text{g °C}}(42.6°C - 24.2°C)} = m$$

$$\frac{2.24 \times 10^4 \text{ J}}{\frac{0.444 \text{ J}}{\text{g °C}}(18.4°C)} = m$$

$$\mathbf{2.74 \times 10^3 \text{ g} = m}$$

Notice two things about the steps to finding the solution to this problem. Some numbers are given in the problem. Sometimes you will have to go to other sources to find appropriate values. In this case the specific heat of iron can be obtained from Table 3.4. The second thing you should notice is the cancellation of units. The only unit remaining is "g" which is in the *denominator of the denominator*, and therefore in the numerator. If this seems confusing, recall that when dividing a fraction by a fraction, that you simply invert the denominator and multiply.

$$\frac{1}{\frac{1}{g}} = 1 \times \frac{g}{1} = g$$

You could also think of this as multiplying both sides by g/g, which leads to the same conclusion.

$$\frac{1}{\frac{1}{g}} \times \frac{g}{g} = g$$

The fourth type of problem is very similar to the third type. This time EQ 3.13 needs to be rearranged to isolate for specific heat. An example follows.

**Example 3.6:** Suppose that $4.52 \times 10^2$ J of heat are removed from an unknown metal with a mass of 40.5 g and its temperature is reduced from 99.7°C to 52.6°C. What is the specific heat (abbreviated *s*) of the substance?

**Solution:** This problem requires some thinking when you set it up. Notice that heat is being removed. That means that the sign on "q" must be negative.

$$\frac{-452 \text{ J}}{40.5 \text{ g}(52.6°C - 99.7°C)} = s$$

$$\frac{-452 \text{ J}}{40.5 \text{ g}(-47.2°C)} = 0.237 \text{ J/g °C}$$

**Example 3.7:** Identify the substance in the above problem.

**Solution:** This question involves examining Table 3.4. The most likely identity of this substance is *silver*. This type of problem shows how you can carry out an experiment in a laboratory and by determining some *physical* property of a substance, like *specific heat* to identify the substance.

The last type of problem necessitates isolating one of the temperature values. Review examples **3.2** and **3.3** to determine how to solve the following problem.

**Example 3.8:** If $8.92 \times 10^2$ J of heat is added to 24.6 g of lead at 22.2°C, what will its new temperature be?

**Solution:** This problem involves identifying the parameter that you need to determine. In this case it is the "final" temperature, and since heat is being added to the lead, this final temperature will be higher than the initial temperature, 22.2°C. The equation you would start with and the subsequent steps are shown below.

$$\frac{892 \text{ J}}{(24.6 \text{ g})(\frac{0.129 \text{ J}}{\text{g °C}})} + 22.2°C = T_2$$

Notice that all the units in the first term have canceled except °C. The final step follows.

$$281°C + 22.2°C = 303°C$$

---

### Practice Problems

**3.46** 7.52 calories equals how many joules?

**3.47** $4.52 \times 10^7$ joules equals how many calories?

**3.48** How much heat is required to raise the temperature of $2.52 \times 10^2$ g of ethanol from 10.5°C to 50.7°C?

**3.49** How much heat must be removed from 54.3 g of ethanol to reduce its temperature from 15.9°C to 12.4°C?

**3.50** The temperature of a piece of lead is raised from -12.3°C to 22.5°C by adding 59.9 J of heat. What is the mass of the lead?

**3.51** The temperature of 45.5 g of an unknown liquid is raised from 24.8°C to 77.4°C by adding 5.82 kJ of heat. What is the specific heat of the substance?

**3.52** If $2.92 \times 10^3$ J of heat is added to 47.7 g of copper at 42.7°C, what would its new temperature be?

---

### 3.8 Law of Conservation of Energy

This states that *energy may be converted from one form to another, but the total quantity of energy remains constant.*

Practically, this means that total energy in the products (both stored and kinetic) of a chemical reaction must equal the total energy originally in the reactants (stored and kinetic). The example given in your textbook regarding fuel cells nicely illustrates this. The combustion of hydrogen in a fuel cell is summarized in EQ 3.14.

$$\text{hydrogen} + \text{oxygen} \rightarrow \text{water} + \text{energy} \qquad \text{(EQ 3.14)}$$

Your textbook noted that burning 1 g of hydrogen in a fuel cell produces 95 kJ of electric energy plus 48 kJ of heat. The total energy produced is thus 143 kJ. If all the energy from this reaction was in the form of heat energy there would be 143 kJ of heat. This is the essence of the **Law of Conservation of Energy**, in that the total is still the same.

If 2 g of hydrogen is burned, 286 kJ of total energy is produced. If this reaction occurs in a fuel cell, how much electric energy and how much heat energy are produced?

To answer this question, you should first of all note that doubling the amount of hydrogen burned (from 1 g to 2 g) doubles the amount of total energy. This should be intuitively obvious. The more fuel you use, the more heat you can produce.

The second thing you need to do to be able to solve this problem is to assume that in a fuel cell the amount of electric energy will be doubled (for 2 g of hydrogen) and the amount of heat energy will be doubled. Thus you would get 190 kJ of electric energy and 96 kJ of heat energy for a total of 286 kJ of energy.

## Practice Problems

**3.53** How much total energy would be produced by burning 3 g of hydrogen?

**3.54** If this reaction was in a fuel cell, how much electric energy and how much heat energy would be produced?

## Integration of Multiple Skills

One of the skills you will need is an ability to compare number sizes with various scales. For example, which is larger, 93 K or -183°C? The solution to this type of problem necessitates comparison of both quantities on the same scale. Here you could do one of two things. Both approaches lead to the correct answer.

- Convert 93 K to -180°C, and then you can determine that this is larger than -183°C.
- Convert -183°C to 90 K, and then you can determine that this is smaller than 93 K.

The second skill you should be continuing to master is the ability to solve chemical story problems. Some examples follow, which will be solved in careful detail.

**Example 3.9:** If you measured heat in Btus (an older unit of heat measurement called the "British thermal unit" which is equal to 252 calories) how many kilojoules of heat would this be? If the substance measured is 34.7 g of copper found to have 982 Btus, how many kilojoules of heat does the substance contain?

**Solution:** Make a mental map of the conversions needed. Some of the conversions are provided by equalities in the problem, and some will need to be looked up elsewhere.

For example, you will need to go back to Table 2.3 if you don't remember that 1 kJ = 1000 J.

$$Btu \rightarrow cal \rightarrow J \rightarrow kJ$$

The final solution to the problem is provided.

$$952 \text{ BTU} \times \frac{252 \text{ cal}}{1 \text{ Btu}} \times \frac{4.184 \text{ J}}{1 \text{ cal}} \times \frac{1 \text{ kJ}}{1000 \text{ J}} = 1000 \text{ kJ} = 1.00 \times 10^3 \text{ kJ}$$

**Example 3.10:** A piece of copper (14.7 g) at 89.7°C is placed in 212 g of water at 25.0°C. At equilibrium (after heat flow from hot to cold object is completed), what is the temperature of the copper in the water?

**Solution:** This problem involves some high-level thinking. The first thing you should recognize is that at equilibrium, both the water and the piece of metal will have the same temperature. Therefore, $T_2$ for both substances is the same. You should also recognize that heat will flow from the hot substance (copper) to the cold substance (water). Now you can apply the law of conservation of energy. Since total energy (in this case heat) is conserved, then $q_{copper} = q_{water}$. Let's look at each heat equation separately.

$$q_{copper} = (14.7 \text{ g})(0.384 \text{ J/g °C})(\Delta T_{copper}) \qquad \text{(EQ 3.15)}$$

$$q_{water} = (212 \text{ g})(4.184 \text{ J/g °C})(\Delta T_{water}) \qquad \text{(EQ 3.16)}$$

Since $q_{copper} = q_{water}$, then the right-hand side of EQ 3.15 equals the right hand side of EQ 3.16, as the following shows:

$$(14.7 \text{ g})(0.384 \text{ J/g °C})(\Delta T_{copper}) = (212 \text{ g})(4.184 \text{ J/g °C})(\Delta T_{water})$$

Now you can substitute initial temperature values and solve for the final temperature, as the following sequence shows.

$$(14.7 \text{ g})(0.384 \text{ J/g °C})(89.7°C - T_f) = (212 \text{ g})(4.184 \text{ J/g °C})(T_f - 25.0°C)$$

$$5.65(89.7°C - T_f) = 887(T_f - 25.0°C)$$

$$506°C - 5.64 T_f = 887 T_f - 22200°C$$

$$22700°C = 893 T_f$$

$$25.4°C = T_f$$

The final thing you must do with this problem is to *assess* the validity of the answer. The final temperature is warmer than the initial temperature of the water and cooler than the initial temperature of the copper, so this is a reasonable answer.

The final skill you should be starting to master involves conceptual challenges. The following is an example of this type of integrated thinking.

**Example 3.11:** Why would steam at 100°C cause a more severe burn than water at 100°C?

**48    Chapter 3**

**Solution:** This first problem involves understanding of what happens to energy when a change of state occurs. You know that heat (energy) must be removed from a vapor when it forms a liquid. If you were to be burned by steam (water vapor), the cooler temperature of your skin would cause the steam to condense (form water), thus releasing heat. This heat would be absorbed by your skin which could lead to a very severe burn. If you were to be burned by water, heat would still flow from the hotter liquid to your cooler skin, but it would not be nearly as much heat as that from the steam condensing.

**Example 3.12:** If you wanted to make a thermometer for measuring the temperature of dry ice (solid carbon dioxide) in an alcohol solution of dry ice which has a temperature in the area of 195 K, what material would you use?

**Solution:** This problem involves using Table 3.1, your ability to convert from one temperature scale to another, and your reasoning skills. If you need to measure a substance at 195 K, you must find a material for the thermometer that is liquid at this temperature.

$$195 \text{ K} = -78°C$$

This temperature is lower than the melting point of mercury; therefore, it would not be a good choice for a thermometer in this range. A better choice would be alcohol, which would still be a liquid at this temperature. Some colored dye could make it easy to see.

## Practice Problems

**3.55** Which is larger, 2110 Cal (that is 2110 kcal) or $9.00 \times 10^6$ J?

**3.56** Which has a higher final temperature, 10.0 g of water at 22.0°C absorbing $4.2 \times 10^3$ J or 20.0 g of ethanol at 22.0°C absorbing 1 kcal of heat?

**3.57** Which is smaller, 250 K or 0°C?

**3.58** If it takes 286 kJ of energy to convert 18 g of water to 2 g of hydrogen and 16 g of oxygen, how much energy is needed to convert 36 g of water to 4 g of hydrogen and 32 g of oxygen?

**3.59** A 10.0–g spoon absorbs 56.7 J of heat. Which would be warmer, a spoon made of copper or a spoon made of silver? (The spoon is initially at 24.0°C.)

**3.60** Why is water the best substance for lakes and oceans (aside from its chemical support of life)?

## Self-Exam

**I. (4 ea) Write T (true) or F (false) in each corresponding blank.**

_____1. The temperature at which a solid melts is called a *boiling point*.

_____2. 5.246 J is larger than 1.1 cal.

_____3. When water stored behind a dam is allowed to flow, potential energy is converted to kinetic energy.

_____ 4. In chemical reactions, total energy is usually lowered.

_____ 5. A vapor has an indefinite volume and an indefinite shape.

_____ 6. The melting point of butanol (an alcohol) is 26°C. At 12.2°C it would be a liquid.

_____ 7. When liquid water is converted to sleet, heat is removed from the water.

_____ 8. Salt water is a heterogeneous mixture.

_____ 9. The specific heat of water is 1 cal/g °C.

_____ 10. It takes more heat to raise the temperature of 10 g of silver than 10 g of water by 20°C.

**II. (4 ea) Multiple Choice. Place the letter of the BEST answer in the corresponding blank.**

_____ 1. Which of the following is **not** an example of a mixture?
   A. Salt water   B. Pure water   C. Air   D. Bronze   E. Milk

_____ 2. Which of the temperatures listed below is the coldest?
   A. -20°C   B. -10°C   C. 200 K   D. 0°C   E. -80°C

_____ 3. Which of the following is a *chemical* property of hydrogen?
   A. It is a gas at room temperature
   B. It is colorless
   C. It is not as dense as air
   D. It cannot exist as a solid
   E. It burns in oxygen to form water

_____ 4. Which of the following is a physical change?
   A. Ice melting
   B. Wood burning
   C. Electrolysis of water to form hydrogen and oxygen
   D. The addition of acid to base to form salt and water
   E. The metabolism of sugar

_____ 5. The burning of hydrogen in oxygen produces 143 kJ of energy per gram of hydrogen. How much energy is produced by 20 g of hydrogen?
   A. 2720 kJ   B. 143 kJ   C. 2860 kJ
   D. 286 kJ   E. 2.86 kJ

_____ 6. Referring to problem 5, the energy produced from burning hydrogen can be converted partially to electric energy and partially to heat energy in a fuel cell. If burning 2 g of hydrogen produces 190 kJ of electric energy, how much heat energy is produced?
   A. 190 kJ   B. 96 kJ   C. 143 kJ
   D. 286 kJ   E. 2860 kJ

_____ 7. Which of the following is an element?
   A. Acetic acid   B. Mercury   C. Brass
   D. Water   E. Carbon dioxide

_____ 8. If 46.0 g of an alcohol contains 24.0 g of carbon, how much carbon is contained in 23.0 g of this same alcohol?
   A. 24.0 g   B. 46.0 g   C. 23.0 g   D. 22.0 g   E. 12.0 g

_____9. Carbon tetrachloride has a boiling point of 77°C and a melting point of -23°C. It will be a liquid at which of the following temperatures?
A. 79°C    B. 72°C    C. -20°C    D. 70 K    E. Both B and C

_____10. How many joules of heat are required to raise the temperature of 28.0 g of aluminum from 21.2°C to 92.3°C? (Specific heat of aluminum = 0.901 J/g °C)
A. 1.82 kJ    B. 2.33 kJ    C. 534 J
D. 2.87 kJ    E. 5.34 kJ

### III. Provide Complete Answers

1. (3 pts) Describe what happens to molecules of water when it boils.

2. (7 pts) Match the following symbols to the elements listed below:
C, Ca, Cd, Cl, Co, Cr, Cu
cadmium, calcium, carbon, chlorine, chromium, cobalt, copper

3. (5 pts) If 480 J of heat is added to a 4.5–g silver spoon at room temperature (22.8°C), what will be the final temperature of the spoon? (The specific heat of silver = 0.237 J/g °C)

4. (5 pts) A 69.7–g piece of copper at 78.9°C is put into a cup of water (424 g) at 24.5°C. What is the new temperature of the water with the copper in it? (The specific heat of copper = 0.384 J/g °C and that of water = 4.184 J/g °C)

## Answers to Practice Problems

| | | | | | |
|---|---|---|---|---|---|
| 3.1 | $\frac{q}{m(T_2 - T_1)} = s$ | 3.2 | $\frac{q}{sm} + T_1 = T_2$ | 3.3 | $T_2 - \frac{q}{sm} = T_1$ |
| 3.4 | 30 | 3.5 | 10 | 3.6 | 10 |
| 3.7 | -118°C | 3.8 | Solid | 3.9 | Liquid |
| 3.10 | Vapor | 3.11 | It would melt. | | |
| 3.12 | It would liquefy (become water). | | | | |
| 3.13 | Physical | 3.14 | Physical | 3.15 | Chemical |
| 3.16 | Physical | 3.17 | Chemical | 3.18 | Chemical |
| 3.19 | Physical | 3.20 | Physical | 3.21 | O |
| 3.22 | Nitrogen | 3.23 | Au | 3.24 | Silver |
| 3.25 | Hg | 3.26 | Iron | 3.27 | I |
| 3.28 | Fluorine | 3.29 | Cr | 3.30 | Calcium |
| 3.31 | U | 3.32 | Plutonium | 3.33 | He |
| 3.34 | Krypton | 3.35 | Cu | 3.36 | Zinc |
| 3.37 | Element | 3.38 | Compound | 3.39 | Compound |
| 3.40 | Element | 3.41 | 33.9 g | 3.42 | 0.4 g |

**3.43** Reactants  **3.44** Products  **3.45** K.E. $= \frac{1}{2} kg \frac{m^2}{s^2}$

**3.46** 31.5 J   **3.47** $1.08 \times 10^7$ cal   **3.48** 24.6 kJ

**3.49** 462 J heat lost or q = -462 J   **3.50** 13.3 g   **3.51** 2.43 J/g °C

**3.52** 202°C   **3.53** 429 kJ

**3.54** 285 kJ electric energy and 144 kJ heat energy

**3.55** $9.00 \times 10^6$ J which = $2.15 \times 10^6$ cal, is larger than 2110 kcal, which = $2.10 \times 10^6$ cal

**3.56** 10.0 g water ($T_f$ = 122.4°C) higher than 20.0 g ethanol (108.1°C)

**3.57** 250 K   **3.58** 572 kJ   **3.59** Silver

**3.60** Water has a high specific heat compared with other substances. Thus it can absorb large amounts of heat without changing temperature very much. Other substances would become too hot to touch on a summer day. Water can also lose lots of heat without becoming cold as rapidly as other substances.

## Answers to Self-Exam

### I. T/F
1. F  2. T  3. T  4. F  5. T  6. F  7. T  8. F  9. T  10. F

### II. MC
1. B  2. E  3. E  4. A  5. C  6. B  7. B  8. E  9. E  10. A

### III. Complete Answers
1. They move so rapidly that they enter the vapor state. These appear as large bubbles of water vapor, which will quickly enter the atmosphere.

2. C: carbon; Ca: calcium; Cd: cadmium; Cl: chlorine; Co: cobalt; Cr: chromium; Cu: copper

3. 473°C

4. 25.3°C

# 4
# Atoms, Molecules, and Ions

This chapter will examine matter at the *very small* level. It will broaden the definition of words such as *element, compound,* and *chemical reaction,* which were described in Chapter 3.

You will learn about the subatomic units and how to express these in a chemical language. You will also learn about the molecular basis for substances, the ionic basis for substances, and the properties of substances in solution.

Keys for success in this chapter:

- Learn new terms and definitions.
- Practice isotope problems.
- Practice ion calculations.
- Continue to learn the names and symbols of the elements.

## Summary of Verbal Knowledge

**atomic theory:** Dalton's explanation of the structure and chemical reactions of matter in terms of atoms.
**atom:** This is the smallest unit of chemistry. It consists of a nucleus and a number of electrons surrounding it.
**element:** A type of matter composed of only one kind of atom, which always has certain specific properties.
**compound:** A type of matter composed of atoms of two or more different elements that are chemically combined in fixed proportions.
**chemical reaction:** A rearrangement of the atoms present in the reacting substances to give new chemical combinations in the substances formed by the reaction.
**electron:** A very light, negatively charged subatomic particle.
**nucleus:** A positively charged core that takes up very little space in the atom. However, it contains most of the atomic mass.
**proton:** A positively charged subatomic particle found in the nucleus of an atom.
**neutron:** An uncharged subatomic particle found in the nucleus of an atom.
**atomic number:** This is the number of protons in an atom. It is unique for each element.
**mass number:** This is the sum of number of protons and neutrons in the nucleus.
**isotopes:** Atoms whose nuclei have the same atomic number but different mass numbers.
**percentage abundance:** This is the percentage of an isotope in a naturally occurring sample of an element.
**atomic mass unit:** This is abbreviated **amu** and equals exactly one-twelfth the mass of a carbon-12 isotope. This is a very small mass unit and approximately equal to $1.661 \times 10^{-24}$ g.

**atomic mass unit:** The weighted average mass (expressed in atomic mass units) of an atom of a naturally occurring element.
**periodic table:** A systematic classification of the elements ordered by increasing atomic number into rows and columns so that the elements in any one column have similar or regularly changing properties.
**period:** This is the grouping of all the elements in a horizontal row of the periodic table.
**group (or family):** This is the grouping of all the elements in a column of the periodic table.
**main group (or representative group):** These are elements in the "A" group of the periodic table.
**transition-metal group:** These are elements in the "B" group of the periodic table.
**metal:** An element that has a characteristic luster or shine and is a relatively good conductor of heat and electricity.
**nonmetal:** An element that has no luster and is nonconductive.
**metalloid:** An element that has properties between metals and nonmetals.
**molecule:** A group of identical or different atoms that are chemically bonded by attractive forces. It is electrically neutral.
**ion:** An electrically charged particle derived by adding or removing electrons from an atom or a chemically bonded group of atoms.
**molecular formula:** A notation that uses atomic symbols with numeric subscripts to convey the exact number of atoms of the different elements that are in a molecule of substance.
**structural formula:** A chemical formula that shows what atoms are bonded to one another.
**anion:** A negatively charged atom or group of atoms. It has more electrons than protons.
**cation:** A positively charged atom or group of atoms. It has more protons than electrons.
**monatomic ion:** An ion having only one atom.
**polyatomic ion:** A group of atoms chemically bonded but having an excess or a deficiency of electrons so that the entire unit has either a positive or negative electric charge.
**formula unit:** The group of atoms or ions explicitly symbolized in the formula of a substance.

## Review of Mathematical and Calculator Skills

There are not many new mathematical skills you will need for this chapter. Many of the calculations in this chapter involve simple arithmetic that won't even require a calculator. However, there are a couple of applied arithmetic skills we will review here.

You will find that you need to apply old skills to incredibly small numbers, so you should make sure that you can handle these. Some examples of handling these numbers will help.

**Example 4.1:** Which is larger, $9.31 \times 10^{-31}$ or $1.67 \times 10^{-27}$? How much larger is the larger number than the smaller number?

> **Solution:** To answer the first question, remember that a *negative* exponent means the number is smaller than 1, or a decimal. The *larger* the *magnitude*, the *smaller* the number. Therefore, $\mathbf{1.67 \times 10^{-27}}$ is the larger number.
>
> To answer the second question you need to know what the words, *how much larger* mean. This could be interpreted in two ways:
> - Larger by a numerical sense (subtract the small number from the larger.)
> - Larger by a multiplication factor (divide the large number by the small number.)

The meaning of the question refers to the second definition. The solution follows:

$$\frac{1.67 \times 10^{-27}}{9.31 \times 10^{-31}} = 1.79 \times 10^3 \text{ times larger}$$

The second skill you need to review involves working with percentages.

**Example 4.2:** What is 75.77% of 34.97?

**Solution:** Remember that the percentage must be changed to a decimal before doing calculations.

As a decimal 75.77% = 0.7577

One of the words that needs to be well understood when trying to solve story problems is a simple one like **of**. Problems involving percents "of" or fractions "of" some quantity, imply **multiplication**.

75.77% of 34.97 means 0.7577 × 34.97 = **26.50**

# Practice Problems

**4.1** How much larger is 1.00 than $1.6605402 \times 10^{-24}$? (*Hint*: Use correct significant figures.)

**4.2** How much larger is $1.6726231 \times 10^{-24}$ than $1.6605402 \times 10^{-24}$?

**4.3** What is 60.2% of 68.9?

## Application of Skills and Concepts

### 4.1 Dalton's Atomic Theory

This theory explains two chemical laws already discussed.

- The law of conservation of mass
- The law of definite proportions

This theory is an explanation of the structure and chemical reactions of matter in terms of atoms. The important highlights are listed.

- All matter is composed of atoms, which are very small, chemically indivisible particles.
- An element is a type of matter composed of only one kind of atom.
- A compound is a type of matter composed of atoms of two or more elements chemically combined to one another in fixed proportions.
- A chemical reaction is a rearrangement of the atoms present in the reacting substances to give new substances with new connections of atoms. The total number of atoms does not change during the course of a reaction.

This section requires some conceptual knowledge. Read the section before you use the postulates of Dalton's atomic theory to answer the following questions.

# Atoms, Molecules, and Ions

> **Practice Problems**
>
> **4.4** If 100 atoms of hydrogen combine with 50 atoms of oxygen to form water molecules, what is the ratio of hydrogen atoms to oxygen atoms in water?
>
> **4.5** A compound of a purple solid decomposes to $1.20 \times 10^{23}$ atoms of nitrogen and $3.60 \times 10^{23}$ atoms of iodine. How many atoms (total) were in the original compound?

## 4.2 Particle Structure of the Atom

This is a very important section for you to comprehend. Your textbook does a nice job of developing the definitions of each subatomic particle and setting these discoveries in historical contexts.

In this *Study Guide* we will concentrate on practicing problems that deal with the numbers of electrons, protons, and neutrons in a given atom.

Table 4.1 summarizes information about the subatomic particles given in your textbook.

**Table 4.1** Some Parameters of Subatomic Particles

| Particle | Atomic Location | Mass (g) | Mass (amu) (to 4 significant figures) | Mass (amu) (to 1 significant figure) | Relative Charge |
|---|---|---|---|---|---|
| Electron | Outside nucleus | $9.31 \times 10^{-28}$ | 0.0005486 | 0 | -1 |
| Proton | Nucleus | $1.67 \times 10^{-24}$ | 1.007 | 1 | +1 |
| Neutron | Nucleus | $1.67 \times 10^{-24}$ | 1.009 | 1 | 0 |

Take some time to compare the relative sizes of each of these particles. Notice that the electron has such a small mass that it can be ignored when calculating the mass of an atom.

This section has some new definitions, which we will summarize, and then we will practice working with the definitions.

The **atomic number** is **always** equal to the number of **protons** in an atom of a given element. We will work with the periodic table in more detail in Section 4.4, but for now, examine the table inside the front cover of your textbook. Notice that the elements are written in order of increasing atomic number. If you know that an atom has 7 protons, then you know that it is N (nitrogen). If you know that you have an atom of U (uranium), then you know that it contains 92 protons.

The **atomic mass number** of an atom is derived by adding the number of **protons** and **neutrons** in the atom. The number of **electrons** has a negligible contribution. This can be summarized in EQ 4.1.

$$\text{mass number} = \text{protons} + \text{neutrons} \qquad \text{(EQ 4.1)}$$

If you know any two numbers in EQ 4.1, you can deduce the third.

**Example 4.3:** An atom that has 17 protons and 18 neutrons would have what mass number? An atom that has 17 protons and a mass of 37 has how many neutrons?

**Solution:** The atoms in both problems can be identified as chlorine, because 17 is the atomic number for this element. The mass number of the first is found by addition.

$$17 + 18 = 35$$

The number of neutrons in the second is found by subtraction.

$$37 - 17 = 20$$

The above example brings us to another important definition. It is that of **isotopes**. These are atoms of the same element that *differ only by the number of neutrons they contain*. The chlorine atom that contains 18 neutrons is an **isotope** of the chlorine atom that contains 20 neutrons.

> An atom of chlorine can only be an isotope to another atom of chlorine; an atom of hydrogen can only be an isotope to another atom of hydrogen; etc.

Isotopes are noted in two ways that you should recognize. These two distinct ways are shown below for the two isotopes of chlorine we have just discussed.

- $^{35}_{17}Cl$ means the mass number is 35; the atomic number is 17; and the number of neutrons is 18 (35 minus 17).

  $^{37}_{17}Cl$ means the mass number is 37; the atomic number is 17; and the number of neutrons is 20 (37 minus 17).

- Chlorine-35 has the same meaning as $^{35}_{17}Cl$.
  Chlorine-37 has the same meaning as $^{37}_{17}Cl$.

Remember that in a neutrally charged atom the number of electrons is equal to the number of protons. Since these two subatomic particles have charges of opposite sign and the same magnitude, the overall charge is 0 when number of electrons equals number of protons. The case where these numbers are not equal will be discussed in Section 4.6.

Table 4.2 below is designed to help you apply the concepts of isotopes and review some nomenclature and symbols for elements.

## Practice Problems

**4.6** An atom has 79 protons. What is it?

**4.7** An atom of sodium (Na) has how many protons?

**4.8** An atom has 35 protons, 35 electrons, and 44 neutrons. What is it? What is its atomic mass number?

**4.9** An atom has 74 neutrons and mass number equal to 127. How many protons does it have? What is it?

**4.10** Which number in $^{35}_{17}Cl$ is unnecessary to completely identify the isotope?

Atoms, Molecules, and Ions        57

**4.11** How many protons and neutrons are present in boron-11?

**4.12** How many protons and neutrons are present in $^{190}_{80}$Hg?

**4.13-4.23** Try to use deductive reasoning to fill in as many blanks as you can in Table 4.2 without consulting anything but a periodic table

Table 4.2 Neutral Isotopes and their Subatomic Particles

| Problem | Element Name | Symbol | Atomic Number | Mass Number | No. of Protons | No. of Electrons | No. of Neutrons |
|---|---|---|---|---|---|---|---|
| 4.13 | Magnesium |  |  | 25 |  |  |  |
| 4.14 |  | Cs |  |  |  |  | 77 |
| 4.15 |  |  | 18 | 40 |  | 18 |  |
| 4.16 |  |  |  | 109 |  |  | 62 |
| 4.17 |  |  |  |  | 26 |  | 30 |
| 4.18 | Lithium |  |  |  |  |  | 4 |
| 4.19 |  |  |  | 3 |  | 1 |  |
| 4.20 |  | Br |  | 81 |  |  |  |
| 4.21 |  |  |  |  |  | 56 | 74 |
| 4.22 |  |  |  | 82 |  |  | 125 |
| 4.23 |  |  |  |  |  | 7 | 8 |

## 4.3 Atomic Weights

This section distinguishes between the terms *atomic mass number* and *atomic weight*. The former refers to one particular *isotope*, while the latter refers to the *average mass* of all naturally occurring isotopes with reference to carbon-12.

Carbon-12 was arbitrarily assigned as the isotope which is *exactly* equal to 12 atomic mass units (amus). Look now at the table of *physical constants* inside the back cover of your textbook. Notice that 1 amu, when converted to grams is also equal to $1.6605402 \times 10^{-24}$ g. Notice that this is nearly equivalent to the mass of a proton ($1.6726231 \times 10^{-24}$ g) or the mass of a neutron ($1.6749286 \times 10^{-24}$ g).

**Example 4.4:** What are the number of protons and neutrons in a copper-63 isotope?

**Solution:** You can easily answer this after completing Section 4.2. There are 29 protons and 34 neutrons. A second isotope of copper is known to exist in nature, copper-65, which contains 29 protons and 36 neutrons. The *atomic mass number* of copper-63 is **63**, and the *atomic mass number* of copper-65 is **65**.

Take a moment to return to the periodic table inside the front of your textbook and examine the *atomic weight* listed for copper. You will see that it is **63.55**. Your textbook explains why this is a decimal, and neither **63** or **65**. The number in the periodic table is **an average of all naturally occurring isotopes** of copper.

## 58  Chapter 4

**Example 4.5:** Test you *number sense* now. Which isotope, copper-63 or copper-65, do you think is more abundant?

**Solution:** Without using a calculator, you should be able to rationalize that more copper atoms are of the copper-63 variety than the copper-65 variety. This should be obvious because 63.55 is much closer to 63 than to 65.

Copper-63 is defined as having a mass of 62.92 amu, and copper-65 has a mass of 64.9278 amu. You may wonder why this is so. It is simply because the copper-63 and copper-65 isotopes are compared with the carbon-12 isotope, which is *exactly* 12 amu.

Scientists have been able to determine (through careful experiments) that 69.09% of all copper atoms are copper-63 isotopes and 30.91% are copper-65 isotopes. This verifies the intuitive conclusion just made. More of the isotopes are copper-63.

Finally, we are ready to determine the *atomic weight* of copper. It is simply the average of all the isotopes. The *atomic weight* contributed from copper-63 can be determined by multiplying the decimal fraction (0.6909) of its abundance times its mass (62.92 amu). A similar approach can be taken to find the *atomic weight* contributed from copper-65. The total *atomic weight* can then be calculated as show in EQ 4.2:

$$(0.6909)(62.93 \text{ amu}) + (0.3091)(64.9278 \text{ amu}) = 63.55 \text{ amu} \qquad (EQ\ 4.2)$$

### Practice Problems

**4.24** The atomic weight of lithium is 6.941. The naturally occurring isotopes for this element are lithium-6 and lithium-7. Which one of these isotopes is more plentiful?

**4.25** The atomic masses of $^{35}_{17}Cl$ and $^{37}_{17}Cl$ are 34.97 amu and 36.956 amu, respectively. The natural abundance of chlorine-35 is 75.77%, and that of chlorine-37 is 24.23%. What is the atomic weight of chlorine?

**4.26** The atomic masses of $^{3}_{1}H$, $^{2}_{1}H$, and $^{1}_{1}H$ are 3.0165 amu, 2.0140 amu, and 1.0078 amu, respectively. The natural abundance of hydrogen-1 is 99.985%, that of hydrogen-2 is 0.01500%, and that of hydrogen-3 is so small that it has a negligible natural abundance. What is the atomic weight of hydrogen?

### 4.4 Periodic Table of the Elements

Before starting this section, locate the modern periodic table inside the front cover of your textbook or Figure 4.9. Notice the following things about the table:

I. The elements are listed in order of **atomic number.**
II. Generally, the **atomic weights** increase with each atomic number (some exceptions).
III. The horizontal rows are called **periods.**
   A. The periods are numbered **1 to 7.**
   B. Not all periods have the same number of elements
   C. **Inner-transition metals** are pulled out of their respective rows.
      This makes it easier to fit the entire table on the page.
IV. The columns are called **groups.**

V. The groups have two types of identification.
  A. **Roman numerals I to VIII** with an **A** or **B**
    1. Group **A** elements are referred to as **main group.**
    2. Group **B** elements are referred to as **transition metals.**
  B. **Arabic numerals 1 to 18**
VI. Elements tend to have properties similar to those in their **group.**
VII. Elements are also grouped by metallic properties.
  A. **Metals** are on the left-hand side of the table (exception hydrogen). Conductors; shiny (usually solids)
  B. **Metalloids** are a diagonal (staircase) going from boron to astatine. Often solids; semiconductors
  C. **Nonmetals** are on the right-hand side of the table (exception hydrogen). Nonconductors; usually gases

The best way to master the material in this section is to practice finding elements in the table and looking for elements that are similarly grouped.

# Practice Problems

**4.27** To what group does oxygen belong? To what period does oxygen belong?

**4.28** To what group does sodium belong? To what period does sodium belong?

**4.29** To what group does helium belong? To what period does helium belong?

**4.30** What are the symbols and numbers of the elements on either side of the lanthanides?

**4.31** Name an element in the same group as tungsten (W).

**4.32** Name an element in the same group as phosphorus (P).

**4.33** Which element (N, Si, or S) will have properties most like P?

**4.34** Find a case where the **atomic weight** of two succeeding elements does not increase when the **atomic number** does.

**4.35** How many elements are in periods, 1? 3? 4? 6?

## 4.5 The Molecular Basis of Substances

This is a short section. Read about the definitions for molecules and ions. The following outline should serve as a good review:

I. Molecules have groups of atoms chemically bonded (connected).
  A. Elements that are molecules
    1. Some elements exist in nature as molecules.
    2. Examples of diatomic molecules
      a. Hydrogen ($H_2$), oxygen ($O_2$), nitrogen ($N_2$)
      b. Fluorine ($F_2$), chlorine ($Cl_2$), bromine ($Br_2$), iodine ($I_2$)
  B. Compounds that are Molecules
    1. Some compounds exist as molecules.
    2. Examples: water, carbon dioxide, carbon tetrachloride, laughing gas, glucose
II. Ions are electrically charged particles.
  A. Formed from atoms

1. Remove or add electrons to an atom.
2. These atoms are electrically charged.
B. Formed from chemically bonded group of atoms
1. Remove or add electrons to the group.
2. These groups are electrically charged.
C. Compounds that are vast arrays of ions
1. Held together by attractions of opposite charges
2. Examples: table salt, baking soda, lye

## Practice Problem

**4.36** What do all the diatomic molecules listed have in common?

### 4.6 Comparing Molecular and Ionic Substances

This section's primary purpose is to help you distinguish between molecular and ionic substances. Table 4.3 summarizes the properties of these two types of substances.

Table 4.3 Summary of Properties of Substances

|  | Types of elements present | Melting Point | Physical State |
|---|---|---|---|
| Molecular | Nonmetal and nonmetal | Low | Gas or liquid |
| Ionic | Metal and nonmetal | High | Solid |

## Practice Problems

**4.37** Sugar contains carbon, hydrogen, and oxygen. Predict if it is molecular or ionic.

**4.38** A common ingredient in antacids is magnesium hydroxide. Predict if it is molecular or ionic.

### 4.7 Molecular Substances and Their Formulas

This section is designed to introduce you to some common molecular substances and how formulas and structures represent them. Tables 4.4 and 4.5 summarize some common molecules.

Table 4.4 Some Molecular Elements

| Name | Molecular Formula |
|---|---|
| Hydrogen | $H_2$ |
| Oxygen | $O_2$ |
| Nitrogen | $N_2$ |
| Fluorine | $F_2$ |
| Chlorine | $Cl_2$ |
| Bromine | $Br_2$ |
| Iodine | $I_2$ |
| Astatine | $At_2$ |
| Phosphorus | $P_4$ |
| Sulfur | $S_8$ |

Table 4.5 Some Molecular Compounds

| Name | Molecular Formula | Molecular Structure |
|---|---|---|
| Water | $H_2O$ | H—O—H |
| Carbon dioxide | $CO_2$ | O=C=O |
| Carbon tetrachloride | $CCl_4$ | Cl—C(Cl)(Cl)—Cl |
| Glucose | $C_6H_{12}O_6$ | H-C(=O); H-C-OH; HO-C-H; H-C-OH; H-C-OH; H-C-OH; H |

Notice that water has 2 atoms of hydrogen and 1 atom of oxygen and thus is written $H_2O$.

You will want to examine these tables and try to see some patterns in each. At this point in the course you should recognize that molecules are held together by attractive forces, called **bonds**. These bonds are represented by lines in molecular structures. You will learn to interpret the lines from an electron standpoint in Chapter 10.

For now, notice that the molecular structure tells you more about the order of connectivity of the atoms than the molecular formula. The latter only tells you what atoms are present in the molecule. The subscript tells you how many of each type of atom are present.

Notice that all the molecular formulas for both elements and compounds contain **only** nonmetals. There are some molecular substances that contain metals, but you will not encounter them in this course.

## Practice Problems

**4.39** Table sugar has 12 atoms of carbon, 22 atoms of hydrogen, and 11 atoms of oxygen in each molecule. What is its molecular formula?

**4.40** How many of each type of atom does $P_2O_{10}$ have?

**4.41** A molecule of vanillin has 8 atoms of carbon, 8 atoms of hydrogen, and 2 atoms of oxygen. What is its molecular formula?

**4.42** Which (there may be several) of the following are molecular substances?
  a. NaCl   b. $PCl_5$   c. CO   d. $N_2O$   e. $N_2O$   f. MgO

## 4.8 Ionic Substances and Their Formulas

There are several important categories for ions.

I. Monatomic ions (contain only **one** atom)
   A. Cations
      1. Have more protons than electrons
      2. Have positive charge
      3. Tend to be metals
   B. Anions
      1. Have more electrons than protons
      2. Have negative charge
      3. Tend to be nonmetals

II. Polyatomic ions (contain a group of atoms chemically bonded together)
   A. Polyatomic cations
      1. Have more protons than electrons
      2. Have positive charge on the group
   B. Polyatomic anions
      1. Have more electrons than protons
      2. Have negative charge on the group

An example of how an atom becomes a cation is shown in EQ 4.3:

$$K \rightarrow K^+ + e^- \quad \text{(EQ 4.3)}$$

A neutral potassium atom has 19 protons and 19 electrons. If it loses one electron, it now has only 18 electrons. That means it has 19 **positive** charges and 18 **negative** charges, and therefore, it has an overall **+1** charge.

An example of how an atom becomes an anion is shown in EQ 4.4.

$$Br + e^- \rightarrow Br^- \quad \text{(EQ 4.4)}$$

A neutral bromine atom has 35 protons and 35 electrons. If it gains an extra electron, it now has 35 protons and 36 electrons. That means it has 35 **positive** charges and 36 **negative** charges, and therefore, it has an overall **-1** charge.

Table 4.6 contains several typical monatomic ions.

For now, you will not be expected to count protons and electrons to determine the charge on polyatomic ions. Table 4.7 does list some common ones you will encounter in the study of chemistry and shows the group which is chemically bonded and the charge on the ion outside the bracket. At this point, it is only important to understand that the group tends to stay together when forming ionic compounds. You will learn to write ionic formulas in Chapter 5.

Like compounds with monatomic ions, the compounds with polyatomic ions exist in vast arrays. Positive and negative charges alternate, and the overall charge of the substance is zero.

## Practice Problems

**4.43-4.48** Try to fill in the number of protons and electrons present in each ion in Table 4.6. Use a periodic table to determine the number of protons for each atom if you do not yet have these memorized.

Table 4.6 Some Monatomic Ions

| Number | Ion | Protons | Electrons |
|---|---|---|---|
| 4.43 | $Li^+$ | | |
| 4.44 | $Al^{3+}$ | | |
| 4.45 | $N^{3-}$ | | |
| 4.46 | $Se^{2-}$ | | |
| 4.47 | $F^-$ | | |
| 4.48 | $Ba^{2+}$ | | |

Table 4.7 Some Polyatomic Ions

| Polyatomic Ion Name | Polyatomic Structure |
|---|---|
| Nitrate ion | 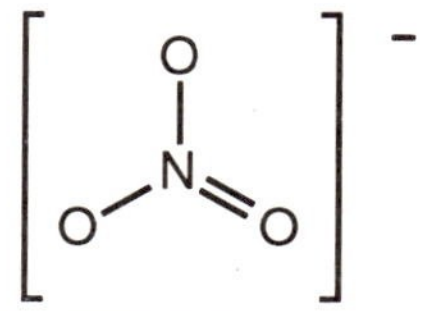 $[O-N(=O)O]^-$ |
| Ammonium ion | 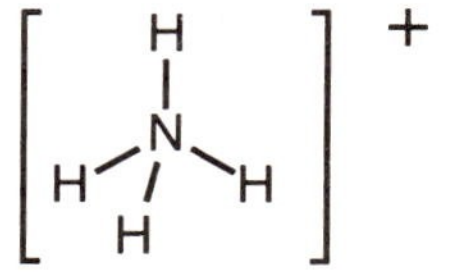 $[NH_4]^+$ |
| Carbonate ion | 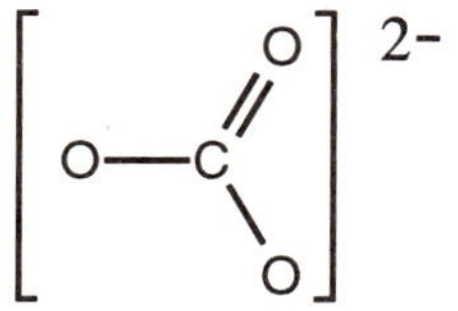 $[O-C(=O)O]^{2-}$ |

## 4.9 Electrical Properties of Substances in Solution

This section explains some natural phenomena of solutions. In many cases, when an ionic substance is dissolved in water (such as table salt), the cations and anions move freely about the solution. When a device such as that shown in Figure 4.17 of *Introductory Chemistry* is inserted into a solution of salt water (contains $Na^+$ and $Cl^-$ ions), a circuit is completed, and the bulb glows.

Sugar, which contains C, H, and O atoms (all nonmetals), also will dissolve in water. This is a molecular substance, and unlike the ionic substances, it does **not** complete the circuit. Thus the bulb does **not** glow.

### Practice Problems

**4.49** Will a solution of dry ice (solid $CO_2$) in pure water ($H_2O$) conduct a current?

**4.50** Will a solution of lye (NaOH) in pure water ($H_2O$) conduct a current?

## Integration of Multiple Skills

One of the skills you will want to develop now and as you proceed through the course is the ability to combine things you learn in earlier chapters with the present one.

**Example 4.6:** How many grams are equal to 132 amu?

**Solution:** Recognize that you can solve this with dimensional analysis. You need to find some equality for amus and grams. One place you can locate this is in Table 2.5

$$132 \text{ amu} \times \frac{1.661 \times 10^{-24} \text{ g}}{1 \text{ amu}} = 2.19 \times 10^{-22} \text{ g}$$

A second multiple skill is the combination of isotopes and ions. These are summarized in Table 4.8.

## Practice Problems

**4.51** How many amus are equal to $1.42 \times 10^{-3}$ g?

**4.52-4.57** Try to fill in the following table using what you learned in Sections 4.3 and 4.8. The first two are completed as a demonstration.

Table 4.8 Subatomic Particles for Ions and Isotopes

| Problem | Ion Name | Symbol | No. of Protons | No. of Electrons | No. of Neutrons | Atomic Number | Mass Number |
|---------|----------|--------|----------------|------------------|-----------------|---------------|-------------|
|         | Fluoride anion | F$^-$ | 9 | 10 | 10 | 9 | 19 |
|         | Calcium cation | Ca$^{2+}$ | 20 | 18 | 20 | 20 | 40 |
| 4.52    | Hydrogen cation |  |  | 0 |  |  | 2 |
| 4.53    |  | Al$^{3+}$ cation |  |  |  |  | 27 |
| 4.54    | anion |  | 8 | 10 | 8 |  |  |
| 4.55    | Lithium cation |  |  | 2 |  |  | 7 |
| 4.56    | cation |  |  | 18 | 20 | 19 |  |
| 4.57    | anion |  | 7 | 10 |  |  | 15 |

The last skill you want to work on is one where you extend what you have learned in this chapter to make predictions about aspects of chemistry which you have not yet studied.

**Atoms, Molecules, and Ions** 65

You learned in this chapter that the overall charge for an ionic substance is neutral. If the compound is table salt (Na⁺ and Cl⁻ ions), then you can deduce that there must be one Na⁺ for each Cl⁻. At one to one, the positive charges cancel out the negative charges.

In this chapter you learned that elements in **groups** have similar properties. You have seen that chlorine forms the anion Cl⁻ in ionic compounds.

**Example 4.7:** Predict the ion that iodine forms.

   **Solution:** Since I and Cl are both in Group VIIA, you might predict that iodine is I⁻ in ionic compounds. This prediction is accurate.

---

## Practice Problems

**4.58** What is the ratio of Mg⁺² to Cl⁻ in an ionic compound?

**4.59** What is the ratio of Mg⁺² to O⁻² in an ionic compound?

**4.60** In ionic compounds, magnesium forms the cation, Mg²⁺. What ion would calcium form?

---

## Self-Exam

**I. (4 ea) Write T (true) or F (false) in each corresponding blank.**

_____ 1. Modern atomic theory is generally credited to Aristotle.

_____ 2. The nucleus of an atom contains one or more protons.

_____ 3. The isotope aluminum-27 has 27 neutrons.

_____ 4. A proton has approximately the same mass as a neutron.

_____ 5. An electron has a negative charge, a proton has a positive charge, and they are equal in mass.

_____ 6. Naturally occurring chlorine consists of ³⁵Cl and ³⁷Cl. The atomic weight of chlorine is 35.45; therefore, ³⁵Cl is most abundant.

_____ 7. The elements in a horizontal row of the periodic table are known as a **period**.

_____ 8. Bromine-81 has 35 protons and 46 electrons.

_____ 9. All molecules contain two or more different elements.

_____ 10. Nonmetals are located on the right-hand side of the periodic table.

**II. (4 ea) Multiple Choice. Place the letter of the BEST answer in the corresponding blank.**

_____ 1. An isotope has 12 electrons, 12 protons, and 10 neutrons is
   A. ¹²₁₂Mg   B. ²²₁₂Mg   C. ²²₁₂Ne   D. ¹²₁₂C   E. ¹²₁₀C

_____ 2. Which of the following compounds is ionic?
   A. CO₂   B. N₂   C. N₂H₄   D. NaOH   E. PCl₃

# 66   Chapter 4

_____ 3. To which group does silver belong?
    A. IB    B. IA    C. 5    D. 11B    E. 5B

_____ 4. An ion with 18 electrons and 16 protons is
    A. Ar    B. $Ar^{2+}$    C. $O^-$    D. $O^{2+}$    E. $O^{2-}$

_____ 5. The elemental identity of an atom can be positively made if you know the number of
    A. Protons    B. Electrons    C. Neutrons
    D. Either A or B    E. All subatomic particles

_____ 6. Which of the following atoms would form a cation in ionic compounds?
    A. C    B. S    C. P    D. Na    E. Ne

_____ 7. How many electrons are present in $N^{3-}$ ion?
    A. 2    B. 7    C. 9    D. 10    E. 3

_____ 8. Which of the following would have properties like calcium?
    A. K    B. Sc    C. Ba    D. Ne    E. Cs

_____ 9. A molecule has two hydrogen atoms, one sulfur atom and four oxygen atoms. Which of the following is the correct molecular formula?
    A. HSO    B. $H_2SO_4$    C. HOS    D. $H_2O_4S$    E. None

_____ 10. Which of the following is a polyatomic ion?
    A. $NO_2$    B. $O^{2-}$    C. $NO_3^-$    D. $Na^+$    E. $Cl_2$

## III. Provide Complete Answers

1. (5 pts) Calculate the atomic weight of copper if 69.2% is isotope copper-63 (62.9296 amu) and 30.8% is isotope copper-65 (64.9278 amu).

2. (5 pts) There are eight elements that exist as diatomic molecules. What are they?

3. (5 pts) The mass of an electron is $9.1094 \times 10^{-31}$ kg. How many amus are equal to one electron? (*Hint*: 1 kg = 1000 g; 1 amu = $1.661 \times 10^{-24}$ g)

4. (5 pts) What are the number of protons, neutrons, and electrons in $^{207}_{82}Pb^{+2}$?

## Answers to Practice Problems

| | | | | | |
|---|---|---|---|---|---|
| 4.1 | $6.02 \times 10^{23}$ | 4.2 | 1.0072765 times | 4.3 | 41.5 |
| 4.4 | 2:1 | 4.5 | $4.80 \times 10^{23}$ atoms. | 4.6 | Gold (Au) |
| 4.7 | 11 | 4.8 | Br (bromine), with atomic mass of 79 | | |
| 4.9 | 53 protons; I (iodine) | 4.10 | 17 (if it is Cl, it must have atomic number 17) | | |
| 4.11 | 5 protons; 6 neutrons | 4.12 | 80 protons; 110 neutrons | | |

## Atoms, Molecules, and Ions

| Problem | Element Name | Symbol | Atomic Number | Mass Number | No. of Protons | No. of Electrons | No. of Neutrons |
|---------|--------------|--------|---------------|-------------|----------------|------------------|-----------------|
| 4.13 | Magnesium | Mg | 12 | 25 | 12 | 12 | 13 |
| 4.14 | Cesium | Cs | 55 | 132 | 55 | 55 | 77 |
| 4.15 | Argon | Ar | 18 | 40 | 18 | 18 | 22 |
| 4.16 | Silver | Ag | 47 | 109 | 47 | 47 | 62 |
| 4.17 | Iron | Fe | 26 | 56 | 26 | 26 | 30 |
| 4.18 | Lithium | Li | 3 | 7 | 3 | 3 | 4 |
| 4.19 | Hydrogen | H | 1 | 3 | 1 | 1 | 2 |
| 4.20 | Bromine | Br | 35 | 81 | 35 | 35 | 46 |
| 4.21 | Barium | Ba | 56 | 130 | 56 | 56 | 74 |
| 4.22 | Lead | Pb | 82 | 207 | 82 | 82 | 125 |
| 4.23 | Nitrogen | N | 7 | 15 | 7 | 7 | 8 |

**4.24** Lithium-7

**4.25** 35.45 amu

**4.26** 1.008 amu

**4.27** Group VIA; Period 2

**4.28** Group IA; Period 3

**4.29** Group VIIIA; Period 1

**4.30** La: 57, Hf: 72

**4.31** Chromium, molybdenum, or rutherfordium

**4.32** Nitrogen, arsenic, antimony, or bismuth

**4.33** N

**4.34** Cobalt, 27, **atomic weight** = 58.3; nickel, 28, **atomic weight** of 58.71.

**4.35** 1: 2; 3: 8; 4: 18; 6: 32

**4.36** They are all nonmetals.

**4.37** Molecular

**4.38** Ionic

**4.39** $C_{12}H_{22}O_{11}$

**4.40** 2 atoms P, 10 atoms O

**4.41** $C_8H_8O_2$

**4.42** b, c, d, e

| Number | Ion | Protons | Electrons |
|--------|-----|---------|-----------|
| 4.43 | $Li^+$ | 3 | 2 |
| 4.44 | $Al^{3+}$ | 13 | 10 |
| 4.45 | $N^{3-}$ | 7 | 10 |
| 4.46 | $Se^{2-}$ | 34 | 36 |
| 4.47 | $F^-$ | 9 | 10 |
| 4.48 | $Ba^{2+}$ | 56 | 54 |

**4.49** No

**4.50** Yes

**4.51** $8.55 \times 10^{20}$ amu

| Problem | Ion Name | Symbol | No. of Protons | No. of Electrons | No. of Neutrons | Atomic Number | Mass Number |
|---------|----------|--------|----------------|------------------|-----------------|---------------|-------------|
| 4.52 | Hydrogen cation | H$^+$ | 1 | 0 | 1 | 1 | 2 |
| 4.53 | Aluminum cation | Al$^{3+}$ | 13 | 10 | 14 | 13 | 27 |
| 4.54 | Oxide anion | O$^{2-}$ | 8 | 10 | 8 | 8 | 16 |
| 4.55 | Lithium cation | Li$^+$ | 3 | 2 | 4 | 3 | 7 |
| 4.56 | Potassium cation | K$^+$ | 19 | 18 | 20 | 19 | 39 |
| 4.57 | Nitride anion | N$^{3-}$ | 7 | 10 | 8 | 7 | 15 |

**4.58** Mg$^{+2}$: Cl$^-$ is 1:2.   **4.59** Mg$^{+2}$: O$^{2-}$ is 1:1.   **4.60** Ca$^{+2}$

## Answers to Self-Exam

**I. T/F**
 1. F   2. T   3. F   4. T   5. F   6. T   7. T   8. F   9. F   10. T

**II. MC**
 1. B   2. D   3. A   4. E   5. A   6. D   7. D   8. C   9. B   10. C

**III. Complete Answers**
 1. 63.5 amu

 2. H$_2$, N$_2$, O$_2$, F$_2$, Cl$_2$, Br$_2$, I$_2$, At$_2$

 3. 5.484 × 10$^{-4}$ amu

 4. 82 protons, 125 neutrons, 80 electrons

# 5
# Chemical Formulas and Names

The skills you need to master in this chapter can be easily summarized. You will need to learn to write and interpret chemical formulas.
    Keys for success in this chapter:

- Learn rules for naming each category of compounds.
- Memorize the basics for nomenclature.
- Use knowledge of periodic table and basic rules of nomenclature to find **patterns** in the naming systems, which will help reduce the amount of memorization needed.

## Summary of Verbal Knowledge

**binary ionic compound:** A compound composed of ions from only two elements.
**oxyanion:** This is a negatively charged polyatomic ion that contains an atom of some element plus one or more oxygen atoms.
**binary molecular compound:** A molecular compound made up of only two elements.
**acid:** This is a compound that produces the hydrogen ion, $H^+$.
**binary acid:** This is a compound that produces a hydrogen ion and contains only hydrogen and one other element.
**oxyacid:** This is a compound that produces a hydrogen ion and contains oxygen and one other nonmetal element.

## Review of Mathematical and Calculator Skills

This chapter is largely dedicated to learning the *language* of chemical formulas. This chapter contains very little numerical manipulation.
    The only math you need is the ability to find positive and negative numbers whose sum equals zero. This concept will be applied when necessary throughout the chapter.

## Application of Skills and Concepts

### 5.1 Formulas for Binary Ionic Compounds

It is very important to understand the meaning of *binary* used in this chapter.

> Binary compounds have **two** different elements, **not** two atoms.

# Chapter 5

There are two major skills you need to master in Section 5.1.

- Learn to predict the charge an element will have when forming ionic compounds.
- Learn to write formulas for ionic compounds.

From Chapter 4 you know that binary ionic compounds consist of cations (positively charged ions) and anions (negatively charged ions) in a vast array, with the overall charge of zero. You also learned about some trends of elements in the periodic table.

- Metals tend to form cations (left-hand side of periodic table).
- Nonmetals tend to form anions (right-hand side of periodic table).

Group A elements *usually* form very predictable ions. There are some simple rules and examples that demonstrate these trends. Group B metals and transition metals *usually* form more than one type of cation, so they are not predictable. If you want to know the charges these metals form, you need to look them up. Take time to examine Figure 5.3 in your textbook. Metals in green form only one cation. Those in orange can form more than one type of cation.

- Group A metals form **cations** with a charge that matches the **Roman numeral** of their group.
    Examples:
    Sodium is in group IA, so it forms **Na⁺**.
    Aluminum is in group IIIA, so it forms **Al³⁺**.

- Group A nonmetals form anions with a charge that matches **8 minus** the **Roman Numeral** of their group.
    Examples:
    Chlorine is in group VIIA, so it forms **8 - 7 = 1**, or **Cl⁻**.
    Phosphorus is in group VA, so it forms **8 - 5 = 3**, or **P³⁻**.

Before trying to write formulas for binary compounds, practice identifying metals that form more than one cation. Also practice predicting the ionic charge on some Group A elements.

## Practice Problems

**5.1** Which of the following form only one cation: K, W, In, Al, Ba?

**5.2** Which of the following form more than one cation: Li, U, Cr, Ga, Mn?

**5.3** Which two Group B elements are exceptional in that they **only** form 2+ cations?

**5.4** Which four Group A elements are exceptional in that they form more than one cation?

**5.5** Use a periodic table to predict the cation of potassium in ionic compounds.

**5.6** Use a periodic table to predict the anion of nitrogen in ionic compounds.

**5.7** Use a periodic table to predict the cation of calcium in ionic compounds.

**5.8** Use a periodic table to predict the cation of indium in ionic compounds.

**5.9** Use a periodic table to predict the anion of sulfur in ionic compounds.

**5.10** Use a periodic table to predict the anion of iodine in ionic compounds.

# Chemical Formulas and Names

Now you have all the tools you need to write ionic formulas. Remember that the overall charge must be neutral. If the size of the *cation* is the same as the size of the *anion*, then the binary formula is simply 1:1. Some illustrations follow.

> Remember: Write the **metal** first, followed by the **nonmetal**.

**Example 5.1:** What is the formula for a compound of potassium and bromine?

  **Solution:** Potassium is $K^+$ and bromine is $Br^-$, so the compound is neutral when there is one cation and one anion. The metal is written first; thus the formula is **KBr**.

**Example 5.2:** What is the formula for a compound of barium and oxygen?

  **Solution:** Barium is $Ba^{2+}$ and oxygen is $O^{2-}$, so the compound is neutral when there is one cation and one anion. Thus the formula is **BaO**.

**Example 5.3:** What is the formula for a compound of gallium and phosphorus?

  **Solution:** Gallium is $Ga^{3+}$ and phosphorus is $P^{3-}$, so the compound is neutral when there is one cation and one anion. Thus the formula is **GaP**.

In cases where the sizes of the cation and anion are unequal, the charges must be neutralized by having a different number of each. Consider the following examples.

**Example 5.4:** What is the formula for a compound of magnesium and chlorine?

  **Solution:** Since magnesium is $Mg^{2+}$ and chlorine is $Cl^-$, there cannot be one cation to one anion. To determine the number of each, you can think of moving the magnitude of each charge to the opposite ion as shown below.

$$Mg^{2+} \, Cl^{-} \longrightarrow MgCl_2$$

  The number of each type of atom is written as a subscript. If the subscript is 1, it is omitted. Thus the formula is **MgCl$_2$**.

**Example 5.5:** What is the formula for a compound of magnesium and nitrogen?

  **Solution:** This can be thought of in the same way as the example for $MgCl_2$. Each of the charge sizes is moved to the opposite atom as a subscript. There is a reason that this works. Since magnesium is $Mg^{2+}$ and nitrogen is $N^{3-}$, there needs to be 3 magnesium atoms ($3 \times +2 = +6$), and 2 nitrogen atoms ($2 \times -3 = -6$) to have an overall charge of 0 ($+6 -6 = 0$). Thus the formula is **Mg$_3$N$_2$**.

## Practice Problems

**5.11** Write formulas for sodium with chlorine, oxygen, and nitrogen.

**5.12** Write formulas for calcium with bromine, sulfur, and nitrogen.

**5.13** Write formulas for aluminum with fluorine, selenium, and phosphorus.

**5.14** Write formulas for sulfur with $Fe^{2+}$ and $Fe^{3+}$.

**5.15** Write formulas for chlorine with $Cu^+$ and $Cu^{2+}$.

## 5.2 Naming Binary Ionic Compounds When the Metal Forms a Single Cation

Binary ionic compounds are the easiest to name. Two simple rules will help.

- Name the metal.
- Name the nonmetal *root* and use the *suffix*, **ide**.

Using these rules:
  NaCl is sodium chlor**ide**.
  $CaBr_2$ is calcium brom**ide**.
  $Al_2O_3$ is aluminum ox**ide**.

## Practice Problems

**5.16-5.21** Try to apply this concept by filling in the blanks in Table 5.1.

Table 5.1 Binary Ionic Formulas and Names

| Number | Name | Formula |
|--------|------|---------|
| 5.16 | Sodium fluoride | |
| 5.17 | | $Na_2S$ |
| 5.18 | Calcium sulfide | |
| 5.19 | | $Li_3N$ |
| 5.20 | Aluminum oxide | |
| 5.21 | | $AlI_3$ |

## 5.3 Naming Binary Ionic Compounds When the Metal Forms Several Cations

The rules for naming these compounds expand on those in Section 5.2. The nonmetal is named in the same way. A new rule stems from the fact that the charge on the metal must be denoted in the name. There are two systems for doing this.

- The stock system
    Use the name of the metal.
    Use a Roman numeral to denote the charge on the cation.

- The classic system
    Use the *suffix* **-ic** for the higher charge.
    Use the *suffix* **-ous** for the lower charge.
    Use the Latin root for the metal.

Iron can form two cations, $Fe^{2+}$ and $Fe^{3+}$. If it forms compounds with chlorine, these would result in $FeCl_2$ and $FeCl_3$, respectively. Under the Stock system, the Roman numerals **II** and **III** would be used for $Fe^{2+}$ and $Fe^{3+}$, respectively. Under the classic system, the suffixes **-ous** and **-ic** would be used for $Fe^{2+}$ and $Fe^{3+}$, respectively. These names are demonstrated in the first two entries of Table 5.2.

Often you will get a formula for binary compound and you will have to deduce the charge it has. An example follows.

**Example 5.6:** What is the name of $Cr_2O_3$?

> **Solution:** You know the name of the metal is *chromium*. You know the nonmetal will be *oxide*. You can deduce the charge on chromium to be +3 by reversing the process for writing subscripts.
>
> $$\underset{\underset{\text{No. of Cr atoms}}{\uparrow}}{2} \times (\overset{\overset{\text{Charge on Cr}}{\downarrow}}{+3}) + \underset{\underset{\text{No. of O atoms}}{\uparrow}}{3} \times (\overset{\overset{\text{Charge on O}}{\downarrow}}{-2}) = 0$$
>
> Since the charge is +3, the Roman numeral is **III**. The name is **chromium(III) oxide**.

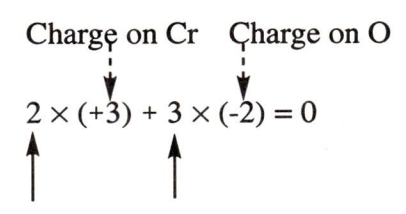

Finally, you will need to be able to write a formula from a name, as the next example shows.

**Example 5.7:** What is the formula for lead(II) bromide?

> **Solution:** The name tells you that the metal is *lead*, the nonmetal is *bromine*, and the charge on the metal is +2. The bromine has a -1 charge (Group VIIA); therefore, the formula is **$PbBr_2$**.

## Chapter 5

> **Practice Problems**
> 5.22-5.28 Table 5.2 has some completed examples and some problems for you to complete.

Table 5.2 Binary Ionic Nomenclature When the Metal Forms Multiple Charges

| Problem | Formula | Stock Name | Classic Name |
|---|---|---|---|
| Example | FeCl$_2$ | Iron(II) chloride | Ferrous chloride |
| Example | FeCl$_3$ | Iron(III) chloride | Ferric chloride |
| 5.22 | | Iron(III) sulfide | |
| 5.23 | | | Ferrous oxide |
| 5.24 | SnCl$_2$ | | Stannous chloride |
| 5.25 | | | Stannic chloride |
| 5.26 | | Copper(I) bromide | |
| 5.27 | | | Cupric oxide |
| 5.28 | CoCl$_2$ | | |

### 5.4 Compounds with Polyatomic Ions

Remember the definition of a polyatomic ion from Chapter 4. The atoms in this group are chemically bonded together, and the overall group has a charge. There are many of these that you need to know, some of which are noted in Table 5.3.

Examine the table closely. Many of these formulas and names you will have to memorize. However, you will be able to reduce your memory work if you can find patterns in the formulas and names assigned to them. Let's look at some examples to demonstrate how recognition of patterns will help you commit these formulas to memory.

**Example 5.8:** Look at the sulfite and sulfate ions. How are they alike? How do they differ?

> **Solution:** The answer to the first question is that they both contain *sulfur* and *oxygen*. Thus the root word *sulf* is in the name of both ions. Both ions also have the same charge, -2. They only differ by the number of oxygen atoms they contain; the *sulfite* has **3**, the *sulfate* has **4**.

Now, without looking at Table 5.3, predict the formulas for the nitrite and nitrate ions. If you tried to make a direct correlation to the sulfite/sulfate ions you might have predicted that nitrite and nitrate were NO$_4^{2-}$ and NO$_3^{2-}$, respectively. Now check your prediction in Table 5.3. You will notice that *nitrite* is **NO$_2^-$** and *nitrate* is **NO$_3^-$**. While the parallels to the *sulfite* and *sulfate* ions are not total, there are some worth noting.

- The oxyanion with fewer oxygens has the suffix **-ous**, and the greater number of oxygens has the suffix **-ic.**
- The oxyanions with the same root have the same charge. Both sulfite and sulfate have a charge of -2; both nitrite and nitrate have a charge of -1.

If you memorize a few of the polyatomic ions, you can deduce many of the others by noting the patterns that are incorporated into their nomenclature.

Table 5.3 Names of Some Common Polyatomic Ions

| Ion Formula | Ion Name | Ion Formula | Ion Name |
|---|---|---|---|
| $NO_2^-$ | Nitrite | $HSO_4^-$ | Hydrogen sulfate (bisulfate) |
| $NO_3^-$ | Nitrate | $PO_4^{3-}$ | Phosphate |
| $SO_3^{2-}$ | Sulfite | $HPO_4^{2-}$ | Hydrogen phosphate |
| $SO_4^{2-}$ | Sulfate | $H_2PO_4^-$ | Dihydrogen phosphate |
| $CO_3^{2-}$ | Carbonate | $NH_4^+$ | Ammonium |
| $HCO_3^-$ | Hydrogen carbonate (bicarbonate) | $OH^-$ | Hydroxide |
| $ClO^-$ | Hypochlorite | $CN^-$ | Cyanide |
| $ClO_2^-$ | Chlorite | $MnO_4^-$ | Permanganate |
| $ClO_3^-$ | Chlorate | $C_2H_3O_2^-$ | Acetate |
| $ClO_4^-$ | Perchlorate | $HSO_3^-$ | Hydrogen sulfite (bisulfite) |

There are several things that you should notice before trying to write formulas.

1. Most polyatomic ions in Table 5.3 are **anions**; the exception is **ammonium ion.**
2. Most *polyatomic* ions have **-ite** or **-ate** for a suffix, while *monatomic* anions have **-ide**.
    There are three notable exceptions.
    Hydroxide has suffix **-ide.**
    Cyanide has suffix **-ide.**
    Ammonium is unique from all others.
3. There are four oxyanions with chlorine (also with bromine and iodine). These require the use of two prefixes, **hypo-** for one less oxygen than the **chlorite** ion and **per-** for one more oxygen than the **chlorate** ion.

Once you know the name of the polyatomic ion, it is easy to name compounds that use these groups. Follow these simple rules:

1. Previous rules for metals still apply. Use Roman numerals in the same cases you did for binary ionic compounds.

2. When naming compounds with a polyatomic ion, just name the group.

    $NaNO_2$ is **sodium nitrite**.

3. Treat the polyatomic ion as a group, and then make sure the total charges equal zero.

    Lithium hydroxide contains $Li^+$ ions and $OH^-$ ions. The compound is **LiOH**.

4. When using more than one polyatomic group, use parenthesis in the formula.

A compound made with calcium and phosphate would have $Ca^{2+}$ ions and $PO_4^{3-}$ ions. Therefore, it would need to contain "3 calcium cations" and "2 phosphate anions" to be neutrally charged. The compound should be written **$Ca_3(PO_4)_2$**.

> ## Practice Problems
>
> **5.29** Use information in Table 5.3 to predict the formulas for hypobromite ion, bromite ion, bromate ion, and perbromate ion.
>
> **5.30** Use information in Table 5.3 to name the following ions: $IO^-$; $IO_2^-$; $IO_3^-$; $IO_4^-$.
>
> **5.31-5.36** Try to fill the missing blanks in Table 5.4 as a way of practicing the nomenclature rules for compounds with polyatomic ions.

Table 5.4 Some Compounds with Polyatomic Ions

| Problem | Formula | Name | Formula | Name |
|---|---|---|---|---|
| 5.31 | $MgCO_3$ |  | $NH_4Cl$ |  |
| 5.32 |  | Sodium hydrogen sulfate |  | Ammonium hydroxide |
| 5.33 | $Al(OH)_3$ |  | $Ca(NO_3)_2$ |  |
| 5.34 |  | Potassium chlorite |  | Barium chlorate |
| 5.35 | $FeSO_4$ |  | $Cu(NO_3)_2$ |  |
| 5.36 |  | Tin(IV) carbonate |  | Copper(I) phosphate |

## 5.5 Binary Molecular Compounds

The major way you will recognize these types of compounds will be to notice that they are made up of two different nonmetals. These are **not ionic**, but **molecular**. There are some rules that will make it easy to write and name these types of compounds.

- The most metallic element is written first.
    Elements in Group IVA are more metallic than Group VA, etc.
    Elements are more metallic going down the periodic table. P is more metallic than N.
- The second element is written with the suffix, **-ide**.
- Use subscripts to denote the number of each type of element in the compound.
- Use Greek prefixes to denote the number of each type of element (see Table 5.5).
    Omit the prefix *mono-* when it applies to the first element in the compound.

Table 5.5 Greek Prefixes

| Prefix | Number | Prefix | Number |
|---|---|---|---|
| mono | 1 | hexa | 6 |
| di | 2 | hepta | 7 |
| tri | 3 | octa | 8 |
| tetra | 4 | nona | 9 |
| penta | 5 | deca | 10 |

Look at two sample problems to illustrate these rules.

**Example 5.9:** Write the formula for a molecular binary compound containing two oxygens and one carbon. Name the compound.

> **Solution:** Since the carbon is more *metallic* than the oxygen, it is written first. The compound is **CO$_2$**.
> Using prefixes, the name is **carbon dioxide.** (Notice that the *mono-* was omitted from the first element.)

**Example 5.10:** Write the formula for a molecular binary compound containing one oxygen and one nitrogen. Name the compound.

> **Solution:** It has the formula **NO**. The compound is called **nitrogen monoxide.**
> Notice that it is **not** "nitrogen monooxide." When two consecutive vowels appear, one is omitted. Thus the name for N$_2$O$_4$ is dinitrogen tetroxide (not "dinitrogen tetraoxide").

---

## Practice Problems

**5.37** What is the correct name for N$_2$O?

**5.38** What is the correct name for N$_2$O$_4$?

**5.39** What is the correct name for NO$_2$?

**5.40** What is the formula for sulfur dioxide?

**5.41** What is the formula for carbon tetrachloride?

**5.42** What is the formula for tetraphosphorus decoxide?

---

### 5.6 Naming Binary Acids

Notice that hydrogen is a nonmetal in Group IA of the periodic table. Hydrogen tends to form +1 cations just like all the metals in Group IA, but the special properties of the H$^+$ cation make it practical to provide a separate category for nomenclature.

A binary acid would be a molecular compound with hydrogen being one element and a nonmetal being the second element. For example, water, H$_2$O, would fit this category. This is so commonly known as water that we will not include it in this category.

Acids with nonmetals in Group VIIA are most notable here. For example, HCl is a molecular compound that is a gas under normal conditions. It would be called *hydrogen monochloride* under the rules in Section 5.5. When you use this compound in the laboratory, it usually is dissolved in water, which means it exists as ions of H$^+$ and Cl$^-$. It is called *hydrochloric acid* in this case. For naming binary acids use the following simple rules:

- Use the *prefix* **hydro-**.
- Use the root of the nonmetal, followed by the *suffix* **-ic**.
- Use the word **acid** following the name.

---

## Practice Problems

**5.43** Name HF.

**5.44** Write the formula for hydroiodic acid.

## 5.7 Naming Oxyacids

Many acidic compounds consist of hydrogen, some other nonmetal, and one or more oxygen atoms. In this case, the hydrogen has a +1 charge, and the nonmetal and oxygen(s) are a polyatomic anion. An example is *sulfuric acid,* which has the formula **$H_2SO_4$**. Some of the following rules may help you determine how this name was assigned. These rules are illustrated in Table 5.6.

- Use the root name of the *nonmetal*.
- Use the *suffix* that corresponds to the *suffix* of the polyatomic anion name (see Table 5.6).
    If the anion *suffix* is **-ate,** the acid *suffix* is **-ic.**
    If the anion *suffix* is **-ite,** the acid *suffix* is **-ous.**
- Use the *prefix* that corresponds to the *prefix* of the polyatomic anion name.
    If the anion *prefix* is **hypo-,** the acid *prefix* is **hypo-.**
    If the anion *prefix* is **per-,** the acid *prefix* is **per-.**
- Use the word **acid** following the name.

Table 5.6 Anions in Ionic Compounds and Corresponding Acids

| Ionic Name | Ionic Formula | Acid Name | Acid Formula |
|---|---|---|---|
| Calcium brom**ite** | $Ca(BrO_2)_2$ | Brom**ous** acid | $HBrO_2$ |
| Sodium **hypo**chlor**ite** | NaClO | **Hypo**chlor**ous** acid | HClO |
| Magnesium nitr**ite** | $Mg(NO_2)_2$ | Nitr**ous** acid | $HNO_2$ |
| Potassium **per**iod**ate** | $KIO_4$ | **Per**iod**ic** acid | $HIO_4$ |
| Barium acet**ate** | $Ba(C_2H_3O_2)_2$ | Acet**ic** acid | $HC_2H_3O_2$ |
| Lithium sulf**ate** | $Li_2SO_4$ | Sulfur**ic** acid | $H_2SO_4$ |
| Sodium io**dide** | NaI | **Hydro**iod**ic** acid | HI |

## Practice Problems

**5.45** Write the formula for sulfurous acid.

**5.46** Write the formula for chloric acid.

**5.47** What is the name of $HNO_3$?

**5.48** What is the name of HClO?

## Integration of Multiple Skills

Each type of compound has been addressed separately throughout this chapter. Now you need to find a way to write a formula or provide a name when the category is not isolated. The following scheme should help you approach nomenclature with a quick system for making sure the name is correct.

When trying to determine the name of a formula, determine which major category (**ionic, molecular, or acid**) it belongs to, and then look at the subcategory, when necessary.

# Chemical Formulas and Names

**Figure 5.1** Summary of Nomenclature

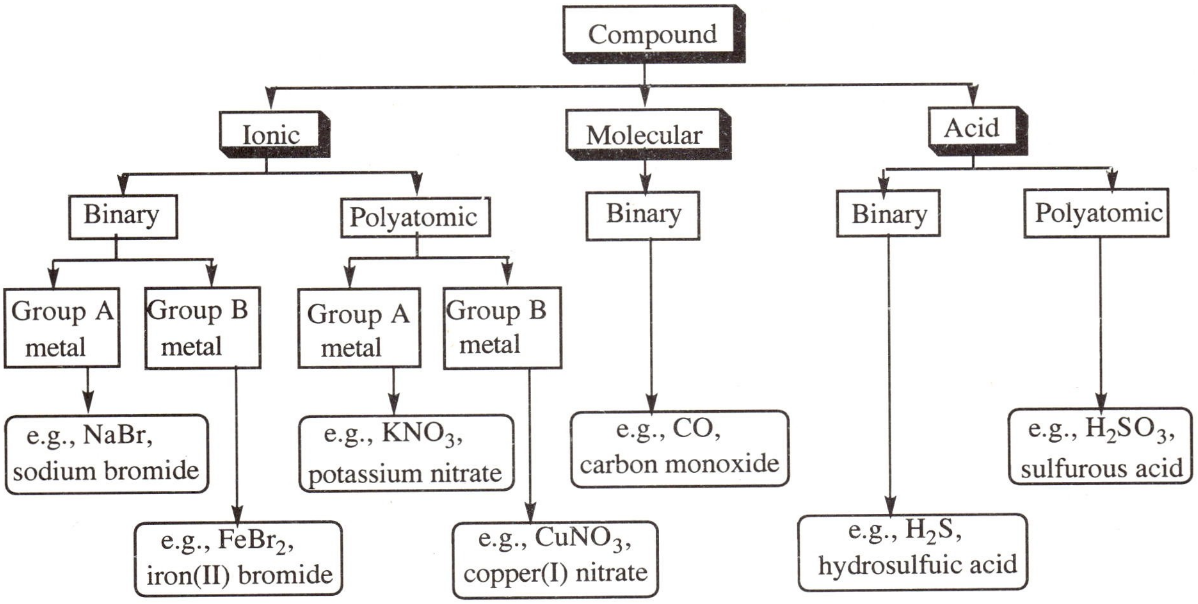

```
                          Compound
            ┌────────────────┼────────────────┐
          Ionic          Molecular           Acid
        ┌───┴───┐            │           ┌────┴────┐
      Binary  Polyatomic   Binary      Binary   Polyatomic
      ┌─┴─┐    ┌─┴─┐         │           │          │
   Group Group Group Group   │           │          │
     A    B    A    B        │           │          │
   metal metal metal metal   │           │          │
     │    │    │    │        │           │          │
   e.g., e.g., e.g., e.g.,  e.g.,       e.g.,     e.g.,
   NaBr, FeBr₂, KNO₃, CuNO₃, CO,         H₂S,      H₂SO₃,
  sodium iron(II) potassium copper(I) carbon    hydrosulfuic sulfurous
  bromide bromide nitrate  nitrate   monoxide    acid        acid
```

## Practice Problems

**5.49-5.66** Using the summary in Figure 5.1, try to fill in the blanks in Table 5.7. This table has many different types of compounds, and you must decide which nomenclature rules to apply, if you are able to see patterns.

**Table 5.7** Names and Formulas of Some Compounds

| Problem | Chemical Name | Chemical Formula | Common Name Usage |
|---------|---------------|------------------|-------------------|
| 5.49 | Sodium hydrogen sulfate | | Sani-flush |
| 5.50 | | MgSO₄ | Epson salts |
| 5.51 | Magnesium hydroxide | | Milk of magnesia |
| 5.52 | | CS₂ | Toxic organic solvent |
| 5.53 | Ammonium hydroxide | | Household cleaner |
| 5.54 | | CaCl₂ | Deicer/wetting agent |
| 5.55 | Sodium chlorate | | Weed killer/charcoal starter |
| 5.56 | | HCl | Stomach acid |
| 5.57 | Tin(II) fluoride (stannous fluoride) | | Toothpaste additive |
| 5.58 | | P₂O₅ | Fertilizer additive |
| 5.59 | Acetic Acid | | Vinegar |
| 5.60 | | NaC₂H₃O₂ | Soluble salt |
| 5.61 | Calcium oxide | | Quicklime |
| 5.62 | | NaNO₂ | Meat preservative |
| 5.63 | Silicon dioxide | | Sand |

80   Chapter 5

| 5.64 |                   | Ca(HCO$_3$)$_2$ | Hard water deposit    |
| 5.65 | diarsenic trioxide |                 | Toxic pesticide       |
| 5.66 |                   | SiC             | Carborundum abrasive  |

# Self-Exam

**I. (4 ea) Write T (true) or F (false) in each corresponding blank.**

_____1. All binary compounds have only two atoms (total).

_____2. Ionic compounds tend to have a metal and a nonmetal.

_____3. SO$_3$ is identical to SO$_3^{2-}$.

_____4. An acid is a compound that forms OH$^-$ ions when dissolved in water.

_____5. Nonmetals tend to form anions in ionic compounds.

_____6. Greek prefixes should be used **only** for naming binary molecular compounds.

_____7. Nickel always forms cations with a +1 charge.

_____8. The compounds in Group VIIIA do not form anions or cations.

_____9. The overall charge of an ionic formula unit is zero.

_____10. The overall charge of a polyatomic anion is negative.

**II. (4 ea) Multiple Choice. Place the letter of the BEST answer in the corresponding blank.**

_____1. What is the name of NaF?
  A. Sodium fluoride      B. Natrium monofluoride    C. Sodium(I) fluoride
  D. Narium fluorine      E. Sodium fluorite

_____2. What is the formula for calcium bromide?
  A. CaBr    B. CAB    C. Ca$_2$Br    D. CaBr$_2$    E. Br$_2$Ca

_____3. What is the name of Fe$_2$S$_3$?
  A. Ferrous sulfate      B. Iron(II) sulfide        C. Ferric sulfite
  D. Iron(III) sulfide    E. Iron(III) sulfate

_____4. What is the formula for lead(IV) acetate?
  A. Pb(C$_2$H$_3$O$_2$)$_4$   B. Pb(C$_2$H$_3$O$_2$)$_2$   C. Pb$_2$Ar   D. Ld(C$_2$H$_3$O$_2$)$_2$   E. PbAr

_____5. What is the name of CS$_2$?
  A. Carbon sulfate       B. Carbon(IV) sulfide      C. Carbon sulfite
  D. Carbon disulfide     E. Carbon sulfide

_____6. What is the formula for bromic acid?
  A. HBr    B. HBrO    C. HBrO$_2$    D. HBrO$_3$    E. HBrO$_4$

_____7. What is the name of Na$_2$SO$_3$?
  A. Sodium sulfate       B. Sodium sulfide          C. Sodium sulfite
  D. Sodium monosulfide   E. Sodium monosulfurtrioxide

# Chemical Formulas and Names 81

_____ 8. What is the formula for copper(II) chlorate?
  A. Cu(ClO)₃   B. CuClO₃   C. Co(ClO₃)₂   D. CoClO₂   E. Cu(ClO₃)₂

_____ 9. Which of the following is a binary molecular compound?
  A. Ferrous sulfide   B. Iron(II) sulfide   C. Carbon disulfide
  D. Bromic acid   E. Ammonium chloride

_____ 10. What is the formula for barium nitride?
  A. BaN   B. Ba₃N   C. Ba₃N₂   D. Ba(NO₃)₂   E. BaNO₂

### III. Provide Complete Answers

1. (5 pts) Write the following in order from most metallic to least metallic (use a periodic table): boron, bromine, carbon, chlorine, fluorine, iodine, nitrogen, oxygen, phosphorus, silicon, sulfur.

2. (5 pts) Write the names of the following acids: HCl, HClO, HClO₂, HClO₃, HClO₄.

3. (10 pts) Provide a formula and name for ionic compounds of the acids in problem 3 with a magnesium cation.

## Answers to Practice Problems

| | | | | | |
|---|---|---|---|---|---|
| 5.1 | K, Al, In, Ba | 5.2 | U, Cr, Mn | 5.3 | Zn, Cd |
| 5.4 | Tl, Sn, Pb, Bi | 5.5 | K⁺ | 5.6 | N³⁻ |
| 5.7 | Ca²⁺ | 5.8 | In³⁺ | 5.9 | S²⁻ |
| 5.10 | I⁻ | 5.11 | NaCl, Na₂O, Na₃N | | |
| 5.12 | CaBr₂, CaS, Ca₃N₂ | 5.13 | AlF₃, Al₂Se₃, AlP | | |
| 5.14 | FeS, Fe₂S₃ | 5.15 | CuCl, CuCl₂ | 5.16 | NaF |
| 5.17 | Sodium sulfide | 5.18 | CaS | 5.19 | Lithium nitride |
| 5.20 | Al₂O₃ | 5.21 | Aluminum iodide | | |

| Problem | Formula | Stock Name | Classic Name |
|---|---|---|---|
| 5.22 | Fe₂S₃ | Iron(III) sulfide | Ferric sulfide |
| 5.23 | FeO | Iron(II) oxide | Ferrous oxide |
| 5.24 | SnCl₂ | Tin(II) chloride | Stannous chloride |
| 5.25 | SnCl₄ | Tin(IV) chloride | Stannic chloride |
| 5.26 | CuBr | Copper(I) bromide | Cuprous bromide |
| 5.27 | CuO | Copper(II) oxide | Cupric oxide |
| 5.28 | CoCl₂ | Cobalt(II) chloride | Cobaltous chloride |

5.29   BrO⁻, BrO₂⁻, BrO₃⁻, BrO₄⁻

5.30   Hypoiodite ion, iodite ion, iodate ion, periodate ion

5.31 Magnesium carbonate, ammonium chloride
5.32 NaHSO$_4$, NH$_4$OH
5.33 Aluminum hydroxide, calcium nitrate
5.34 KClO$_2$, Ba(ClO$_3$)$_2$
5.35 Iron(II) sulfate, copper(II) nitrate
5.36 Sn(CO$_3$)$_2$, Cu$_3$PO$_4$
5.37 Dinitrogen monoxide
5.38 Dinitrogen tetroxide
5.39 Nitrogen dioxide
5.40 SO$_2$
5.41 CCl$_4$
5.42 P$_4$O$_{10}$
5.43 Hydrofluoric acid
5.44 HI
5.45 H$_2$SO$_3$
5.46 HClO$_3$
5.47 Nitric acid
5.48 Hypochlorous acid
5.49 NaHSO$_4$
5.50 Magnesium sulfate
5.51 Mg(OH)$_2$
5.52 Carbon disulfide
5.53 NH$_4$OH
5.54 Calcium chloride
5.55 NaClO$_3$
5.56 Hydrochloric acid
5.57 SnF$_2$
5.58 Diphosphorus pentoxide
5.59 HC$_2$H$_3$O$_2$
5.60 Sodium acetate
5.61 CaO
5.62 Sodium nitrite
5.63 SiO$_2$
5.64 Calcium hydrogen carbonate
5.65 As$_2$O$_3$
5.66 Silicon carbide

## Answers to Self-Exam

**I. T/F**
1. F  2. T  3. F  4. F  5. T  6. T  7. F  8. T  9. T  10. T

**II. MC**
1. A  2. D  3. D  4. A  5. D  6. D  7. C  8. E  9. C  10. C

**III. Complete Answers**
1. B, Si, C, P, N, S, O, I, Br, Cl, F

2. HCl: hydrochloric acid
   HClO: hypochlorous acid
   HClO$_2$: chlorous acid
   HClO$_3$: chloric acid
   HClO$_4$: perchloric acid

3. MgCl$_2$: magnesium chloride
   Mg(ClO)$_2$: magnesium hypochlorite
   Mg(ClO$_2$)$_2$: magnesium chlorite
   Mg(ClO$_3$)$_2$: magnesium chlorate
   Mg(ClO$_4$)$_2$: magnesium perchlorate

# 6
# Chemical Reactions and Equations

This chapter introduces the *language* of *chemical reactions*. You will learn to write chemical reactions, apply the law of conservation of mass, and recognize types of reactions.

Keys for success in this chapter:

- Learn the symbolism for describing a chemical reaction.
- Practice writing reactions using the correct symbolism.
- Study the categories for types of chemical reactions.
- Master the art of balancing equations by practicing.
  Spend 60% of your study time on this last category.

## Summary of Verbal Knowledge

**reactants:** Chemical substance(s) involved at the start of a chemical reaction.
**products:** Chemical substance(s) that result(s) from a chemical reaction.
**chemical equation:** This is a symbolic way of expressing a chemical reaction.
**coefficients:** The numbers in front of the formulas in the chemical equation.
**catalyst:** A substance that it is not changed by the reaction but causes it to speed up.
**balanced chemical equation:** This is a symbolic representation of a chemical reaction where the number of atoms of each element on the left side of the arrow IS equal to the number of atoms of each element on the right side.
**decomposition reaction:** One in which a single compound breaks up into two or more other substances.
**combination reaction:** One in which two substances chemically combine to form a third.
**single-replacement reaction:** One in which one element reacts by replacing another element in a compound.
**activity series:** A list of metallic elements (plus hydrogen) ordered by their relative activities in single-replacement reactions.
**double-replacement reaction:** One in which two compounds exchange parts to form two new compounds. These are also known as *metathesis reactions*.
**precipitate:** This is a solid that will not dissolve appreciably in a liquid such as water.
**insoluble:** The inability to dissolve.
**soluble:** The ability to dissolve.
**spectator ions:** These are ions that remain unchanged in a double-replacement reaction.
**precipitation reaction:** A double-replacement reaction that results in an insoluble solid formation.
**acid:** A compound that produces hydrogen ions ($H^+$) when dissolved in water.
**base:** A compound that produces hydroxide ions ($OH^-$) when dissolved in water.
**neutralization reaction:** A double-replacement reaction where a base reacts with an acid to produce water.

**salt:** This is an ionic compound formed during a neutralization reaction.
**combustion reaction:** One in which a substance reacts with either pure oxygen or oxygen in the air with the rapid release of heat and the appearance of a flame.

## Review of Mathematical and Calculator Skills

You need to review the use of fractions. For example, $\frac{7}{2} \times 2 = 7$.

This will be used when multiplying a coefficient times a subscript in a chemical formula. The following example illustrates this point.

**Example 6.1:** How many oxygen atoms are present in $\frac{7}{2}$ oxygen molecules?

> **Solution:** You always multiply the number in the numerator (in this case, 7) of the fraction times each subscript (in this case, 2) and divide by the number in the denominator (in this case, 2).
>
> $\frac{7}{2} O_2$ has **7 O atoms** $(7 \times 2 \div 2 = 7)$

A second example illustrates the distribution of a coefficient throughout all subscripts in a chemical formula.

**Example 6.2:** How many atoms of each type of element are present in 8 molecules of $CO_2$?

> **Solution:** $8CO_2$ has: 8 carbon atoms $(8 \times 1)$
> 16 oxygen atoms $(8 \times 2)$.

Another time you will need to apply this distribution is in a case where there is more than one polyatomic ion group in a formula. This is best illustrated by the following example.

**Example 6.3:** How many atoms of each type of element are in a molecule of calcium phosphate?

> **Solution:** The first step to solving this problem requires the correct formula (Chapter 5).
>
> $Ca_3(PO_4)_2$ has: 3 calcium atoms (3)
> 2 phosphorus atoms $(2 \times 1)$
> 8 oxygen atoms $(2 \times 4)$

A final problem illustrates the combination of the last two types.

**Example 6.4:** How many atoms of each type of element are in 5 molecules of $Ca_3(PO_4)_2$?

> **Solution:** $5Ca_3(PO_4)_2$ has: 15 calcium atoms $(5 \times 3)$
> 10 phosphorus atoms $(5 \times 2 \times 1)$
> 40 oxygen atoms $(5 \times 2 \times 4)$

# Chemical Reactions and Equations

## Practice Problems

**6.1** How oxygen atoms are present in 6 molecules of oxygen?

**6.2** How many atoms of each type are present in 3 formula units of hydrochloric acid?

**6.3** How many atoms of each type are present in a formula unit of copper(II) nitrate?

**6.4** How many atoms of each type are present in 4 formula units of aluminum sulfate?

## Application of Skills and Concepts

### 6.1 Recognizing Chemical Reactions

This first section serves as an introduction. It allows you to think about experiences with various chemical reactions. In any given system, one or more of the following observations might indicate that a chemical reaction has taken place:

- Heat evolution (reaction vessel feels warm)
- Heat loss (reaction vessel feels cool)
- Light emission or absorption
- Formation of an insoluble solid in a previously clear solution
- Evolution of a gas
- A change in color

While the above might indicate a reaction has occurred, they do not necessarily describe the reaction at the molecular level. The subsequent sections will look at reactions in more detail.

### 6.2 Chemical Equations

This section will address several items of information that can be obtained from a *chemical equation*.

- The identity of the reacting substance(s) and product substance(s)
- The relative amounts of each substance in the reaction
- The states of each substance (gas, liquid, solid, dissolved)
- The reaction conditions (heat added, catalyst used, solvents used, etc.)

The generic chemical reaction might be summarized by EQ 6.1:

$$A + B \rightarrow C + D \qquad \text{(EQ 6.1)}$$

This equation would be interpreted as *substance A reacts with substance B to form substances C and D.*

One thing EQ 6.1 does not tell is the relative amounts of each substance. Most chemical equations provide this information by using **coefficients.** Look at the chemical equation (EQ 6.2) symbolizing the following word description of a reaction:

> *1 ionic formula unit of calcium carbonate reacting with 2 formula units of hydrochloric acid to form 1 ionic formula unit of calcium chloride plus 1 molecule of water plus 1 molecule of carbon dioxide.*

$$CaCO_3 + 2HCl \rightarrow CaCl_2 + H_2O + CO_2 \qquad (EQ\ 6.2)$$

The **coefficients** for $CaCO_3$, $CaCl_2$, $H_2O$, and $CO_2$ are all **1**. This number is *implied,* so it does not appear in the equation. The **coefficient** for HCl is **2**. Coefficients multiply each atom in the equation times its **subscript**.

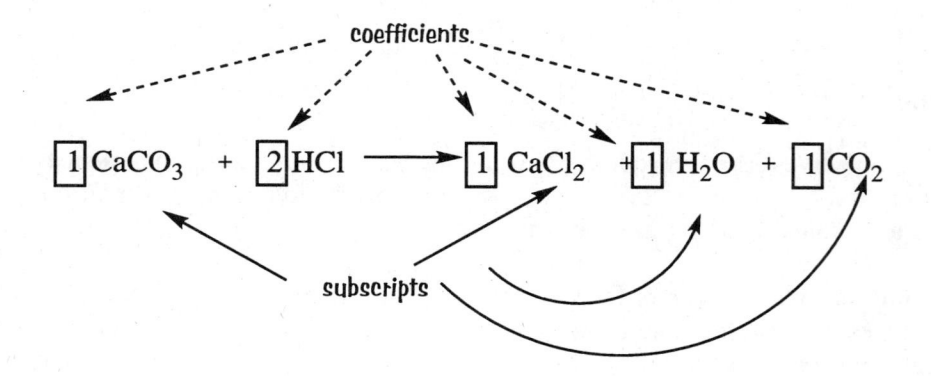

$$\boxed{1}CaCO_3 + \boxed{2}HCl \longrightarrow \boxed{1}CaCl_2 + \boxed{1}H_2O + \boxed{1}CO_2$$

(coefficients point to the boxed numbers; subscripts point to the numbers within the formulas)

Some equations will be written so that additional information can be provided. This may include one or more of the following:

- The *state* of the substance
    - (s): solid
    - (l): liquid
    - (g): gas
    - (aq): aqueous
- The presence of a catalyst
    - Written over or under the arrow
- The addition of heat
    - Written as the Greek letter delta (Δ)

Notice that EQ 6.2 could be written with some of the additional information.

$$CaCO_3(s) + 2HCl(aq) \longrightarrow CaCl_2(aq) + H_2O(l) + CO_2(g)$$

This could be described as with the following words:

> *1 ionic formula unit of solid calcium carbonate reacts with 2 formula units of aqueous hydrochloric acid to form 1 ionic formula unit of aqueous calcium chloride plus 1 molecule of liquid water plus 1 molecule of carbon dioxide gas.*

## Practice Problems

**6.5** State in words what the following equation means:

$4Fe + 3O_2 \rightarrow 2Fe_2O_3$

**6.6** State in words what the following equation means:

$2H_2O \rightarrow 2H_2 + O_2$

**6.7** What are the reaction conditions for the following equation?

$2KClO_3(s) \xrightarrow{\Delta} 2KCl(s) + 3O_2(g)$

**6.8** What is the state of each substance in the following equation?

$2\,Na(s) + 2H_2O(l) \rightarrow 2NaOH(aq) + H_2(g)$

**6.9** Write an equation from the following description of a chemical reaction: Two formula units of aqueous hydrochloric acid react with one formula unit of solid calcium hydroxide to form two molecules of water and one formula unit of aqueous calcium chloride.

**6.10** Write an equation from the following description of a chemical reaction:
One molecule of solid carbon reacts with one molecule of oxygen gas to produce one molecule of carbon dioxide gas.

## 6.3 Balancing Chemical Equations

The chemical equations in Section 6.2 have all been *balanced* equations. That is, they are written to show that the *law of conservation of mass* is obeyed during a chemical reaction. Practically, this means that the number of atoms of each type of element on the left side of the arrow (in the reactants) must equal the number of atoms of each type of element on the right side of the arrow (in the products).

Balancing equations can almost be described as an *art*. However, even if you aren't artistically inclined, you can still be successful in this endeavor.

When writing balanced equations, do the following:

- Write the formulas of the reactants and products **correctly**. Review Chapter 5 if you have forgotten how to do this.
- Calculate the number of atoms of each type of element on the *left* and *right* sides of the equation.
- Use coefficients in front of substances to make the number of atoms of each type of element on the *left* equal to that on the *right*.
- Recalculate the number of atoms of each type of element on the *left* and *right* sides of the equation to make sure that they are correct.

> **Never** adjust the number of atoms on a side of the equation by altering the **subscripts** in a chemical formula.

The following example problem and stepwise solution will show one way to arrive at a *balanced equation*.

**Example 6.5:** Sulfuric acid reacts with sodium hydroxide to produce water and sodium sulfate. Write a balanced equation for this reaction.

**Solution:** The *unbalanced* equation with correctly written chemical formulas is shown below.

$$H_2SO_4 + NaOH \rightarrow H_2O + Na_2SO_4$$

Before going on to the next steps, stop to make sure you can write the correct chemical formulas (on your own) for all the substances in this equation.
Below is a system of bookkeeping for all the atoms on each side of the arrow in.

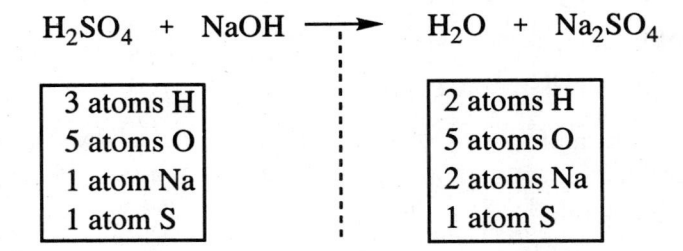

$$H_2SO_4 + NaOH \rightarrow H_2O + Na_2SO_4$$

| 3 atoms H | 2 atoms H |
| 5 atoms O | 5 atoms O |
| 1 atom Na | 2 atoms Na |
| 1 atom S | 1 atom S |

A quick examination of the atoms tells you that the right-hand side is short of hydrogen atoms and the left-hand side is short of sodium atoms.

> Remember that the **only** way to balance this equation is with **coefficients** in front of substances.

Try to put a 2 in front of the sodium hydroxide on the left. This also affects the number of oxygen and hydrogen atoms. A new count now reveals the following.

$$H_2SO_4 + 2NaOH \rightarrow H_2O + Na_2SO_4$$

| 4 atoms H | 2 atoms H |
| 6 atoms O | 5 atoms O |
| 2 atom Na | 2 atoms Na |
| 1 atom S | 1 atom S |

Now the sodium and sulfur atoms *balance*, but the right side of the equation is short 2 hydrogens and 1 oxygen. This can be solved by placing a "2" in front of the water to yield a correctly balanced equation.

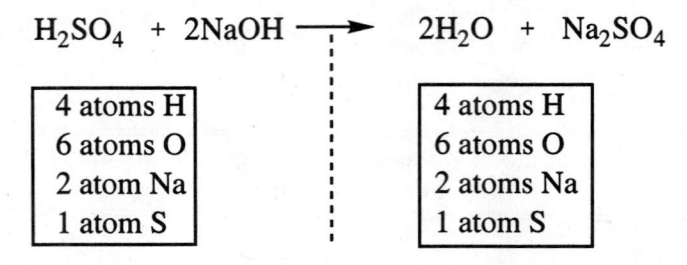

$$H_2SO_4 + 2NaOH \rightarrow 2H_2O + Na_2SO_4$$

| 4 atoms H | 4 atoms H |
| 6 atoms O | 6 atoms O |
| 2 atom Na | 2 atoms Na |
| 1 atom S | 1 atom S |

**Chemical Reactions and Equations**     89

The usual goal in balancing equations is to obtain the *smallest* coefficients possible, where they are still *whole numbers*. The next example reaction is an equation that requires some special insights for balancing.

**Example 6.6:** Balance this combustion reaction $C_4H_{10} + O_2 \rightarrow CO_2 + H_2O$.

**Solution:** The original bookkeeping for atoms looks as follows.

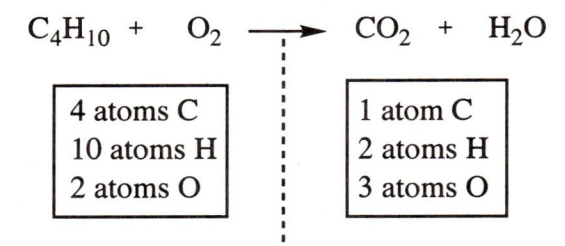

$$C_4H_{10} + O_2 \longrightarrow CO_2 + H_2O$$

| 4 atoms C | 1 atom C |
| 10 atoms H | 2 atoms H |
| 2 atoms O | 3 atoms O |

> A hint on this type of equation: *It usually is easiest to try to balance the pure element last.*

Balancing the carbon and the hydrogen leads to the following:

$$C_4H_{10} + O_2 \longrightarrow 4CO_2 + 5H_2O$$

| 4 atoms C | 4 atom C |
| 10 atoms H | 10 atoms H |
| 2 atoms O | 13 atoms O |

Now the only problem is to balance the oxygen atoms, but the subscript 2 makes this difficult. Clearly, 6.5 or 13/2 as a coefficient of oxygen would result in 13 atoms of oxygen on the right side of the equation. This is *technically balanced*.

$$C_4H_{10} + \frac{13}{2} O_2 \rightarrow 4CO_2 + 5H_2O$$

All *whole* numbers can be achieved by multiplying each coefficient by two to obtain the preferable *balanced* equation below.

$$2C_4H_{10} + 13O_2 \rightarrow 8CO_2 + 10H_2O$$

Check to make sure that you can account for the following atoms on both sides:
         8 carbon atoms
         20 hydrogen atoms
         26 oxygen atoms

Now you are ready to practice the skills of balancing combined with word descriptions of reactions. Try to put as much time as possible on this section of problems.

Remember: **Active** learning is better than **Passive** learning. Try to master these problems without the answer page. If you need the answer (this is **passive**) redo the problems, until you can solve them with knowledge and logical thought (this is **active**).

## Practice Problems

**6.11** What is the molecular formula for carbon dioxide? How many atoms of each type are in 5 molecules of carbon dioxide?

**6.12** What is the formula unit for barium phosphate? How many atoms of each type are in 2 molecules of barium phosphate?

Use the following unbalanced equation to answer problems **6.13-6.16**

$$C_2H_6 + O_2 \rightarrow CO_2 + H_2O$$

**6.13** How many atoms of each type of element are on the left?

**6.14** How many atoms of each type of element are on the right?

**6.15** How many atoms of each type of element needs to be added to each side?

**6.16** Write the balanced equation using the smallest whole-number coefficients.

For the equations written out in problems **6.17-6.23,** balance by writing the correct coefficients in the box. You can omit any coefficients which are 1.

**6.17** ☐ Al + ☐ $O_2$ → ☐ $Al_2O_3$

**6.18** ☐ $AgNO_3$ + ☐ $K_2CrO_4$ → ☐ $Ag_2CrO_4$ + ☐ $KNO_3$

**6.19** ☐ $C_2H_6O$ + ☐ $O_2$ → ☐ $CO_2$ + ☐ $H_2O$

**6.20** ☐ $KClO_3$ → ☐ KCl + ☐ $O_2$

**6.21** ☐ $C_2H_6$ + ☐ $Br_2$ → ☐ $C_2H_5Br$ + ☐ HBr

**6.22** ☐ Zn + ☐ NaOH → ☐ $Na_2ZnO_2$ + ☐ $H_2$

**6.23** ☐ $CH_4$ + ☐ $O_2$ → ☐ $CO_2$ + ☐ $H_2O$

For problems **6.24-6.27**, write a balanced equation from the word description.

**6.24** Water decomposes to hydrogen and oxygen.

**6.25** Nitrogen and hydrogen combine to form ammonia ($NH_3$).

**6.26** Aluminum metal reacts with iron(III) oxide, a magnesium catalyst, and heat to produce iron metal and aluminum oxide.

**6.27** Aqueous barium chloride reacts with aqueous potassium sulfate to form aqueous potassium chloride and solid barium sulfate.

The second part of Chapter 6 deals with one classification system for chemical reactions. These include the following:

- Combination reactions
- Decomposition reactions

- Single-replacement reactions
- Double-replacement reactions
    - Solid formation
    - Gas formation
    - Water formation
- Combustion reactions

The details of each type of reaction and some hints for identification will follow in Sections 6.4-6.10.

## 6.4 Combination and Decomposition Reactions

These two types of reactions are very self-descriptive, and should be quite easy to recognize. Combination reactions are of the generic type in EQ 6.3.

$$A + B \rightarrow C \qquad \text{(EQ 6.3)}$$

Decomposition reactions are just the opposite, as in EQ 6.4.

$$C \rightarrow A + B \qquad \text{(EQ 6.4)}$$

---

**Practice Problems**

Look at the reactions in problems **6.28-6.29**, and identify them as either *combination* or *decomposition* reactions. For problems **6.30** and **6.31**, write the balanced equation.

**6.28** $2NI_3(s) \rightarrow 3I_2(g) + N_2(g)$

**6.29** $2Na(s) + Cl_2(s) \rightarrow 2NaCl(s)$

**6.30** An electric current is applied to molten sodium chloride to produce solid sodium metal and chlorine gas.

**6.31** Hydrogen gas and oxygen gas form water vapor.

---

## 6.5 Single Replacement Reactions

The secret for recognizing these reactions is to *look for an element alone that replaces an element in a compound, leaving a new element isolated*. This is demonstrated by the generic reaction in EQ 6.5.

$$A + BX \rightarrow AX + B \qquad \text{(EQ 6.5)}$$

Not all elements will replace another in a compound. If fact, there is a *pecking order* for these reactions. The theory behind these observations will be addressed in Chapter 16. Table 6.1 is an activity series for some metals (including hydrogen) and a second series of Group VII nonmetals. To interpret the series, notice that if A (from EQ 6.5) in Table 6.1 is listed before B, then it will replace B in a compound. If A follows B in Table 6.1, no reaction will occur. Equations 6.6 to 6.8 illustrate these situations.

$$Mg + ZnCl_2 \rightarrow MgCl_2 + Zn \qquad \text{(EQ 6.6)}$$

$$\text{Ag} + \text{ZnCl}_2 \rightarrow \text{NO RXN} \qquad \text{(EQ 6.7)}$$

$$\text{F}_2 + \text{KCl} \rightarrow \text{KF} + \text{Cl}_2 \qquad \text{(EQ 6.8)}$$

**Table 6.1** Activity Series of Some Elements

| Metal (Including Hydrogen) | Metal Ion | Metal (Including Hydrogen) | Metal Ion | Group VII Element | Group VII Ion |
|---|---|---|---|---|---|
| Li | Li$^+$ | Cd | Cd$^{2+}$ | F$_2$ | F$^-$ |
| K | K$^+$ | Co | Co$^{2+}$ | Cl$_2$ | Cl$^-$ |
| Ba | Ba$^{2+}$ | Ni | Ni$^{2+}$ | Br$_2$ | Br$^-$ |
| Ca | Ca$^{2+}$ | Sn | Sn$^{2+}$ | I$_2$ | I$^-$ |
| Na | Na$^+$ | Pb | Pb$^{2+}$ | | |
| Mg | Mg$^{2+}$ | H$_2$ | H$^+$ | | |
| Al | Al$^{3+}$ | Cu | Cu$^{2+}$ | | |
| Zn | Zn$^{2+}$ | Hg | Hg$^{2+}$ | | |
| Cr | Cr$^{2+}$ | Ag | Ag$^+$ | | |
| Fe | Fe$^{2+}$ | Pt | Pt$^{2+}$ | | |
| Cd | Cd$^{2+}$ | Au | Au$^{3+}$ | | |

## Practice Problems

For the following problems, write the products and balance the reaction. If no reaction occurs, write "NO RXN." Use the activity series to determine if a reaction occurs.

**6.32** Mg + HCl $\rightarrow$

**6.33** Mg + CaCl$_2$ $\rightarrow$

**6.34** Ag + Zn(NO$_3$)$_2$ $\rightarrow$

**6.35** Cl$_2$ + NaBr $\rightarrow$

**6.36** Metallic copper is placed into a solution of silver nitrate

**6.37** Metallic calcium is placed into a solution of cobalt(II) nitrate

## 6.6 Double-Replacement Reactions

These all have the generic equation shown in EQ 6.9.

$$\text{AX} + \text{BY} \rightarrow \text{AY} + \text{BX} \qquad \text{(EQ 6.9)}$$

Notice that the A and B in EQ 6.9 have *changed* places with the X and Y. For all these reactions, the two cations switch places and become paired with the other anion.

- Cations are metals or polyatomic cations.
- Anions are nonmetals or polyatomic anions.

> **Practice Problems**
> Which of the following are double-replacement reactions? If not, identify the type.
>
> **6.38** HCl + NaOH → H$_2$O + NaCl
>
> **6.39** 2AgNO$_3$ + Zn → Zn(NO$_3$)$_2$ + 2Ag
>
> **6.40** MgCl$_2$ + 2AgNO$_3$ → 2AgCl + Mg(NO$_3$)$_2$
>
> **6.41** Ag + O$_2$ → Ag$_2$O

## 6.7 Double Replacement: A Solid Forms (Precipitation)

The way to identify this type of reaction is to look for the formation of a solid from two aqueous substances. If the equation does not tell you which one of the products (if any) forms a solid, you will need to determine this by knowing some rules of solubility.

### Rules for Solubility

- **Always soluble** ions
    - Group IA cations and NH$_4^+$ cation
    - Acetate, nitrate, hydrogen carbonate anions
- **Usually soluble** ions
    - Fluorides (except CaF$_2$)
    - Chlorides, bromides and iodides (except with Pb$^{2+}$, Ag$^+$, Hg$^+$)
    - Sulfates (except Ca$^{2+}$, Pb$^{2+}$, Ba$^{2+}$, Sr$^{2+}$)
    - Oxalates, C$_2$O$_4^{2+}$ (except Ca$^{2+}$)
- **Usually insoluble** ions
    - Carbonates, hydroxides, sulfides (except with the **always soluble** ions)

> The **most** important things to memorize are the **always soluble ions.**

When trying to determine which (if any) product is a solid precipitate, write the substances resulting from the reaction. Then use your knowledge of solubility rules to identify the solid. Two examples will demonstrate how to do this.

**Example 6.7:** Write a balanced equation for AgNO$_3$(aq) + KCl(aq) →

> **Solution:** The first way to solve the above is to write the **correct** formulas for the double-replacement reaction.
>
> AgNO$_3$(aq) + KCl(aq) → AgCl + KNO$_3$
>
> The next step is to balance the equation. The above equation is balanced as written. Finally, use the solubility rules to determine which (if any) product is an insoluble solid (precipitate). An examination of the above rules shows that *silver chloride* is the *solid*. The solution follows.

$$AgNO_3(aq) + KCl(aq) \rightarrow AgCl(s) + KNO_3(aq)$$

Another example demonstrates a situation where no solid forms.

**Example 6.8:** $NH_4NO_3(aq) + CaCl_2(aq) \rightarrow$

> **Solution:** For this equation, when the double replacement substances are written and the equation is correctly balanced, it looks as follows:
>
> $$2NH_4NO_3(aq) + CaCl_2(aq) \rightarrow Ca(NO_3)_2 + 2NH_4Cl$$
>
> However, if you check the solubility rules, you will notice that neither of the products will form an insoluble solid (all ammonium cations and nitrate anions are soluble). Therefore, this should be written as "NO RXN" because there is no chemical change.
>
> $$NH_4NO_3(aq) + CaCl_2(aq) \rightarrow NO\ RXN$$

The best way to learn how to write these reactions is to practice.

## Practice Problems

**6.42** Which of the following compounds are soluble?
$NaC_2H_3O_2$, $BaCO_3$, $SrCl_2$, $(NH_4)_2SO_4$, $AgBr$

**6.43** Which of the following compounds are insoluble?
potassium sulfate, calcium carbonate, barium hydrogen carbonate, magnesium hydroxide

For problems **6.44-6.50,** write a balanced equation complete with (s) for compounds forming an insoluble precipitate. If no solid forms, write "NO RXN."

**6.44** $KBr(aq) + Pb(NO_3)_2(aq) \rightarrow$

**6.45** $NiSO_4(aq) + NaOH(aq) \rightarrow$

**6.46** $NiSO_4(aq) + Ba(OH)_2(aq) \rightarrow$

**6.47** $Na_2SO_4(aq) + NH_4NO_3(aq) \rightarrow$

**6.48** An aqueous solution of silver nitrate is mixed with an aqueous solution of $K_2CrO_4$. The result is the formation of a red solid.

**6.49** Aqueous cobalt(II) acetate is mixed with aqueous sodium carbonate, and a blue solid forms.

**6.50** Aqueous lead(II) nitrate is mixed with aqueous sodium iodide, and a yellow solid forms.

## 6.8 Double Replacement: A Gas Forms

There are two ways that a gas can form. The first is shown in EQ 6.10. In this case, one of the products is a gas.

$$AX + BY \rightarrow AY + BX(g) \qquad (EQ\ 6.10)$$

Sometimes gas formation is actually two reactions. The first is a double replacement, followed by the decomposition of one of the products. The generic forms are shown in EQ 6.11 and EQ 6.12.

$$AX + BY \rightarrow AY + BX \qquad \text{(EQ 6.11)}$$

$$BX \rightarrow H_2O + BO_2(g) \qquad \text{(EQ 6.12)}$$

For the purposes of this course, anytime one of the following compounds is a product of a double-replacement reaction, a gas will be produced.

- $H_2S$ (is a gas)
- $H_2SO_3$ (becomes $SO_2(g)$ and water)
- $H_2CO_3$ (becomes $CO_2(g)$ and water)

Apply the same rules for double-replacement used in Section 6.7 to the next problem set. If sulfurous or carbonic acid forms, the products are a gas ($SO_2$ and $CO_2$, respectively) and water.

---

**Practice Problems**

Write the products for the following reactions and balance the equations.

**6.51** $CaCO_3 + H_2SO_4 \rightarrow$

**6.52** $MgS + HCl \rightarrow$

**6.53** Sodium bicarbonate (sodium hydrogen carbonate) reacts with acetic acid.

---

### 6.9 Double Replacement: Water Forms (Neutralization)

Review the definitions of acid, base, and salt at the beginning of this chapter. The generic equation for this type of reaction is shown in EQ 6.13:

$$AX + BY \rightarrow H_2O + BX \qquad \text{(EQ 6.13)}$$

Where AX is an acid, BY is a base, and BX is a salt.

One of these types of reactions can be demonstrated in the word description below.

**Example 6.9:** Nitric acid reacts with magnesium hydroxide. What are the products? Write a balanced equation.

**Solution:** The base (magnesium hydroxide) reacts with the acid to produce water and the corresponding salt. The salt will be made of the *cation from the base* ($Mg^{2+}$) and the *anion from the acid* (nitrate ion). The unbalanced equation follows:

$$HNO_3 + Mg(OH)_2 \rightarrow H_2O + Mg(NO_3)_2$$

Balance the equation **only after** the correct formulas for the products are written.

$$2HNO_3 + Mg(OH)_2 \rightarrow 2H_2O + Mg(NO_3)_2$$

## Practice Problems

Write products for the following reactions, and balance the equations.

**6.54** $H_2SO_4 + NaOH \rightarrow$

**6.55** $HC_2H_3O_2 + KOH \rightarrow$

**6.56** Calcium hydroxide reacts with phosphoric acid.

**6.57** Hydrobromic acid reacts with lithium hydroxide.

## 6.10 Combustion Reactions

These reactions have a separate category (even though they often could be included in combination reactions) because they are so numerous. To identify these reactions, look for one reactant to be oxygen. An example is shown in EQ 6.14.

$$C_2H_6O + 2O_2 \rightarrow 2CO_2 + 3H_2O \qquad (EQ\ 6.14)$$

Try balancing the combustion reactions below. While doing so, notice that each of these reactions contains oxygen as one of the reactants.

## Practice Problems

**6.58** $C_3H_8 + O_2 \rightarrow CO_2 + H_2O$

**6.59** $Ca + O_2 \rightarrow CaO$

**6.60** Elemental carbon reacts with oxygen to form carbon monoxide.

**6.61** Gasoline ($C_8H_{18}$) reacts with oxygen to form carbon dioxide and water.

## Integration of Multiple Skills

This section requires that, given reactants or a word description, you can write a balanced equation, and classify it. This involves the use of several skills.

- Writing the correct formulas for molecules or ionic compounds
- Recognizing the type of reaction taking place
- Providing the correct products
- Applying the *law of conservation of mass,* to balance the equation

## Practice Problems

Write a balanced equation for each of the following descriptions, and classify the reaction.

**6.62** Carbon reacts with oxygen to form carbon dioxide.

**6.63** Nitrogen tribromide forms nitrogen and bromine.

> **6.64** Nitrous acid reacts with calcium hydroxide.
>
> **6.65** Potassium iodide reacts with silver nitrate.
>
> **6.66** Copper metal reacts with silver nitrate.

# Self-Exam

**I. (4 ea) Write T (true) or F (false) in each corresponding blank.**

_____1. The Greek symbol delta (Δ) indicates *heat* in chemical reactions.

_____2. Reactants are written on the left-hand side of a chemical equation.

_____3. A *precipitation* reaction and a *single-replacement* reaction are identical.

_____4. A *metathesis* reaction and a *single-replacement* reaction are identical.

_____5. Carbon tetrachloride is an acid.

_____6. Calcium hydroxide is a base.

_____7. Sodium bromide is a salt.

_____8. Glucose plus oxygen react to produce $CO_2$ and $H_2O$. This is a *combustion* reaction.

_____9. Use the symbol (**aq**) to indicate that a substance in an equation is dissolved.

_____10. Five molecules of potassium nitrite contain ten atoms of oxygen.

**II. (4 ea) Multiple Choice. Place the letter of the BEST answer in the corresponding blank.**

_____1. Which of the following **does not** indicate that a chemical reaction has taken place?
   A. Heat is released.   B. Gas is Evolved.   C. Water is evaporated.
   D. The color changes.   E. Light is emitted.

_____2. How many atoms of oxygen are present in $Ca_3(PO_4)_2$?
   A. 1   B. 2   C. 4   D. 8   E. 24

_____3. The following equation is an example of what type of reaction?

$$ZnS + 2HI \rightarrow ZnI_2 + H_2S(g)$$

   A. Combustion   B. Combinations   C. Double replacement
   D. Decomposition   E. Single replacement

_____4. The following equation is an example of what type of reaction?

$$2C_4H_{10} + 13O_2 \rightarrow 8CO_2 + 10H_2O$$

   A. Combustion   B. Combination   C. Double replacement
   D. Decomposition   E. Single replacement

_____ 5. Which of the following is a single replacement reaction?
  A. $2HNO_3 + Mg(OH)_2 \rightarrow 2H_2O + Mg(NO_3)_2$
  B. $2C_8H_{18} + 25O_2 \rightarrow 16CO_2 + 18H_2O$
  C. $CaCO_3 + 2HCl \rightarrow CaCl_2 + H_2O + CO_2$
  D. $2Na(s) + Cl_2(s) \rightarrow 2NaCl(s)$
  E. $Zn + Cd(NO_3)_2 \rightarrow Cd + Zn(NO_3)_2$

_____ 6. Which of the following results is a double-replacement reaction?
  A. Potassium chlorate is heated.      B. Acetic acid is added to $NaHCO_3$.
  C. Chloric acid is added to Mg.       D. Magnesium is added to silver nitrate.
  E. Oxygen is added to hydrogen.

_____ 7. Which of the following results is a decomposition reaction?
  A. Potassium chlorate is heated.      B. Acetic acid is added to $NaHCO_3$.
  C. Chloric acid is added to Mg.       D. Magnesium is added to silver nitrate.
  E. Oxygen is added to hydrogen.

_____ 8. What is(are) the solid precipitate when KCl is added to $Pb(NO)_3$?
  A. $Cl_2$    B. $KNO_3$    C. $PbCl_2$    D. $PbCl$    E. Both B and C

_____ 9. What are the products when zinc metal is added to $Cu(NO)_3$?
  A. Copper metal and zinc metal        B. Copper metal and zinc nitrate
  C. NO RXN                             D. Zinc cuprate
  E. Zinc and copper oxide

_____ 10. What are the products when chlorine gas is added to a solution of KI?
  A. Iodine and KCl    B. ICl    C. KICl    D. $Cl_2$ and $I_2$    E. NO RXN

### III. Provide Complete Answers

1. (4 pts) Balance the following equations.
   A. $KClO_3 \rightarrow KCl + O_2$
   B. $H_2 + N_2 \rightarrow NH_3$

2. (8 pts) Complete and balance the following equations
   A. $H_2SO_4 + Na_2CO_3 \rightarrow$
   B. $C_4H_{10} + O_2 \rightarrow$
   C. $Cu + AgNO_3 \rightarrow$

3. (8 pts) Write balanced equations for the following descriptions.
   A. Aqueous phosphoric acid is added to solid magnesium hydroxide.
   B. Solid nitrogen triiodide decomposes to nitrogen gas and iodine.

## Answers to Practice Problems

| | | | |
|---|---|---|---|
| 6.1 | 12 atoms O | 6.2 | 3 atoms H; 3 atoms Cl |
| 6.3 | 1 atom Cu; 2 atoms N; 6 atoms O | 6.4 | 8 atoms Al; 12 atoms S; 48 atoms O |

**6.5** Four atoms of iron react with three molecules of oxygen to produce two formula units of iron(III) oxide.

**6.6** Two molecules of water form two molecules of hydrogen and two molecules of oxygen.

**6.7** Two formula units of solid potassium chlorate are heated to from two formula units of solid potassium chloride and three molecules of oxygen gas.

**6.8** Sodium: solid; water: liquid; sodium hydroxide: aqueous; hydrogen: gas

**6.9** $2HCl(aq) + Ca(OH)_2(s) \rightarrow 2H_2O + CaCl_2(aq)$

**6.10** $C(s) + O_2(g) \rightarrow CO_2(g)$

**6.11** $CO_2$; 5 atoms C, 10 atoms O in 5 molecules

**6.12** $Ba_3(PO_4)_2$; 6 atoms Ba, 4 atoms P, 16 atoms O in 2 molecules

**6.13** 2C, 6H, 2O

**6.14** 1C, 2H, 3O

**6.15** Left needs 1 oxygen, right needs 1 carbon, 4 hydrogen

**6.16** $2C_2H_6 + 7O_2 \rightarrow 4CO_2 + 6H_2O$

**6.17** $4Al + 3O_2 \rightarrow 2Al_2O_3$

**6.18** $2AgNO_3 + K_2CrO_4 \rightarrow Ag_2CrO_4 + 2KNO_3$

**6.19** $C_2H_6O + 3O_2 \rightarrow 2CO_2 + 3H_2O$   **6.20** $2KClO_3 \rightarrow 2KCl + 3O_2$

**6.21** $C_2H_6 + Br_2 \rightarrow C_2H_5Br + HBr$   **6.22** $Zn + 2NaOH \rightarrow Na_2ZnO_2 + H_2$

**6.23** $CH_4 + 2O_2 \rightarrow CO_2 + 2H_2O$

**6.24** $2H_2O \rightarrow 2H_2 + O_2$   **6.25** $N_2 + 3H_2 \rightarrow 2NH_3$

**6.26** $2Al(s) + Fe_2O_3 \xrightarrow[Mg]{\Delta} Al_2O_3 + 2Fe(s)$

**6.27** $BaCl_2(aq) + K_2SO_4(aq) \rightarrow 2KCl(aq) + BaSO_4(s)$

**6.28** Decomposition

**6.29** Combination

**6.30** $2NaCl \rightarrow 2Na(s) + Cl_2(g)$, decomposition

**6.31** $2H_2(g) + O_2(g) \rightarrow 2H_2O(g)$, combination

**6.32** $Mg + 2HCl \rightarrow MgCl_2 + H_2$   **6.33** $Mg + CaCl_2 \rightarrow$ NO RXN

**6.34** $Ag + Zn(NO_3)_2 \rightarrow$ NO RXN   **6.35** $Cl_2 + 2NaBr \rightarrow 2NaCl + Br_2$

**6.36** Cu + 2AgNO$_3$ → 2Ag + Cu(NO$_3$)$_2$     **6.37** Ca + Co(NO$_3$)$_2$ → Co + Ca(NO$_3$)$_2$

**6.38** Double replacement     **6.39** Single replacement

**6.40** Double replacement     **6.41** Combination

**6.42** NaC$_2$H$_3$O$_2$, SrCl$_2$, (NH$_4$)$_2$SO$_4$

**6.43** Calcium carbonate, magnesium hydroxide

**6.44** 2KBr(aq) + Pb(NO$_3$)$_2$(aq) → 2KNO$_3$(aq) + PbBr$_2$(s)

**6.45** NiSO$_4$(aq) + 2NaOH(aq) → Ni(OH)$_2$(s) + Na$_2$SO$_4$(aq)

**6.46** NiSO$_4$(aq) + Ba(OH)$_2$(aq) → Ni(OH)$_2$(s) + BaSO$_2$(s)

**6.47** Na$_2$SO$_4$(aq) + NH$_4$NO$_3$(aq) → NO RXN

**6.48** 2AgNO$_3$(aq) + K$_2$CrO$_4$(aq) → 2KNO$_3$(aq) + Ag$_2$CrO$_4$(s)

**6.49** Co(C$_2$H$_3$O$_2$)$_2$(aq) + Na$_2$CO$_3$(aq) → CoCO$_3$(s) + 2NaC$_2$H$_3$O$_2$(aq)

**6.50** Pb(NO$_3$)$_2$(aq) + 2NaI(aq) → PbI$_2$(s) + 2NaNO$_3$(aq)

**6.51** CaCO$_3$(s) + H$_2$SO$_4$(aq) → CaSO$_4$(aq) + H$_2$O(l) + CO$_2$(g)

**6.52** MgS(s) + 2HCl(aq) → H$_2$S(g) + MgCl$_2$

**6.53** NaHCO$_3$(aq) + HC$_2$H$_3$O$_2$(aq) → H$_2$O(l) + CO$_2$(g) + NaC$_2$H$_3$O$_2$(aq)

**6.54** H$_2$SO$_4$(aq) + 2NaOH(aq) → 2H$_2$O(l) + Na$_2$SO$_4$(aq)

**6.55** HC$_2$H$_3$O$_2$(aq) + KOH(aq) → KC$_2$H$_3$O$_2$(aq) + H$_2$O(l)

**6.56** 3Ca(OH)$_2$(s) + 2H$_3$PO$_4$(aq) → 6H$_2$O(l) + Ca$_3$(PO$_4$)$_2$(s)

**6.57** HBr(aq) + LiOH(aq) → H$_2$O(l) + LiBr(aq)

**6.58** C$_3$H$_8$ + 5O$_2$ → 3CO$_2$ + 4H$_2$O

**6.59** 2Ca + O$_2$ → 2CaO

**6.60** 2C + O$_2$ → 2CO

**6.61** 2C$_8$H$_{18}$ + 25O$_2$ → 16CO$_2$ + 18H$_2$O

**6.62** C + O$_2$ → CO$_2$, combustion

**6.63** 2NBr$_3$ → N$_2$ + 3Br$_2$, decompositiion

**6.64** 2HNO$_2$ + Ca(OH)$_2$ → Ca(NO$_2$)$_2$ + 2H$_2$O, double replacement (neutralization)

**6.65** KI + AgNO$_3$ → AgI(s) + KNO$_3$, double replacement (precipitation)

**6.66** Cu + 2AgNO$_3$ → Cu(NO$_3$)$_2$ + 2Ag, single replacement

# Answers to Self-Exam

**I. T/F**
1. T  2. T  3. F  4. T  5. F  6. T  7. T  8. T  9. T  10. T

**II. MC**
1. C  2. D  3. C  4. A  5. E  6. B  7. A  8. C  9. B  10. A

**III. Complete Answers**
1. A. 2KClO$_3$ → 2KCl + 3O$_2$
   B. 3H$_2$ + N$_2$ → 2NH$_3$

2. A. H$_2$SO$_4$ + Na$_2$CO$_3$ → Na$_2$SO$_4$ + CO$_2$ + H$_2$O
   B. 2C$_4$H$_{10}$ + 13O$_2$ → 8CO$_2$ + 10H$_2$O
   C. Cu + 2AgNO$_3$ → 2Ag + Cu(NO$_3$)$_2$

3. A. 2H$_3$PO$_4$ + 3Mg(OH)$_2$ → 6H$_2$O + Mg$_3$(PO$_4$)$_2$
   B. 2NI$_3$ → N$_2$ + 3I$_2$

# 7
# Chemical Composition

This chapter will look at measurements of masses at the atomic (molecular) level, which use the atomic mass unit (amu), and at the practical level, which use grams (g). It will also relate the two types of measurements by Avogadro's number. This chapter is very heavy on calculational chemistry.

Keys for success in this chapter:

- Learn terminology and try to comprehend how small atoms are.
- Review *dimensional analysis.*
- Examine example problems and understand how the solutions were obtained.
- Practice as many problems as possible.

> **Active** learning is very important in this section. Before moving on, make sure you can do calculations without the answer already provided.

## Summary of Verbal Knowledge

**molecular weight:** The sum of the atomic weights of the atoms in the compound.
**formula weight:** The sum of the atomic weights of all the atoms in the formula unit of the compound.
**Avogadro's number:** The number of atoms in exactly 12 g of carbon-12. Rounded to four figures, there are $6.022 \times 10^{23}$ atoms in 12 g of carbon-12.
**mole:** A quantity of a substance that contains Avogadro's number of items (usually refers to atoms, molecules, or formula units).
**molar mass:** This is the mass of one mole of a substance.
**percentage composition:** This means the percent of each element (by mass) in a compound.
**mass percentage:** This refers to the number of grams of a given element in 100 g of a compound.
**empirical formula:** This is a ratio of elements in a compound using the smallest whole numbers.
**molecular formula:** This is the actual formula of each element in a compound.

## Review of Mathematical and Calculator Skills

There are several math skills that need to be reviewed.

- Working with percentages
- Finding smallest whole number ratios
- Applying more dimensional analysis

**Chemical Composition** 103

Lets look at each of these skills through several example problems. When dealing with percentages there are two types of problems you will encounter.

- The first involves finding the percent of a known quantity.
- The second involves determining what percent quantity A is of quantity B.

The first type may be illustrated by the following:

**Example 7.1:** What is 28% of 42 g?

  **Solution:** To solve this problem, convert the 28% to a decimal, 0.28, and then multiple.

$$0.28 \times 42 \text{ g} = 12 \text{ g}$$

**Example 7.2:** 12 g is what percent of 46 g?

  **Solution:** To solve this problem, divide 12 g by 46 g (this leads to a decimal), and multiply by 100 to get the percent.

$$\frac{12 \text{ g}}{46 \text{ g}} = 0.26 \times 100 = 26\%$$

The second type of problem involves finding smallest whole number ratios. An example will demonstrate this.

**Example 7.3:** What is the ratio of 2:6:4?

  **Solution:** The answer can be found by noticing that all the numbers are divisible by 2. Therefor, the actual ratio is **1:3:2**.

The final skill to review for this chapter is that of dimensional analysis, which needs to be used to solve the following problems.

**Example 7.4:** $6.02 \times 10^{23}$ molecules of oxygen has how many oxygen atoms?

  **Solution:** The answer using dimensional analysis follows. Remember to use *unit labels* for this type of problem.

$$6.02 \times 10^{23} \text{ molecules O}_2 \times \frac{2 \text{ atoms O}}{1 \text{ molecule O}_2} = 1.20 \times 10^{24} \text{ atoms O}$$

**Example 7.5:** How many formula units of magnesium nitrate are necessary to provide $4.27 \times 10^{23}$ nitrate ions?

  **Solution:** $4.27 \times 10^{23}$ nitrate ions $\times \dfrac{1 \text{ Mg(NO}_3)_2}{2 \text{ nitrate ions}} = 2.14 \times 10^{23}$ Mg(NO$_3$)$_2$

Follow each step of the solution, and verify that each unit cancels until the correct answer is achieved. Check your mastery of these skills by trying the practice problems below.

## Practice Problems

**7.1** What is 52% of 46 g?

**7.2** 12 g is what percent of 42 g?

**7.3** 11.3 g is what percent of 15.8 g?

**7.4** Find the smallest whole number ratio of 6:9.

**7.5** Find the smallest whole number ratio of 5:3:2.

**7.6** Find the smallest whole number ratio of 8:8:2.

**7.7** How many atoms of hydrogen are in $2.52 \times 10^{22}$ molecules of hydrogen molecules?

**7.8** How many atoms of nitrogen are in $1.67 \times 10^{24}$ molecules of dinitrogen tetroxide?

**7.9** How many ionic formula units of calcium phosphate are needed to obtain $5.46 \times 10^{21}$ calcium cations?

## Application of Skills and Concepts

### 7.1 Molecular Weight and Formula Weight

This section involves using the periodic table to determine atomic, molecular, and formula weights in amus. Remember from Chapter 4 that the weights for each atom listed in the periodic table are the average weights of all naturally occurring isotopes of that element, using carbon-12 as a reference.

There are three easy steps to arrive at the molecular or formula weights.

- Find the atomic weight of each element in the periodic table.
- Multiply the weight by the number of atoms of that element.
- Sum all the weights.

The following examples illustrate the application of this process.

**Example 7.6:** Find the atomic weight of gold.

  **Solution:** To solve the above, just look up the average weight of a gold atom in the periodic table. To the nearest tenth of an amu, it is **197.0 amu** per atom.

**Example 7.7:** Find the molecular weight of water.

  **Solution:** To solve the above, consider the molecular formula for water, $H_2O$. Using the scheme above, the molecular weight of water to the nearest tenth can be obtained:

  $$\begin{array}{ll} 2 \text{ atoms H} \times \;\; 1.0 \text{ amu/atom} = & 2.0 \text{ amu} \\ \underline{1 \text{ atoms O} \times 16.0 \text{ amu/atom} =} & \underline{16.0 \text{ amu}} \\ \text{Sum} & 18.0 \text{ amu} \end{array}$$

**Example 7.8:** Find the formula weight of potassium sulfide.

**Solution:** This final example requires writing the correct formula ($K_2S$) and then solving just as you did with the previous problem:

$$\begin{array}{l} 2 \text{ atoms K} \times 39.1 \text{ amu/atom} = 78.2 \text{ amu} \\ \underline{1 \text{ atoms S} \times 32.1 \text{ amu/atom} = 32.1 \text{ amu}} \\ \text{Sum} \hspace{4.5cm} 110.3 \text{ amu} \end{array}$$

## Practice Problems

**7.10** What is the atomic weight of platinum?

**7.11** What is the molecular weight of dinitrogen monoxide?

**7.12** What is the formula weight of iron(II) phosphate?

## 7.2 The Mole

The **mole** is a unit which specifies a certain number of objects, much the same as a unit like "dozen." You know that a dozen roses means 12 roses; a dozen eggs means 12 eggs; etc.

A mole contains $6.02 \times 10^{23}$ objects.

This is a very large number; thus in most cases it refers to very small items, such as atoms, molecules, or ions. You might have a mole of atoms, which means $6.02 \times 10^{23}$ atoms, but you could have a mole of roses, which means $6.02 \times 10^{23}$ roses.

The number $6.02 \times 10^{23}$ is actually rounded off from the more precise $6.0221367 \times 10^{23}$. Some important things to remember about this number:

- The number of objects in a mole, $6.02 \times 10^{23}$, is also known as *Avogadro's number*.
- This number is based on the number of carbon-12 atoms in exactly 12 g of carbon.
- The abbreviation for a mole is **mol.**

Before going on to some problems, lets look at Avogadro's number in its relation to carbon-12. Recall from Chapter 4 that carbon-12 contains 6 protons and 6 neutrons, each having a mass of 1 amu. Thus the atomic mass of carbon-12 is 12 amu. The following dimensional analysis conversion demonstrates how the mass of an amu can be converted to grams.

$$\frac{12 \text{ g } ^{12}C}{6.0221367 \times 10^{23} \text{ atoms } ^{12}C} \times \frac{1 \text{ atom } ^{12}C}{12 \text{ amu}} = \frac{1.6605402 \times 10^{-24} \text{ g}}{1 \text{ amu}}$$

Check the mass of 1amu in the back cover of your textbook. You will notice that it matches the determination just made. Use your *number sense* to realize that an *amu* is very **small** and Avogadro's number is very **large**.

## 106   Chapter 7

There are four types of problems that use Avogadro's number in the calculations.

- Convert moles to molecules (or atoms or formula units)
- Convert molecules (or atoms or formula units) to moles
- Convert moles to atoms in a chemical formula
- Convert atoms in a chemical formula to moles

Four sample problems and their solutions follow. They demonstrate how dimensional analysis can be used to solve these conversions. After examining the sample problems, try the practice problems.

**Example 7.9:** How many molecules of hydrogen are in 2.32 moles of hydrogen?

$$\text{Solution: } 2.32 \text{ mols H}_2 \times \frac{6.02 \times 10^{23} \text{ molecules}}{1 \text{ mol H}_2} = 1.40 \times 10^{24} \text{ molecules H}_2$$

Notice that the units cancel to lead to the desired units. When you do dimensional analysis problems, **always write the labels**. When writing a complete label is too time consuming, use abbreviations. For example, the following abbreviations will be used in subsequent problems.

- Moles are mol.
- Molecules are mcs.
- Formula units are FU.
- Molecular formulas are MF.
- Empirical formulas are EF.

**Example 7.10:** $7.21 \times 10^{21}$ atoms of silver equal how many moles?

$$\text{Solution: } 7.21 \times 10^{21} \text{ atoms Ag} \times \frac{1 \text{ mol Ag}}{6.02 \times 10^{23} \text{ atoms Ag}} = 1.20 \times 10^{-2} \text{ mol Ag}$$

**Example 7.11:** $5.61 \times 10^{-2}$ moles of silver nitrate contain how many atoms of oxygen?

**Solution:**

$$5.61 \times 10^{-2} \text{ mol AgNO}_3 \times \frac{6.02 \times 10^{23} \text{ FU AgNO}_3}{1 \text{ mol AgNO}_3} \times \frac{3 \text{ atoms O}}{1 \text{ FU AgNO}_3} = 1.01 \times 10^{23} \text{ atom O}$$

**Example 7.12:** How many moles of carbon tetrachloride are needed to obtain $7.89 \times 10^{24}$ atoms Cl?

**Solution:**

$$7.89 \times 10^{24} \text{ atoms Cl} \times \frac{1 \text{ mcs CCl}_4}{4 \text{ atom Cl}} \times \frac{1 \text{ mol CCl}_4}{6.02 \times 10^{23} \text{ mcs CCl}_4} = 3.28 \text{ mol CCl}_4$$

# Chemical Composition

## Practice Problems

**7.13** How many atoms are in 4.20 moles of calcium?

**7.14** $2.12 \times 10^{24}$ formula units of potassium chlorate equal how many moles?

**7.15** How many atoms of oxygen are in $1.25 \times 10^{-2}$ moles of sulfuric acid?

**7.16** $2.52 \times 10^{21}$ hydrogen atoms are present in how many moles of hydrogen molecules?

**7.17** How many pounds would $2.34 \times 10^{-4}$ moles of pennies weigh? (1 penny = 2.45 g)

**7.18** How many moles of roses contain a million petals? (15 petals on the average rose)

## 7.3 Molar Mass

The **molar mass** is the mass (in grams) of *one mole of a substance*. It is calculated in the same way as the molecular (formula) mass by using the periodic table. The only difference is that the unit is now **grams** instead of **amus**.

## Practice Problems

**7.19-7.23** Table 7.1 lists some molecules, formula units, molecular (formula) masses, and molar masses. The first entries are complete. The rest should be filled in as a way to practice the calculations for masses. This table also provides a chance to review nomenclature.

**Table 7.1** Molecular (Formula) Weights and Molar Masses

| Problem | Formula | Name | Molecular (Formula) Weight | Molar Mass |
|---------|---------|------|----------------------------|------------|
|         | $Pb(SO_4)_2$ | Lead(II) sulfate | 399.4 amu | 399.4 g |
|         | NaOH | Sodium hydroxide | 40.0 amu | 40.0 g |
| 7.19 | $BaI_2$ | | | |
| 7.20 | | Carbon monoxide | | |
| 7.21 | $(NH_4)_3PO_4$ | | | |
| 7.22 | | Cobalt(II) chloride | | |
| 7.23 | $SO_2$ | | | |

## 7.4 Molar Masses in Calculations: Grams to Moles

This section deals with the conversion of grams to moles. Before this can be done, you need to calculate the molar mass of the substance of interest. An example follows.

**Example 7.13:** 425 g of copper(I) chloride is equal to how many moles of this compound?

**Solution:** Using calculations from the periodic table, 1 mol CuCl = 99.0 g.

$$425 \text{ g CuCl} \times \frac{1 \text{ mol CuCl}}{99.0 \text{ g CuCl}} = 4.29 \text{ mol CuCl}$$

## Practice Problems

**7.24** How many moles are there in 27.2 g of sulfurous acid?

**7.25** $2.16 \times 10^{-2}$ g of potassium carbonate is equal to how many moles of this compound?

**7.26** How many moles are there in $4.77 \times 10^4$ g of sodium hydroxide?

### 7.5 Molar Masses in Calculations: Moles to Grams

This section is the reverse of the previous one. Again, a calculation of the molar mass is necessary to perform this calculation. The following example demonstrates this type of problem.

**Example 7.14:** $1.27 \times 10^{-1}$ mol of silver nitrate is equal to how many grams of this compound?

**Solution:** Using calculations from the periodic table, 1 mol $AgNO_3$ = 169.9 g.

$$1.27 \times 10^{-1} \text{ mol } AgNO_3 \times \frac{169.9 \text{ g } AgNO_3}{\text{mol } AgNO_3} = 21.6 \text{ g } AgNO_3$$

## Practice Problems

**7.27** How many grams are contained in 2.22 moles of calcium fluoride?

**7.28** $9.76 \times 10^{-2}$ moles of uranium(IV) oxide equal how many grams of this compound?

**7.29** What is the mass of 43.1 moles of nitrogen gas?

### 7.6 Percentage Composition

The **percentage** in this case is a reference to mass. A compound contains at least two different atoms, and they will each contribute to a percent of the overall mass of the compound. The sum of the mass percentages of each elemental percentage will equal 100%. The following example illustrates how to solve these problems.

**Example 7.15:** What is the percentage composition of each element in carbon dioxide?

**Solution:** A molecule of $CO_2$ contains:
1 C atom × 12.0 amu/atom = 12.0 amu
2 O atoms × 16.0 amu/atom = 32.0 amu
Total molecular weight = 44.0 amu

The percentage by mass for carbon is 12.0 amu/44.0 amu × 100 = 27.3%.
The percentage by mass for oxygen is 32.0 amu/44.0 amu × 100 = 72.7%.
Total                                                           100.0%

Notice that this problem could have been solved using a mole of carbon dioxide.

A mole of $CO_2$ contains:
1 mol C atoms × 12.0 g/mol = 12.0 g
2 mol O atoms × 16.0 g/mol = 32.0 g
Total molar mass           = 44.0 g

The percentage by mass for carbon is 12.0 g/44.0 g × 100 = 27.3%.
The percentage by mass for oxygen is 32.0 g/44.0 g × 100 = 72.7%.
Total                                                    100%

Units cancel in these types of percentage problems, so making the calculation at either the molecular or molar level leads to the same correct answer.

## Practice Problems

**7.30-7.32** Try one or both of these approaches when solving the problems in Table 7.2.

**Table 7.2** Calculating Mass Percents

| Number | Formula | Name | Molar Mass | Elemental Percents |
|--------|---------|------|------------|--------------------|
|        | $CCl_4$ | Carbon tetrachloride | 154.0 g | 7.79% C<br>92.2% Cl |
| 7.30   | $Pb(SO_4)_2$ |    |            | Pb<br>S<br>O |
| 7.31   |         | Aluminum oxide |    | Al<br>O |
| 7.32   | NaOH    |      |            | Na<br>O<br>H |

## 7.7 Chemical Analysis and Mass Percentages

This section is very similar to Section 7.6. It shows you how to compute mass percent by using experimental techniques. This is often referred to as **elemental analysis**. In this case a measured amount of reactant is decomposed or combusted with oxygen. The masses of the products are carefully measured, and the mass percent of each element in the original reactant can be carefully obtained.

**Example 7.16:** Suppose you have 15.0 g of some unknown compound which contained only carbon, hydrogen and oxygen. By chemical analysis you are able to conclude that the original compound contained:

6.00 g of carbon
1.00 g of hydrogen
8.00 g of oxygen

Use this information to determine the mass percent of each element in the unknown compound.

**Solution:** This problem in very similar to the problems in Section 7.6. In this case you don't know the molar mass of the compound, but the total of mass percent of each element should still equal 100%. The solution follows.

## 110 Chapter 7

$$6.00\text{g C}/15.0\text{g total} \times 100 = 40.0\% \text{ C}$$
$$1.00\text{g H}/15.0\text{g total} \times 100 = 6.7\% \text{ H}$$
$$\underline{8.00\text{g O}/15.0\text{g total} \times 100 = 53.3\% \text{ O}}$$
$$\text{Total} \qquad\qquad\qquad\qquad 100.0\%$$

Use information in the subsequent problems to obtain similar solutions about mass percent. Make the same assumptions that were made in the example problem.

---

**Practice Problems**

Determine the mass percent of each element in the following.

**7.33** A 12.6-g sample of a compound is found to contain 8.18 g C, 1.70 g H, and 2.79 g O.

**7.34** A 1.63-g sample of a compound is found to contain 0.520 g K, 0.473 g Cl, and 0.637 g O.

**7.35** A 52.7-g sample of a compound thought to be water is found to contain 5.86 g H, and 46.8 g O. Do these results indicate that the compound is correctly identified?

**7.36** A 17.8-g sample of a compound is found to contain 13.6 g C and 1.14 g H. If you assume that the remaining mass is from oxygen, what is the mass percent of each?

---

## 7.8 Empirical Formulas

These formulas are derived from experimental evidence; thus the name **empirical**. In Section 7.7 you were introduced to elemental analysis, which provides a method for determining the percent of each element in a compound.

Mathematical treatment of this evidence leads to the *empirical formula (EF)* of the compound. This differs from the *molecular formula (MF)* in that it represents the ratio of each element only in the lowest whole number comparisons.

The determination of empirical formulas is quite easy if the following rules are used.

### Rules for Determination of Empirical Formulas

1. Write each mass percent assuming 100 g total sample.
   For example, 45.4% becomes 45.4 g.
2. Convert each elemental mass to moles.
3. Divide each mole quantity by the smallest quantity and assign as respective subscripts.
4. If numbers in the preceding step are not whole numbers (or nearly so), multiply the subscripts times a small whole number (e.g., 2 or 3) until whole numbers are achieved.

Two examples should demonstrate the application of these rules.

**Example 7.17:** A sample consists of 40% C, 6.67% H, and 53.3% O. What is its empirical formula?

**Solution:** The following conversions [each element to moles (assuming 100-g sample)] are the first steps necessary.

**Chemical Composition** 111

$$40.0 \text{ g C} \times \frac{1 \text{ mol C}}{12.0 \text{ g C}} = 3.33 \text{ mol C}$$

$$6.67 \text{ g H} \times \frac{1 \text{ mol H}}{1.0 \text{ g H}} = 6.67 \text{ mol H}$$

$$53.3 \text{ g O} \times \frac{1 \text{ mol O}}{16.0 \text{ g O}} = 3.33 \text{ mol O}$$

If these numbers are used for subscripts, the empirical formula would be $C_{3.33}H_{6.67}O_{3.33}$. This is not an acceptable form because the numbers are not integers. The second step involves dividing each mole number by the smallest so that whole number subscripts can be obtained.

3.33 mol C ÷ 3.33 = **1** mol C
6.67 mol H ÷ 3.33 = **2** mol H
3.33 mol O ÷ 3.33 = **1** mol O

The numbers obtained from this calculation are whole numbers. Now the empirical formula can be written as **CH$_2$O.**

A second example illustrates a case where **rule 4** must be applied.

**Example 7.18:** What is the empirical formula of a substance that is 48.6% C, 8.11% H, and 43.2% O?

**Solution:** Start solving this problem just as in the case above. The first two steps are shown without all the details. Check your understanding by applying the math, as in the first example.

48.6 g C = 4.05 mol C ÷ 2.70 = 1.5 mol C
8.11 g H = 8.11 mol H ÷ 2.70 = 3.0 mol H
43.2 g O = 2.70 mol O ÷ 2.70 = 1.0 mol O

If the numbers obtained were applied as subscripts, the empirical formula would be $C_{1.5}H_3O$. The is not acceptable because there is a decimal included. By trial and error you can find a small integer that will be used to multiply each subscript. In this case it is 2.

$(C_{1.5}H_3O) \times 2 = C_3H_6O_2 =$ **empirical formula**

---

**Practice Problems**

**7.37** A compound is 87.5% nitrogen and 12.5% hydrogen. What's its empirical formula?

**7.38** A compound is 84.2% carbon and 15.8% hydrogen. What's its empirical formula?

**7.39** A compound is 70.6% C, 5.9% H, and 23.5% O. Calculate its empirical formula.

**7.40** A compound is 52.9% Al and 47.1% O. What is its empirical formula?

## 7.9 Molecular Formulas

This can be calculated if you know the elements present, the percent of each, and the molar mass (see Integrated Skills of Chapter 7). They can also be calculated from the empirical formula (EF) and the molar mass.

This is done by calculating the mass of one empirical formula unit and dividing that mass into the molar mass. This tells you the whole number that you must use to reach the MF. Each element in the EF must be multiplied by the number of EF mass units in the molar mass.

**Example 7.19:** An empirical formula (EF) is determined to be $CH_2O$. The molar mass is 60.0 g. What is the molecular formula (MF)?

**Solution:** Mass of $CH_2O$ = 30.0 g
60.0 g ÷ 30.0 g = 2
$(CH_2O) \times 2 = C_2H_4O_2$ = molecular formula

## Practice Problems

**7.41** A compound has an EF of $NH_2$ and a molar mass of 32.0 g. What is its MF?

**7.42** A compound has an EF of $C_4H_9$ and a molar mass of 114.0 g. What is its MF?

**7.43** A compound has an EF of $C_4H_4O$ and a molar mass of 136.0 g. What is its MF?

**7.44** A compound has an EF of $Al_2O_3$ and a molar mass of 102.0 g. What is its MF?

## Integration of Multiple Skills

There are several problems that you will be able to solve by using more than one skill from this chapter and from previous chapters. Some examples follow.

The first thing you want to be able to do is compare the amounts of an element in various compounds.

**Example 7.20:** Which has more oxygen (by mass), 0.438 mol of dinitrogen monoxide, 12.7 g of nitrogen monoxide, or $2.45 \times 10^{23}$ molecules of nitrogen dioxide?

**Solution:** First, you have to write correct molecular formulas for each compound and convert the amount of oxygen in each to one common unit. The common unit that might be useful in this case is grams. The conversion for each follows:

$$N_2O: 0.438 \text{ mol } N_2O \times \frac{44.0 \text{g } N_2O}{1 \text{ mol } N_2O} \times \frac{16.0 \text{g O}}{44.0 \text{g } N_2O} = 7.01 \text{g O}$$

$$NO: 12.7 \text{g NO} \times \frac{16.0 \text{g O}}{30 \text{g NO}} = 6.77 \text{g O}$$

$$NO_2: 1.23 \times 10^{23} \text{ mcs } NO_2 \times \frac{1 \text{ mol } NO_2}{6.02 \times 10^{23} \text{ mcs } NO_2} \times \frac{46.0 \text{ g } NO_2}{1 \text{ mol } NO_2} \times \frac{32.0 \text{ g O}}{46.0 \text{ g } NO_2} = 6.54 \text{ g O}$$

A comparison of all three solutions reveals that the largest amount of oxygen is in 0.438 mol of dinitrogen monoxide.

The second skill you want to have is the ability to use percent by mass and molar mass to find molecular formula and empirical formula.

**Example 7.21:** The molar mass of a compound is 110 g. It is 65.5% C, 5.45% H, and 29.1% O. What are the molecular formula and empirical formula?

**Solution:** Like many problems in chemistry, this can be solved in several ways. One easy way involves using the mass percent of each element and multiplying it times the molar mass to find the mass contribution of each element.

$$0.655 \times 110 = 72.0 \text{ g C}$$
$$0.0545 \times 110 = 6.00 \text{ g H}$$
$$0.291 \times 110 = 32.0 \text{ g O}$$

To obtain the number of moles of each of these elements in the molecular formula, you just need to multiply each by the molar mass of each element.

$$72.0 \text{ C} \times 1 \text{ mol C}/12.0 \text{ g} = 6 \text{ mol C}$$
$$6.0 \text{ H} \times 1 \text{ mol H}/1.0 \text{ g} = 6 \text{ mol H}$$
$$32.0 \text{ O} \times 1 \text{ mol O}/16.0 \text{ g} = 2 \text{ mol O}$$

The molecular formula is thus $\mathbf{C_6H_6O_2}$.

Remember that by definition, the empirical formula is just the lowest ratio of elements in the molecular formula. By inspection, you should notice that if you divide each subscript in the molecular formula by 2, you will obtain the empirical formula, $\mathbf{C_3H_3O}$.

The third skill is the ability to convert molecules to grams. This requires a series of multiplications using dimensional analysis.

**Example 7.22:** Convert $7.35 \times 10^{21}$ molecules of sulfur dioxide to grams.

**Solution:** Notice that there is no direct conversion for this. However, notice that two different **true** equalities involving the mole can be employed in the solution to this problem.

$$1 \text{ mol} = 6.02 \times 10^{23} \text{ molecules of } SO_2$$
$$1 \text{ mol} = 64.1 \text{ g of } SO_2$$

Notice how both these equalities are used in the solution.

$$7.35 \times 10^{21} \text{ mcs } SO_2 \times \frac{1 \text{ mol } SO_2}{6.02 \times 10^{23} \text{ mcs } SO_2} \times \frac{64.1 \text{ g } SO_2}{1 \text{ mol } SO_2} = 0.783 \text{ g } SO_2$$

The last integrated skill is the ability to convert grams to molecules or atoms. This is the reverse of the previous problem with an additional step. Follow the steps, and cross-off the units that cancel, so that you can understand the logic in the solution to this type of problem.

**Example 7.23:** How many sodium ions are present in 78.5 g of sodium sulfate?

Solution:

$$78.5 \text{ g Na}_2\text{SO}_4 \times \frac{1 \text{ mol Na}_2\text{SO}_4}{142.1 \text{ g Na}_2\text{SO}_4} \times \frac{6.02 \times 10^{23} \text{ mcs Na}_2\text{SO}_4}{1 \text{ mol Na}_2\text{SO}_4} \times \frac{2 \text{ ions Na}^+}{1 \text{ mcs Na}_2\text{SO}_4} =$$

$$6.65 \times 10^{23} \text{ ions Na}^+$$

## Practice Problems

**7.45** Which is larger, $7.21 \times 10^{19}$ atoms or silver or 0.00100 mol of silver?

**7.46** Which is larger, 2.0 mol of $C_6H_{12}O_6$ or 350 g of $C_6H_{12}O_6$?

**7.47** Which contains more oxygen (total mass)?
  A. 22.0 g of $CO_2$    B. 0.938 mol of CO    C. $5.50 \times 10^{23}$ water molecules

**7.48** A compound is 65.5% C and 5.5% H. The rest of the mass is assumed to be oxygen. What is its EF? It has a molar mass of 110 g. What is its MF?

**7.49** How many grams are in $6.25 \times 10^{22}$ molecules of potassium chlorite?

**7.50** How many carbon atoms are in 90.0 g of $C_6H_{12}O_6$?

## Self-Exam

**I. (4 ea) Write T (true) or F (false) in each corresponding blank.**

_____ 1. The molecular weight of one mole of silver is 107.9 amus.

_____ 2. There are $6.02 \times 10^{23}$ oxygen atoms in 1 mole of oxygen molecules.

_____ 3. It is **impossible** to have $1.0 \times 10^{-2}$ atoms of a substance.

_____ 4. It is **possible** to have $1.0 \times 10^{-2}$ moles of a substance.

_____ 5. 1 mole of carbon-12 atoms has a mass of exactly 12 g.

_____ 6. 1 atom of carbon-12 has a mass of exactly 12 amu.

_____ 7. One mole of eggs is smaller than 2 dozen eggs.

_____ 8. The subscripts on a molecular formula are always the same or larger than those of the corresponding empirical formula.

_____ 9. The percent of oxygen in $NO_2$ is greater than in $CO_2$.

_____ 10. Aluminum nitride (AlN) is 50% aluminum and 50% nitrogen by mass.

**II. (4 ea) Multiple Choice. Place the letter of the BEST answer in the corresponding blank.**

_____1. What is the atomic mass of sodium?
    A. 23.0 g    B. 46.0 g    C. 23.0 amu    D. 46.0 amu    E. 11.0 amu

_____2. What is the molar mass of iodine?
    A. 126.9 g    B. 253.8 g    C. 126.9 amu    D. 253.8 amu    E. 53.0 amu

_____3. What is the molecular mass of $NH_3$?
    A. 10.0 g    B. 17.0 g    C. 10.0 amu    D. 15.0 amu    E. 17.0 amu

_____4. What is the molar mass of $Al_2(SO_4)_3$?
    A. 214.3 g    B. 198.8 amu    C. 198.8 g    D. 342.2 amu    E. 342.2 g

_____5. What is the percent sulfur in $Al_2(SO_4)_3$?
    A. 28.1%    B. 17.6%    C. 48.4%    D. 42.7%    E. 2.81%

_____6. Which is the largest amount of oxygen?
    A. 2.0 mol $CO_2$    B. $1.18 \times 10^{24}$ molecules $CO_2$    C. 4400 mg $CO_2$
    D. 80 g $CO_2$    E. $1 \times 10^6$ dozen molecules $CO_2$

_____7. How many grams are in 6.25 mol of $Al_2(SO_4)_3$?
    A. 214 g    B. 2.14 kg    C. 54.8 g    D. 0.548 kg    E. 329 g

_____8. How many molecules are in 6.25 mol of $Al_2(SO_4)_3$?
    A. 214    B. $2.14 \times 10^{23}$    C. $35.5 \times 10^{23}$
    D. $9.04 \times 10^{23}$    E. $3.76 \times 10^{24}$

_____9. $6.25 \times 10^{22}$ molecules of $Na_2CO_3$ is how many grams?
    A. 1020    B. $6.63 \times 10^{22}$    C. 104
    D. 11.0    E. 1.04

_____10. What is the empirical formula of $C_{10}H_{12}O_4$?
    A. $C_{10}H_{12}O_4$    B. CHO    C. $C_{20}H_{24}O_8$    D. $C_5H_6O_2$    E. $C_2H_3O$

**III. Provide Complete Answers**

1. (5 pts) Calculate the molecular mass and the molar mass of carbon tetrabromide.

2. (5 pts) Calculate the number of oxygen atoms in 6.24 mol of glucose ($C_6H_{12}O_6$).

3. (5 pts) Calculate the empirical formula of a substance that is 40.00% carbon, 6.67% hydrogen and 53.33% oxygen.

4. (5 pts) Find the molecular formula for a substance that is 49.0% carbon, 2.7% hydrogen, and 48.3% chlorine with a molar mass of 147 g.

# Answers to Practice Problems

| | | | | | |
|---|---|---|---|---|---|
| **7.1** | 24 g | **7.2** | 29% | **7.3** | 71.5% |
| **7.4** | 2:3 | **7.5** | 5:3:2 | **7.6** | 4:4:1 |

**116   Chapter 7**

**7.7**   $5.04 \times 10^{22}$ atoms H in $2.52 \times 10^{22}$ molecules of $H_2$

**7.8**   $3.34 \times 10^{24}$ atoms N in $1.67 \times 10^{24}$ molecules of $N_2O_4$

**7.9**   $1.82 \times 10^{21}$ $Ca_3(PO_4)_2$

**7.10**   195.1 amu                     **7.11**   44.0 amu

**7.12**   357.4 amu                     **7.13**   $2.53 \times 10^{24}$ atoms

**7.14**   3.52 mol $KClO_3$             **7.15**   $3.01 \times 10^{22}$ atoms

**7.16**   $2.09 \times 10^{-3}$ mol

**7.17**   $2.34 \times 10^{-4}$ mol pennies $\times \dfrac{6.02 \times 10^{23} \text{ pennies}}{\text{mol}} \times \dfrac{2.45 \text{ g}}{\text{penny}} \times \dfrac{1 \text{ lb}}{454 \text{ g}} = 7.60 \times 10^{17}$ lb

**7.18**   $1 \times 10^6$ petals $\times \dfrac{1 \text{ rose}}{15 \text{ petals}} \times \dfrac{1 \text{ mol}}{6.02 \times 10^{23} \text{ roses}} = 1.11 \times 10^{-19}$ mol roses

| Problem | Formula | Name | Molecular (Formula) Weight | Molar Mass |
|---|---|---|---|---|
| 7.19 | $BaI_2$ | Barium iodide | 391.1 amu | 391.1 g |
| 7.20 | CO | Carbon monoxide | 28.0 amu | 28.0 g |
| 7.21 | $(NH_4)_3PO_4$ | Ammonium phosphate | 149.0 amu | 149.0 g |
| 7.22 | $CoCl_2$ | Cobalt(II) chloride | 129.9 amu | 129.9 g |
| 7.23 | $SO_2$ | Sulfur dioxide | 64.0 amu | 64.0 g |

**7.24**   0.332 mol $H_2SO_3$

**7.25**   $1.56 \times 10^{-4}$ mol $K_2CO_3$

**7.26**   1190 mol NaOH                **7.27**   173 g $CaF_2$

**7.28**   26.3 g $UO_2$                **7.29**   1.21 kg $N_2$

| Problem | Formula | Name | Molar Mass | Elemental Percents |
|---|---|---|---|---|
| 7.30 | $Pb(SO_4)_2$ | Lead(IV) sulfate | 399.4 g | 51.9% Pb<br>16.0% S<br>32.1% O |
| 7.31 | $Al_2O_3$ | Aluminum oxide | 102.0 g | 52.9% Al<br>47.1% O |
| 7.32 | NaOH | Sodium hydroxide | 40.0 g | 57.5% Na<br>40.0% O<br>2.5% H |

**7.33**   64.9% C, 13.4% H, 22.1% O

**7.34**   31.9% K, 29.0% Cl, 39.1% O

**7.35**   11.1% H, 88.9% O; yes these agree with $H_2O$

**7.36** 76.4% C, 6.4% H, 17.2% O    **7.37** $NH_2$

**7.38** $C_4H_9$    **7.39** $C_4H_4O$

**7.40** $Al_2O_3$    **7.41** $N_2H_4$

**7.42** $C_8H_{18}$    **7.43** $C_8H_8O_2$

**7.44** $Al_2O_3$

**7.45** 0.00100 mol ($6.02 \times 10^{20}$ atoms Ag) [larger than $7.21 \times 10^{19}$ atoms of Ag]

**7.46** 2.0 mol $C_6H_{12}O_6$ (360 g) [larger than 350 g of $C_6H_{12}O_6$]

**7.47** 22.0 g $CO_2$ (16.0 g O); [larger than 0.938 mol CO (15.0 g O) or $5.50 \times 10^{23}$ $H_2O$ molecules (14.6 g O)]

**7.48** EF = $C_3H_3O$; MF = $C_6H_6O_2$

**7.49** 11.1 g $KClO_2$

**7.50** $1.81 \times 10^{24}$ atoms C

## Answers to Self-Exam

**I. T/F**
1. F    2. F    3. T    4. T    5. T    6. T    7. F    8. T    9. F    10. F

**II. MC**
1. C    2. B    3. E    4. E    5. A    6. A    7. B    8. E    9. D    10. D

**III. Complete Answers**

1. 1 molecule $CBr_4$ = 331.6 amu; 1 mole $CBr_4$ = 331.6 g

2. $2.25 \times 10^{25}$ atoms oxygen

3. $CH_2O$

4. $C_6H_4Cl_2$

# 8
# Quantities in Chemical Reactions

This chapter uses the balanced chemical equation to design strategies for precise outcomes. You will learn to convert the mass of one substance into a proportional amount of another substance in the equation. Like Chapter 7, this chapter is very heavy on calculational chemistry.

Keys for success in this chapter:

- Learn terminology.
- Review *dimensional analysis* and *molar mass* calculations.
- Examine example problems and understand how the solutions were obtained.
- Practice as many problems as possible. Spend 60% to 70% of your time on practice.

## Summary of Verbal Knowledge

**stoichiometry:** This is the system of using a balanced chemical equation to calculate quantities of the reactants and products of a reaction.
**limiting reactant:** The reactant that is used up when a reaction goes to completion.
**stoichiometric quantity:** This is a case when the amounts of reactants exactly match those dictated by the balanced equation.
**theoretical yield:** The calculated maximum mass of a product that can be obtained from the given amounts of reactants.
**actual yield:** The actual amount of product that is obtained from the given amounts of reactants.
**percentage yield:** The actual yield divided by the theoretical yield times 100.

## Review of Mathematical and Calculator Skills

All the skills you practiced in Chapters 6 and 7 will come in handy in this chapter. Check your skills on the first couple of problems.

### Practice Problems

**8.1** Write a balanced equation for the reaction of lead(II) nitrate with potassium iodide.

**8.2** What is the molar mass of $Na_3PO_4$?

**8.3** How many grams are there in $3.49 \times 10^{-2}$ mol of $Na_3PO_4$?

**8.4** How many moles are contained in 28.9 g of $Na_3PO_4$?

# Quantities in Chemical Reactions

## Application of Skills and Concepts

### 8.1 Interpretation of the Balanced Chemical Equation

In Chapter 6, the equation below could be described in the following words:

$$2Na + Cl_2 \rightarrow 2NaCl$$

> *Two atoms of sodium react with one molecule chlorine to form two ionic formula units of sodium chloride.*

The above equation could be described on the molar level. [This can be done because we know that Avogadro's number can be used to convert numbers of molecules or atoms to grams (molar mass).]

> *Two moles of sodium react with one mole of chlorine to form two moles of sodium chloride.*

In practice, most calculations and descriptions of reactions are made in the latter format.

---

**Practice Problems**

**8.5** Describe the equation in words (at the molar level): $2H_2O \rightarrow 2H_2 + O_2$

**8.6** Describe the equation in words (at the molar level): $3H_2 + N_2 \rightarrow 2NH_3$

**8.7** Write an equation for "one mole of phosphoric acid reacts with three moles of potassium hydroxide to produce three moles of water and one mole of potassium phosphate."

**8.8** Write an equation for "one mole of chlorine reacts with two moles of lithium bromide to produce two moles of lithium chloride and one mole of bromine."

---

### 8.2 Mole Calculations from Chemical Equations

This section shows how to use balanced equations and their set ratios of substances to predict relative amounts of other substances in the equations. Examples are the best way to illustrate this.

**Example 8.1:** The balanced equation for the formation of water from hydrogen and oxygen is shown below. If 4 moles of hydrogen are used, how many moles of oxygen are needed?

$$2H_2 + O_2 \rightarrow 2H_2O$$

**Solution:** By inspection and use of your *number sense*, you can see that 4 is twice the coefficient in front of hydrogen, which would mean that the coefficient in front of oxygen must also be doubled, thus 2 moles of oxygen would be needed. Often the numbers of real-life situations are not this easy to manipulate in your head. Dimensional analysis combined with the coefficients in the balanced equation can be used to make sure that the correct solutions is obtained. For the above equation, three equalities involving the substances can be written.

2 mol hydrogen = 1 mol oxygen
2 mol hydrogen = 2 mol water
1 mol oxygen = 2 mol water

Notice that all the equalities are derived from the coefficients in the balanced equation. To solve the above problem, formally, use the first equality. To decide which substance to put in the denominator, use the usual rules for dimensional analysis. The denominator should contain the same units as the previous numerator. The solution follows and matches the previous answer obtained mentally.

$$4 \text{ mol } H_2 \times \frac{1 \text{ mol } O_2}{2 \text{ mol } H_2} = 2 \text{ mol } O_2$$

**Example 8.2:** How many mols of each reactant are needed to produce 17 mol water?

**Solution:** This solution uses the other two equalities listed above. Again, choose the equality based on the two substances in the balanced equations that need to be compared.

$$17 \text{ mol } H_2O \times \frac{2 \text{ mol } H_2}{2 \text{ mol } H_2O} = 17 \text{ mol } H_2$$

$$17 \text{ mol } H_2O \times \frac{1 \text{ mol } O_2}{2 \text{ mol } H_2O} = 8.5 \text{ mol } O_2$$

### Practice Problems
Use the equation below to answer problems 8.9-8.11:

$$2NaOH + H_2SO_4 \rightarrow 2H_2O + Na_2SO_4$$

**8.9** How many moles of sulfuric acid are needed to completely react with 0.525 mol NaOH?

**8.10** How many moles of each reactant are needed to produce 2.39 mol $Na_2SO_4$?

**8.11** If 0.786 mol of water is produced, how many moles of $Na_2SO_4$ are produced?

## 8.3 Mass Calculations from Chemical Equations

This section uses the same equalities of the balanced equations as the problems in Section 8.2, but it expands on the information. In this section you will learn to use dimensional analysis, equalities from the balanced equation, and molar masses to convert the following:

- Grams of substance **A** to moles of substance **B**.
- Moles of substance **A** to grams of substance **B**.
- Grams of substance **A** to grams of substance **B**.

For these problems, **A** is designated as the substance which you have information about and **B** as the substance that needs to be determined. The following problems illustrate typical questions that will often be encountered. The solutions use Figure 8.1 as a scheme which helps decide where to start on a problem and where to finish.

# Quantities in Chemical Reactions 121

**Figure 8.1** Scheme for Conversions Within a Balanced Equation

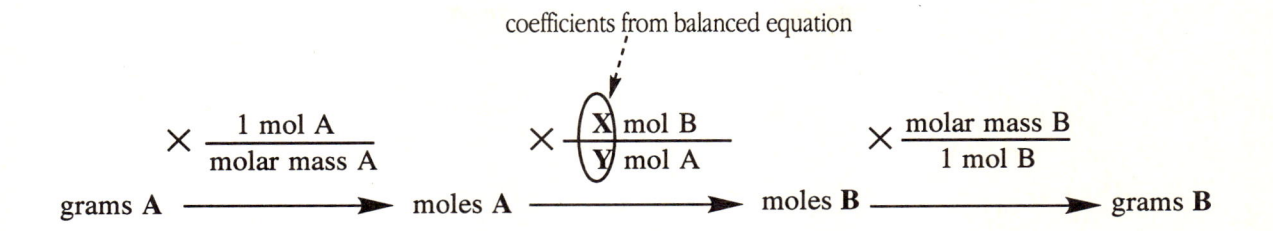

coefficients from balanced equation

grams A $\xrightarrow{\times \dfrac{1 \text{ mol A}}{\text{molar mass A}}}$ moles A $\xrightarrow{\times \dfrac{X \text{ mol B}}{Y \text{ mol A}}}$ moles B $\xrightarrow{\times \dfrac{\text{molar mass B}}{1 \text{ mol B}}}$ grams B

**Example 8.3:** The balanced equation for the formation of silver and copper(II) nitrate from copper and silver nitrate is shown below. If 5.24 g of copper is used, how many moles of silver nitrate are needed for this reaction to occur?

$$2AgNO_3 + Cu \rightarrow 2Ag + Cu(NO_3)_2$$

**Solution:** Notice that the scheme in Figure 8.1 shows that you can convert grams of **A** (the substance you have information about) to grams of **B** (the substances you want to know about). The solution in this process always goes through an equality that involves the coefficients of **A** and **B** from the balanced equation.

To solve the above problem, you need to start at grams of **A** (copper) and stop at moles of **B** (silver nitrate). Carefully follow the steps shown below. Notice that the coefficients of the balanced equations in the solutions below are highlighted in **bold**.

$$5.24 \text{ g Cu} \times \frac{1 \text{ mol Cu}}{63.6 \text{ g Cu}} \times \frac{\mathbf{2} \text{ mol AgNO}_3}{\mathbf{1} \text{ mol Cu}} = 0.165 \text{ mol AgNO}_3$$

**Example 8.4:** How many grams of copper are needed to produce 1.79 mol silver?

**Solution:** This problem involves going from moles of **A** (silver) to grams of **B** (copper).

$$1.79 \text{ mol Ag} \times \frac{\mathbf{1} \text{ mol Cu}}{\mathbf{2} \text{ mol Ag}} \times \frac{63.6 \text{ g Cu}}{1 \text{ mol Cu}} = 56.9 \text{ g Cu}$$

**Example 8.5:** How many grams of silver nitrate are needed to produce 0.578 g silver?

**Solution:** This problem involves going from grams of **A** (silver nitrate) to grams of **B** (silver).

$$0.578 \text{ g Ag} \times \frac{1 \text{ mol Ag}}{107.9 \text{ g Ag}} \times \frac{\mathbf{1} \text{ mol AgNO}_3}{\mathbf{1} \text{ mol Ag}} \times \frac{169.9 \text{ g AgNO}_3}{1 \text{ mol AgNO}_3} = 0.910 \text{ g AgNO}_3$$

Use the scheme in Figure 8.1 to design solutions for problems when amounts of one substance in a balanced equation need to be converted to amounts of another. Make sure to select the correct starting and stopping points. Remember that coefficients from the balanced equation will always be used at some point in these solutions.

> **Practice Problems**
> Use the following balanced equation to answer problems **8.12-8.15**:
>
> $$2KClO_3 \rightarrow 2KCl + 3O_2$$
>
> **8.12** How many moles of oxygen are produced from 28.7 g of potassium chlorate?
>
> **8.13** How many grams of KCl are produced from 0.456 mol of $KClO_3$?
>
> **8.14** How many grams of KCl are produced from enough $KClO_3$ to form 350 g $O_2$?
>
> **8.15** How many grams of $KClO_3$ are needed to form 22.2 g of KCl?

## 8.4 Identifying Limiting Reactants

In a chemical reaction, if exact ratios of reactants match those in the balanced equation, the reaction is said to occur with **stoichiometric** amounts of reactants. In reality, this is often not the case. For example, when you burn gasoline from your car, it reacts with oxygen from the air. There is usually an unlimited amount of oxygen, but once all the gas in the tank has been converted to products, the reaction is over.

For reactions where the conditions do not have stoichiometric amounts, the reactant that is totally consumed is known as the **limiting reactant**, and that which is partially left over is called the **excess reagent**.

The purpose of this section is to learn how to identify each. The following simple rules serve as a great guide.

1. Use amount of reactant **A** (select either) to calculate stoichiometric amount of **B**.
2. Compare calculated amount of **B** to given amount of **B**, and select one of the following:
   If given **B** equals calculated **B**, there is no limiting reagent.
   If given **B** is larger than calculated **B**, then **A** is the limiting reagent.
   If given **B** is smaller than calculated **B**, then **B** is the limiting reagent.

Some sample problems with EQ 8.1 will demonstrate how these problems can be solved. Remember, it doesn't matter which reactant you select for **A** or **B**.

$$2AgNO_3 + CaCl_2 \rightarrow 2AgCl(s) + Ca(NO_3)_2 \qquad \text{(EQ 8.1)}$$

**Example 8.6:** If 0.40 mol of silver nitrate and 0.20 mol of calcium chloride are combined, which is the limiting reagent?

**Solution:** Try selecting silver nitrate as reagent **A**. The conversion below shows that 0.40 moles of this would require 0.20 moles of calcium chloride.

$$0.40 \text{ mol } AgNO_3 \times \frac{1 \text{ mol } CaCl_2}{2 \text{ mol } AgNO_3} = 0.20 \text{ mol } CaCl_2$$

Since this is the precise amount of calcium chloride in the reaction, neither reagent is limiting, and this reaction occurs under stoichiometric amounts of reagents.

**Example 8.7:** If 7.28 mol of silver nitrate and 3.80 mol of calcium chloride are combined, which is the limiting reagent?

**Solution:** Try using silver nitrate again as **A**. The calculation shows that 7.28 mol of silver nitrate requires

$$7.28 \text{ mol AgNO}_3 \times \frac{1 \text{ mol CaCl}_2}{2 \text{ mol AgNO}_3} = 3.64 \text{ mol CaCl}_2$$

In this case calculated **B** (CaCl$_2$) is smaller than actual **B**, so **A** (CaCl$_2$) is **limiting.**

If CaCl$_2$ had been selected as reactant **A**, the following calculation would still have led to CaCl$_2$ as the limiting reagent. In this case calculated **B** (AgNO$_3$) is larger than actual **B**, so **A** (CaCl$_2$) is **limiting.**

$$3.80 \text{ mol CaCl}_2 \times \frac{2 \text{ mol AgNO}_3}{1 \text{ mol CaCl}_2} = 7.60 \text{ mol AgNO}_3$$

**Example 8.8:** If 10.3 g of silver nitrate and 3.78 g of calcium chloride are combined, which is the limiting reagent?

**Solution:** Select silver nitrate as **A**, and solve for **B**.

$$10.3 \text{ g AgNO}_3 \times \frac{1 \text{ mol AgNO}_3}{169.9 \text{ g AgNO}_3} \times \frac{1 \text{ mol CaCl}_2}{2 \text{ mol AgNO}_3} \times \frac{111.1 \text{ g CaCl}_2}{1 \text{ mol CaCl}_2} = 3.37 \text{ g CaCl}_2$$

Since 10.3 g of silver nitrate completely reacts with 3.37 g of calcium chloride and 3.78 g of calcium chloride was used in the reaction, CaCl$_2$ is in **excess** and AgNO$_3$ is **limiting.**

---

## Practice Problems

Use the following to answer which (if any) is the limiting reagent in each case below.

$$2HCl(aq) + Na_2CO_3(aq) \rightarrow 2NaCl(aq) + H_2O(l) + CO_2(g)$$

**8.16** 0.167 mol HCl and 0.100 mol Na$_2$CO$_3$

**8.17** 5.0 mol HCl and 2.5 mol Na$_2$CO$_3$

**8.18** 7.3 g HCl and 9.9 g Na$_2$CO$_3$

**8.19** 0.15 g HCl and 0.25 g Na$_2$CO$_3$

**8.20** 5.40 × 10$^{-3}$ mol HCl and 0.213 g Na$_2$CO$_3$

---

## 8.5 Calculations with Limiting Reactants

The calculations with limiting reactants usually involve predicting the amount of one or more of the products from given information about amounts of reactants.

**Example 8.10:** Using the reaction below, calculate the amount of solid calcium hydroxide that will be formed from 9.6 g of each reactant:

$$CaCl_2(aq) + 2NaOH(aq) \rightarrow 2NaCl(aq) + Ca(OH)_2(s)$$

**Solution:** In this case, you can use stoichiometry to convert each of the starting reagents to a theoretical amount of product.

$$9.6 \text{ g CaCl}_2 \times \frac{1 \text{ mol CaCl}_2}{111.1 \text{ g CaCl}_2} \times \frac{1 \text{ mol Ca(OH)}_2}{1 \text{ mol CaCl}_2} \times \frac{74.1 \text{ g Ca(OH)}_2}{1 \text{ mol Ca(OH)}_2} = 6.4 \text{ g Ca(OH)}_2$$

$$9.6 \text{ g NaOH} \times \frac{1 \text{ mol NaOH}}{40.0 \text{ g NaOH}} \times \frac{1 \text{ mol Ca(OH)}_2}{2 \text{ mol NaOH}} \times \frac{74.1 \text{ g Ca(OH)}_2}{1 \text{ mol Ca(OH)}_2} = 8.9 \text{ g Ca(OH)}_2$$

In the above calculations, the calcium chloride is converted to a **smaller** amount of product. This is the **real theoretical yield,** because once that amount of product is produced, the limiting reagent is totally consumed, and no more product can be produced.

When trying to calculate theoretical yield given the amounts of reactants, do the following:

> *Convert each amount of starting material to an amount of one of the products. The **smaller** amount is **always** the theoretical yield.*

**Example 8.11:** For the above reaction, calculate the amount of the excess reagent that will remain.

**Solution:** Use stoichiometry and dimensional analysis to see how much of the reagent in excess is needed to react with all the limiting reagent.

$$9.6 \text{ g CaCl}_2 \times \frac{1 \text{ mol CaCl}_2}{111.1 \text{ g CaCl}_2} \times \frac{2 \text{ mol NaOH}}{1 \text{ mol CaCl}_2} \times \frac{40.0 \text{ g NaOH}}{1 \text{ mol NaOH}} = 6.9 \text{ g NaOH}$$

Notice that the amount of NaOH used is smaller than the amount used in the reaction. The amount left is simply

9.6 g - 6.9 g = **3.3 g NaOH left** after the reaction is complete.

When trying to calculate the amount of excess reactant remaining given amounts of reactants and the limiting reactant, do the following:

> *Convert the amount of limiting reactant to the amount of excess reactant needed. Subtract the amount of excess reactant needed from the amount used in the reaction.*

## Practice Problems

Use the following balanced equation to answer problems **8.21-8.24**:

$$3H_2 + N_2 \rightarrow 2NH_3$$

**8.21** How many moles of $NH_3$ can be produced from 2.6 mol $H_2$ and 0.90 mol $N_2$?

**8.22** How much of the reagent in excess is left?

**8.23** How many grams of $NH_3$ are produced from 62.5 g $H_2$ and 315 g of $N_2$?

**8.24** How much of the reagent in excess is left?

## 8.6 Percentage Yields

These types of problems compare the **actual yield** to the **theoretical yield**. The former is always **smaller or equal to** the latter. The actual yield often is not as high as expected from stoichiometric calculations. This could be the result of several factors, such as product left on the side of the reaction vessel, incomplete reaction, or side reactions.

To solve these problems, do the following:

- Calculate the theoretical yield.
- Divide the actual yield by the theoretical yield and multiply by 100.

**Example 8.12:** For the reaction below, if 1.57 g of $C_7H_6O_2$ was used with an excess of $SOCl_2$ and 1.71 g of $C_7H_5ClO$ was obtained, what is the percent yield?

$$C_7H_6O_2 + SOCl_2 \rightarrow HCl + SO_2 + C_7H_5ClO$$

**Solution:** The theoretical yield can be calculated from the limiting reagent, $C_7H_5ClO$. Convert it to mass of product.

$$1.57 \text{ g } C_7H_6O_2 \times \frac{1 \text{ mol } C_7H_6O_2}{122.0 \text{ g } C_7H_6O_2} \times \frac{1 \text{ mol } C_7H_5ClO}{1 \text{ mol } C_7H_6O_2} \times \frac{140.5 \text{ g } C_7H_5ClO}{1 \text{ mol } C_7H_5ClO} = 1.81 \text{ g } C_7H_5ClO$$

The actual yield of this product from the lab is 1.71g (less than theoretical). The percent yield is solved as follows:

$$\text{percent yield} = \frac{\text{A.Y.}}{\text{T.Y.}} \times 100 = \frac{1.71 \text{ g}}{1.81 \text{ g}} \times 100 = 94.5\%$$

## Practice Problems

Use the balanced equation that follows to solve problems **8.25-8.30**.

$$C_7H_5ClO + C_2H_6O \rightarrow HCl + C_9H_{10}O_2$$

**8.25** If the theoretical yield of $C_9H_{10}O_2$ is 25.6 g, and the actual yield is 21.2 g, what is the percentage yield?

**8.26** If you need 315 g of $C_9H_{10}O_2$ and you know the reaction can lead to 82.8% yield, how much does the theoretical yield need to be?

**8.27** If you need 105.0 g of $C_9H_{10}O_2$ and the reaction gives 82.8% yield, how much of each reactant should be used?

**8.28** If 125 g of $C_7H_5ClO$ and 65.5 g of $C_2H_6O$ are combined, what is the theoretical yield of $C_9H_{10}O_2$? What is the percentage yield if 112 g is actually obtained?

## Integration of Multiple Skills

Real reaction conditions often require the use of stoichiometry as described in this chapter with previous skills for unit conversion. The following example shows one of these instances.

**Example 8.11:** For the reaction below, if $1.00 \times 10^6$ gallons of gasoline ($C_8H_{18}$) is burned in an excess of oxygen, how many pounds of carbon dioxide are produced? (Density of gasoline is 0.703 g/mL.)

$$2C_8H_{18} + 25O_2 \rightarrow 18H_2O + 16CO_2$$

**Solution:** This problem can be solved by the following conversions. Notice that some information must be obtained from earlier tables.

gal gas → L gas → mL gas → g gas → mol gas → mol $CO_2$ → g $CO_2$ → lb $CO_2$

(Note that the solution is continued on the second line because of space limitations.)

$$1.00 \times 10^6 \text{ gal } C_8H_{18} \times \frac{3.784 \text{ L } C_8H_{18}}{1 \text{ gal } C_8H_{18}} \times \frac{1000 \text{ mL } C_8H_{18}}{1 \text{ L } C_8H_{18}} \times \frac{0.703 \text{ g } C_8H_{18}}{1 \text{ mL } C_8H_{18}} \times \frac{1 \text{ mol } C_8H_{18}}{114 \text{ g } C_8H_{18}} \times$$

$$\frac{16 \text{ mol } CO_2}{2 \text{ mol } C_8H_{18}} \times \frac{44.0 \text{ g } CO_2}{1 \text{ mol } CO_2} \times \frac{1 \text{ lb } CO_2}{454 \text{ g } CO_2} = 1.81 \times 10^7 \text{ lb } CO_2 = 18 \text{ million lb } CO_2$$

## Practice Problems

Due to their difficulty, the problems below are shown in some detail in the answer section.

**8.29** Using the reaction for the combustion of gasoline, how many pounds of water are produced from burning $2.00 \times 10^6$ gallons of $C_8H_{18}$?

**8.30** If a city has 12,500 drivers who average 678 miles a week at an average rate of 27.2 mi/gal, how many pounds of carbon dioxide does this city produce in a week?

# Self-Exam

**I. (4 ea) Write T (true) or F (false) in each corresponding blank.**

_____1. 2.0 mol of water is larger than 20.0 g of water.

_____2. The coefficients in a balanced equation can refer to the substances at the molecular level or the molar level.

_____3. Mass calculations can be made on an *unbalanced equation*.

_____4. The limiting reagent is the one that remains once a reaction is complete.

_____5. The actual yield cannot be higher than the theoretical yield.

Use the following balanced equation to answer questions 6-10.

$$CH_4 + 2O_2 \rightarrow CO_2 + 2H_2O$$

_____6. If 1 mol of oxygen is used, 0.5 mol of $CH_4$ is needed.

_____7. If 4 mol of $CH_4$ is reacted with an excess of oxygen, 2 mol of $CO_2$ is the most that can be produced.

_____8. If there are 32 g of $CH_4$ and 32 g of $O_2$, oxygen is the limiting reagent.

_____9. If 88 g of carbon dioxide is produced from the reaction, 88 g of water is produced.

_____10. The mole ratio of $CH_4$ to $O_2$ in the reaction is 1:2.

**II. (4 ea) Multiple Choice. Place the letter of the BEST answer in the corresponding blank.**

_____1. Sulfur dioxide reacts with oxygen to form sulfur trioxide. The correct balanced equation for this reaction is
   A. $SO_2 + O \rightarrow SO_3$ 
   B. $SO_2 + O_2 \rightarrow SO_3$
   C. $SO_2 + O_2 \rightarrow 2SO_3$ 
   D. $2SO_2 + O \rightarrow 2SO_3$
   E. $2SO_2 + O_2 \rightarrow 2SO_3$

_____2. For the reaction $Ba(OH)_2 + Na_2SO_4 \rightarrow BaSO_4(s) + 2NaOH$, how many moles of solid barium sulfate will be produced from 1.2 mol of each reactant?
   A. 2.4 mol   B. 1.2 mol   C. 0.6 mol   D. 233.4 mol   E. 116.7 mol

_____3. For the reaction in question 2, how many grams of solid barium sulfate will be produced from 6.79 g of barium hydroxide with an excess of sodium sulfate?
   A. 9.25 g   B. 6.79 g   C. 4.98 g   D. 4.65 g   E. 2.33 g

_____4. For the reaction in question 2, how many grams of each reactant are needed to produce 3.56 g of barium sulfate?
   A. 3.56 g $Ba(OH)_2$ and 3.56 g $Na_2SO_4$
   B. 1.78 g $Ba(OH)_2$ and 1.78 g $Na_2SO_4$
   C. 171 g $Ba(OH)_2$ and 142 g $Na_2SO_4$
   D. 2.61 g $Ba(OH)_2$ and 2.17 g $Na_2SO_4$
   E. 4.85 g $Ba(OH)_2$ and 5.85 g $Na_2SO_4$

## 128  Chapter 8

_____5. For the reaction in question 2, if the theoretical yield is 25.7 g of barium sulfate and 22.6 g of barium sulfate is actually recovered, what is the percentage yield?
   A. 22.6%   B. 25.7%   C. 87.9%   D. 95.5%   E. 114%

Use the following balanced equation to answer questions 6-10.

$$2C_4H_{10} + 9O_2 \rightarrow 8CO_2 + 10H_2O$$

_____6. If 5 mol of water is produced, how many moles of oxygen are required?
   A. 5.0 mol   B. 4.5 mol   C. 10.0 mol   D. 7.5 mol   E. 1.0 mol

_____7. This equation represents the burning of butane in atmospheric oxygen. In a real-life situation, what is usually the limiting reagent?
   A. $C_4H_{10}$   B. $O_2$   C. $CO_2$   D. $H_2O$   E. Atmosphere

_____8. What is the maximum amount of water that can be produced from 6 mol of hydrogen and 27 mol of oxygen?
   A. 33 mol   B. 180 g   C. 540 g   D. 1.21 kg   E. 10 mol

_____9. If a 79.8% yield of carbon dioxide is obtained from the reaction, where the theoretical yield was 27.7 g, how many grams of carbon dioxide were obtained?
   A. 52.1 g   B. 34.7 g   C. 27.7 g   D. 22.1 g   E. 2.77 g

_____10. What mass of carbon dioxide can be produced from burning 1.00 L of butane (density = 0.579 g/mL)?
   A. 35.1 g   B. 17.6 g   C. 878 g   D. 3.51 kg   E. 1.76 kg

### III. Provide Complete Answers

1. (5 pts) For the following equation, what is the limiting reagent if 16.5 g of copper is reacted with 68.9 g of silver nitrate?

$$2AgNO_3 + Cu \rightarrow 2Ag + Cu(NO_3)_2$$

2. (5 pts) For the reaction in question 1, what is the mass of silver that can be recovered?

3. (5 pts) For the reaction in question 1, how much of the reagent in excess remains after the reaction is completed?

4. (5 pts) What is the name of the other product in this reaction?

## Answers to Practice Problems

**8.1**  $Pb(NO_3)_2 + 2KI \rightarrow PbI_2(s) + 2KNO_3$   **8.2**   164 g $Na_3PO_4$

**8.3**  5.72 g $Na_3PO_4$   **8.4**   0.176 mol $Na_3PO_4$

**8.5**  Two moles of water decompose to two moles of hydrogen and one mole of oxygen.

**8.6**  Three moles of hydrogen combine with one mole of nitrogen to produce two moles of $NH_3$.

8.7 $H_3PO_4 + 3KOH \rightarrow 3H_2O + K_3PO_4$

8.8 $Cl_2 + 2LiBr \rightarrow 2LiCl + Br_2$

8.9 0.263 mol $H_2SO_4$

8.10 4.78 mol NaOH, 2.39 mol $H_2SO_4$

8.11 0.393 mol $Na_2SO_4$

8.12 0.351 mol $O_2$

8.13 34.4 g KCl

8.14 550 g KCl

8.15 36.1 g $KClO_3$

8.16 0.167 mol HCl

8.17 Neither is limiting.

8.18 9.9 g $Na_2CO_3$

8.19 0.15 g HCl

8.20 0.213 g $Na_2CO_3$

8.21 1.7 mol $NH_3$ ($H_2$ is limiting.)

8.22 0.03 mol $N_2$ left (0.90 mol minus 0.87 mol)

8.23 354 g of $NH_3$ ($H_2$ is limiting)

8.24 23 g of $N_2$ left (315 g minus 292 g)

8.25 82.8%

8.26 380 g $C_9H_{10}O_2$

8.27 118 g $C_7H_5ClO$, 38.9 g $C_2H_6O$

8.28 133 g of $C_9H_{10}O_2$, 84.2%

8.29 gal gas → L gas → mL gas → g gas → mol gas → mol $H_2O$ → g $H_2O$ → lb $H_2O$ = $1.67 \times 10^7$ lb of water

8.30 $12{,}500 \text{ drivers} \times \dfrac{678 \text{ mi}}{\text{drivers-week}} \times \dfrac{1 \text{ gal } C_8H_{18}}{27.2 \text{ mi}} \times \dfrac{3.784 \text{ L } C_8H_{18}}{1 \text{ gal } C_8H_{18}} \times \dfrac{1000 \text{ mL } C_8H_{18}}{1 \text{ L } C_8H_{18}} \times$

$\dfrac{0.703 \text{ g } C_8H_{18}}{1 \text{ mL } C_8H_{18}} \times \dfrac{1 \text{ mol } C_8H_{18}}{114 \text{ g } C_8H_{18}} \times \dfrac{16 \text{ mol } CO_2}{2 \text{ mol } C_8H_{18}} \times \dfrac{44.0 \text{ g } CO_2}{1 \text{ mol } CO_2} \times \dfrac{1 \text{ lb } CO_2}{454 \text{ g } CO_2} =$

$5.54 \times 10^6$ lb $CO_2$ = 5.54 million lb $CO_2$

## Answers to Self-Exam

### I. T/F
1. F  2. T  3. F  4. F  5. T  6. T  7. F  8. T  9. F  10. T

### II. MC
1. E  2. B  3. A  4. D  5. C  6. B  7. A  8. C  9. D  10. E

### III. Complete Answers
1. Silver nitrate

2. 43.8 g Ag

3. 3.6 g Cu

4. Copper(II) nitrate

# 9
# Electron Structure of Atoms

This chapter takes a break from the high stress on computational chemistry that you encountered in Chapters 7 and 8. Chapter 9 will have a minimum amount of calculations and very many new conceptual ideas. These concepts will be applied at the atomic level, so they are too small to see. The textbook presents models of what is happening at the atomic level to make it easier to understand these abstract ideas.

Keys for success in this chapter:

- Read the chapter to understand the new concepts.
- Learn the definition of the new terms.
- Practice the problems at the end of each section.
- Use the periodic table for problem solutions.
- Look for patterns of behavior in the periodic table to decrease memory work.

## Summary of Verbal Knowledge

**wavelength:** The distance between two peaks (or the distance between two troughs) of a wave.
**photon:** A particle of light consisting of an extremely small packet of energy.
**energy level:** This is one of the allowed energy values that an electron in an atom can have.
**orbital:** This is a mathematical function (called a **wave function**) describing a region in space where an electron is likely to be found.
**electron shell:** This is an energy level describing a region of space where electrons can be found. It is numbered from 1 to infinity ($n = 1 \cdots n = \infty$)
**subshell:** This is a subset of orbitals of an electron shell. All subshells (of a given shell, n) have exactly the same energy and the same shape.
**electron configuration:** A particular distribution of electrons among the different subshells of an atom.
**Pauli exclusion principle:** This says that an orbital can hold no more than two electrons.
**Hund's rule:** This says that electrons do not pair up in an orbital unless all orbitals in a subshell already contain one electron.
**valence electrons:** These are the electrons in the outer shell of an atom.
**periodic law:** This says that when you arrange the elements by atomic number, their chemical and physical properties vary periodically.
**isoelectronic:** Atoms with the same number of electrons. Cl⁻ is isoelectronic with Ar, since both have 10 electrons.
**ionization energy:** The energy needed to remove an electron from its nucleus (to infinity).
**electron affinity:** The ease with which an atom will accept an extra electron. A *strong* electron affinity means the atom easily accepts an extra electron.

**atomic radius:** The distance from the center of an atom to the outside of the area occupied by its electrons.
**ionic radium:** The distance from the center of an ion to the outside of the area occupied by its electrons.
**metallic character:** The ability of an atom to behave like a metal (i.e., easily lose electrons).

## Review of Mathematical and Calculator Skills

This chapter is not too heavy on the mathematical skills. One thing you need to review is how to square numbers. Look at equation, EQ 9.1 and Example 9.1.

$$n^2 = n \times n \qquad (EQ\ 9.1)$$

**Example 9.1:** What is $3^2$?

**Solution:** $3^2 = 3 \times 3 = 9$

Conversions using dimensional analysis also should be reviewed. You may need to go back to Chapter 2 and check some of the equalities in earlier tables.

**Example 9.2:** How many meters are in 540 nm?

**Solution:** $540\ nm \times \dfrac{1\ m}{1 \times 10^9\ nm} = 5.4 \times 10^{-7}\ m$

---

**Practice Problems**

**9.1** What is $1^2$?

**9.2** What is $6^2$?

**9.3** How many meters are in 720 nm?

**9.4** How many nanometers are in $7.10 \times 10^{-7}$ m?

---

## Application of Skills and Concepts

### 9.1 Light and Other Forms of Electromagnetic Radiation

Just as the title to this section implies, **light** is a form of **electromagnetic radiation**. Some common forms of this type of energy include the following:

- Visible light
- Ultraviolet light
- Infrared light
- Microwaves
- Radio waves

## Chapter 9

- X-rays
- Gamma rays

All these forms of radiation share many things in common.

- They travel at the speed of light (3.00 × 10⁸ m/s) in a vacuum.
- They all have characteristic wavelengths and frequencies.
- They all have an electric component and a magnetic component, hence the name.
- They all have characteristic energy which is directly proportional to the frequency and indirectly proportional to the wavelength.

This course will not treat the energy involved in a quantitative manner, but you should develop a qualitative intuition for the size of the numbers used. The speed of light is usually symbolized by the letter "c," as shown in equation, EQ 9.2. It is a product of the wavelength times the frequency. You should realize that this is a very high rate of speed. Under normal distances it cannot be detected. Rather, when a light is turned on, it seems to travel from the source to your eye instantaneously.

$$c = \lambda \nu \qquad (EQ\ 9.2)$$

$c$ = speed of light (electromagnetic radiation), $3.00 \times 10^8$ m/s

$\lambda$ = wavelength of light in meters (represented by the Greek letter lambda)

$\nu$ = frequency of light in cycles per second (s⁻¹), also known as hertz (Hz).

The wavelength can be understood by examining the wave in Figure 9.1. All forms of electromagnetic radiation travel in a wave (sine wave) like that shown in Figure 9.1. The wavelength can be calculated either by measuring from crest to crest or trough to trough.

**Figure 9.1 Wave of EM radiation**

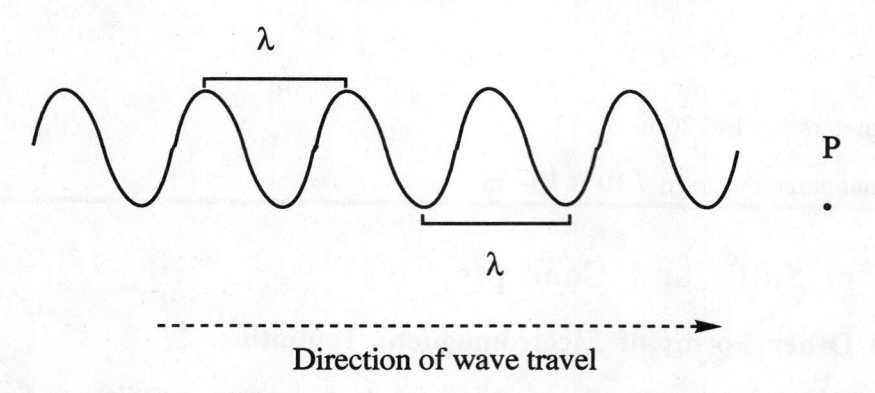

Direction of wave travel

The frequency can be measured by counting the number of waves that go past a given location (like point P) in a second.

Even though we will not do any specific calculations with the wavelength and frequency of electromagnetic radiation, you should recognize that the waves are traveling very fast, and the frequencies can be very large (such as for x-rays and gamma rays, which are in the range of $10^{20}$ to $10^{15}$ cycles per second) to quite small (such as for some TV and radio signals, which might be a million or even hundreds of cycles per second).

You should also recognize that since wavelength times frequency always equals "c," if one gets smaller, the other must get larger. Look at the sample of some examples of electromagnetic radiation in Figure 9.2 to verify that this is true. Frequencies (ν) decrease from left to right, and wavelengths (λ) increase from left to right.

**Figure 9.2** Electromagnetic Spectrum

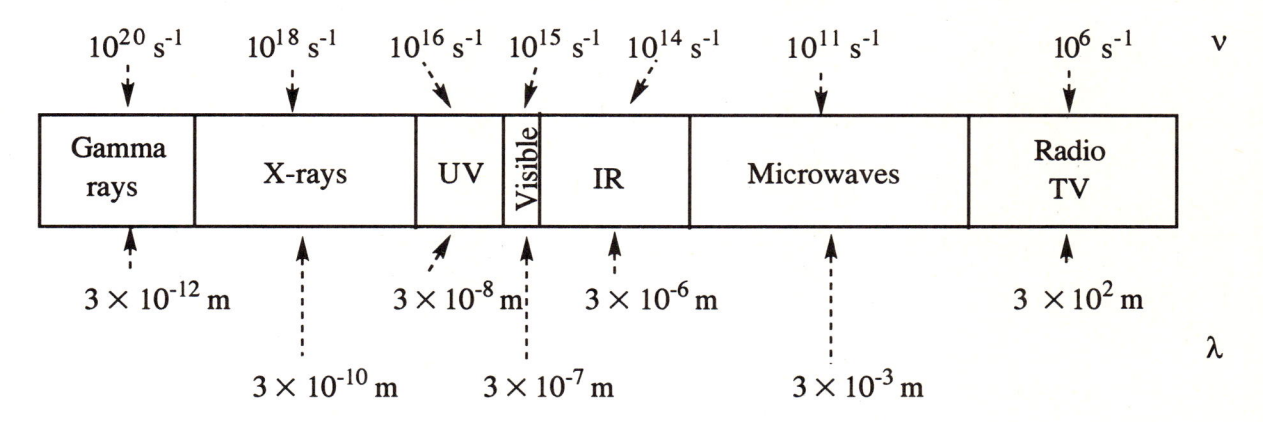

| Gamma rays | X-rays | UV | Visible | IR | Microwaves | Radio TV |

Frequencies (top): $10^{20}$ s$^{-1}$, $10^{18}$ s$^{-1}$, $10^{16}$ s$^{-1}$, $10^{15}$ s$^{-1}$, $10^{14}$ s$^{-1}$, $10^{11}$ s$^{-1}$, $10^{6}$ s$^{-1}$   ν

Wavelengths (bottom): $3 \times 10^{-12}$ m, $3 \times 10^{-10}$ m, $3 \times 10^{-8}$ m, $3 \times 10^{-7}$ m, $3 \times 10^{-6}$ m, $3 \times 10^{-3}$ m, $3 \times 10^{2}$ m   λ

The above are just examples. Electromagnetic radiation runs the whole continuum of the spectrum. Visible light can be seen as part of this continuum when looking at a rainbow or white light separated by a prism.

Each type of radiation may consist of one or many moving waves. These waves may also be thought of as particles. One wave (particle) has a very small amount of energy. It is known as a **photon**. All waves have different energy. Visible light has different energy than UV light. The energy is *directly* proportional to frequency as shown in EQ 9.3 below. Conversely, it is *indirectly* proportional to wavelength.

$$E = h\nu = h\frac{c}{\lambda}$$   (EQ 9.3)

where  E = energy
h = constant
ν = frequency
λ = wavelength
c = the speed of light

Examine EQ 9.3 enough to convince yourself that if you are comparing two different photons, the one with the higher frequency will have the higher energy. Conversely, the one with the lower wavelength will have a higher frequency.

**Example 9.3:** Compare microwaves (λ = $10^{-2}$ m) and UV radiation (λ = $10^{-8}$ m). Which has the larger frequency? Which has the larger energy?

**Solution:** Recognize that microwaves are **longer** than UV waves. That means that microwaves have **lower** frequency that UV waves.
The energy is directly proportional to the frequency. Thus **UV waves**, which have higher frequency, also have **higher energy**.

# 134 Chapter 9

> **Practice Problems**
>
> **9.5** Which has larger frequency, radio waves ($\lambda = 10$ mm) or x-rays ($\lambda = 1$ nm)?
>
> **9.6** Which has the highest energy yellow light ($\lambda = 602$ nm) or violet light ($\lambda = 395$ nm)?

## 9.2 Bohr's Theory of the Atom

The reason for looking at photons of electromagnetic energy is to try to understand what happens when a photon particle interacts with an atom (or how photons react with matter).

Bohr noticed that when a sample of hydrogen gas was excited with energy, the energy emitted from the excited sample did not exhibit a continuous spectrum (see Figure 9.7 in your textbook, *Introductory Chemistry*) when passed through a grating (prism).

Bohr proposed that these lines were the results of selective photons of energy being emitted from excited electrons dropping back to lower energy levels. Bohr helped develop **quantum mechanics** from his work. This theory says that only certain energy levels are possible for the electron.

Think of possible energy levels as steps (not evenly placed apart). An electron is in the lowest energy level, or **ground state,** when it is in the energy level closest to the nucleus.

Look at Figure 9.3 which is a pictorial representation of **quanta** or energy levels that the electron may occupy.

**Figure 9.3** Quantum Levels for the Hydrogen Atom.
a) Levels (not to scale); b) The one electron of H in the lowest energy level (ground state);
c) Electron that has been promoted to a higher energy level (excited state); d) electron falls from $n = 4$ to $n = 2$ and emits a photon.

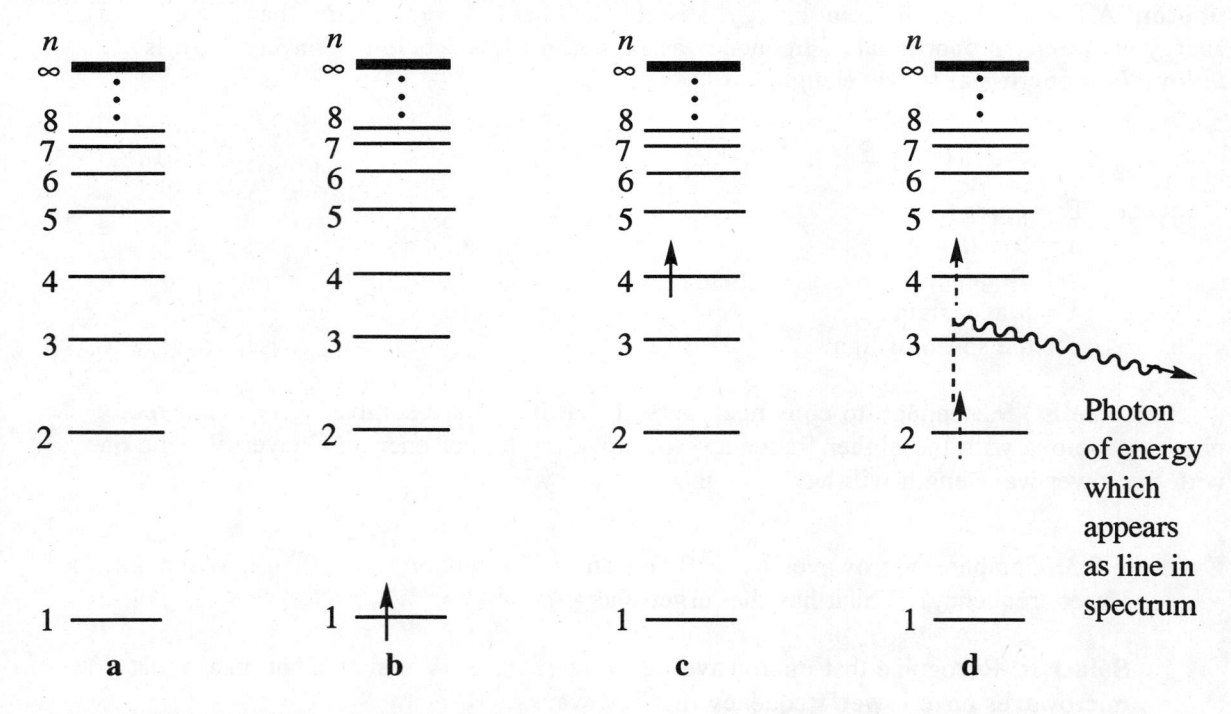

Notice that the energy levels are referred to as $n$ and the electron is represented by an arrow.

Once you understand the energy levels of the hydrogen atom, you are ready to try some problems.

**Example 9.4:** An electron going from energy level $n = 6$ to a lower allowed level could emit how many possible different photons? Which photon has the highest energy?

**Solution:** Use Figure 9.3 to assist in answering this problem. From $n = 6$ the photon could fall to one of five levels below. Each of these transitions would emit a different photon. The transitions are listed below from least energy to highest energy. (Think of the electron falling to lower energy levels like a marble falling down stairs).

$n = 6 \rightarrow n = 5$ + photon (lowest energy)
$n = 6 \rightarrow n = 4$ + photon
$n = 6 \rightarrow n = 3$ + photon
$n = 6 \rightarrow n = 2$ + photon
$n = 6 \rightarrow n = 1$ + photon (highest energy)

# Practice Problems

**9.7** Which electron is in the higher energy state, one in $n = 5$ shell or one in $n = 4$ shell?

**9.8** Does it take energy or release energy for an electron to go from $n = 2$ to $n = 3$?

**9.9** Does it take energy or release energy for an electron to go from $n = 4$ to $n = 1$?

**9.10** How many lines from photon emissions can excited electrons at $n = 5$ level have? Use Figure 9.3 to order these by energy.

## 9.3 Orbitals, Electron Shells, and Subshells

Before trying to go too far in this section, review the definitions of terms at the beginning of the chapter.

In multielectron atoms, the transitions from excited electrons are more complicated than the levels proposed by Bohr.

It was soon discovered that electrons could have certain *major energy* levels called **shells,** and within each shell there were **subenergy levels** called **subshells.**

The place where an electron is most likely to be found is called an **orbital**. This is a wave function (a mathematical definition of the electron's likely locality). This differs slightly from Bohr's theory, in that this space does not look like a planet's orbit about the sun.

Before looking at the shape of some orbitals lets look at how the location of an electron can be described. There are four quantum numbers that uniquely describe an electron's location.

1. Shell level (Same as Bohr's $n$, where $n$ goes from 1 to $\infty$.)
2. Subshell (shape of orbital, named *s, p, d, f, g,* ···).
3. Subshell direction (all orbitals of a subshell have the same or degenerate energies but have different arrangements in space.
4. Electron spin (+1/2 or -1/2).

At any given shell level, $n$, the number of subshells varies. As $n$ gets larger, the number of subshells gets larger.

## 136  Chapter 9

Each additional new subshell also has more degenerate energy levels. Look at Table 9.1, which summarizes these observations.

**Table 9.1** Summary of First Four Major Energy Shells

| $n$ | Total Orbitals $n^2$ | Number of Subshells $n$ | Subshell Shapes | Number of Degenerate Subshells |
|---|---|---|---|---|
| 1 | 1 | 1 | $s$ | 1 |
| 2 | 4 | 2 | $s$ | 1 |
|   |   |   | $p$ | 3 |
| 3 | 9 | 3 | $s$ | 1 |
|   |   |   | $p$ | 3 |
|   |   |   | $d$ | 5 |
| 4 | 16 | 4 | $s$ | 1 |
|   |   |   | $p$ | 3 |
|   |   |   | $d$ | 5 |
|   |   |   | $f$ | 7 |

Examine the information in the table carefully. Look for patterns. Test your ability to see these patterns by doing the practice problems at the end of this section.

**Example 9.5:** How many orbitals are in the 4p subshell? How many orbitals are in the entire $n = 4$ shell?

**Solution:** To determine the number of orbitals in the 4p shell, notice that **all** $p$ subshells have three orbitals, thus the answer is **3**.

To determine the number of orbitals totally in the $n = 4$ shell, just square the number. The answer is $4^2 =$ **16** orbitals total.

Figures 9.4 and 9.5 demonstrate some pictorial representations of orbitals and their shapes. For both figures, energy increases from bottom to top. Also notice that as $n$ increases, the size of the orbital increases. Even though these are not drawn totally to scale, you should note that a 1s orbital is **smaller** than a 2s, which is **smaller** than a 3s, etc. The same can be said for $p$ orbitals and other subshell orbitals.

Note the shapes of the orbitals.

- $s$ orbitals are always spherical
- $p$ orbitals have a dumbbell shape.
  Notice that they can be oriented in the x, y, or z direction, but all have the same energy.
- $d$ orbitals have a shape similar to a four-leaf clover (note the exception in Figure 9.5)
  Even though we can't perceive of five different directions, these all have the same energy.
- $f$ orbitals can be projected as models with computer programming.

Figure 9.6 illustrates energy levels (again energy increases from bottom to top) without the shapes. Each horizontal line represents the energy level of an orbital. Compare Figure 9.6 with information in Table 9.1. The shading on the orbitals represents a change in the sign on the wavefunction.

**Figure 9.4** Energy Levels of Shells and Subshells (Including Shapes) for $n = 1$ and 2

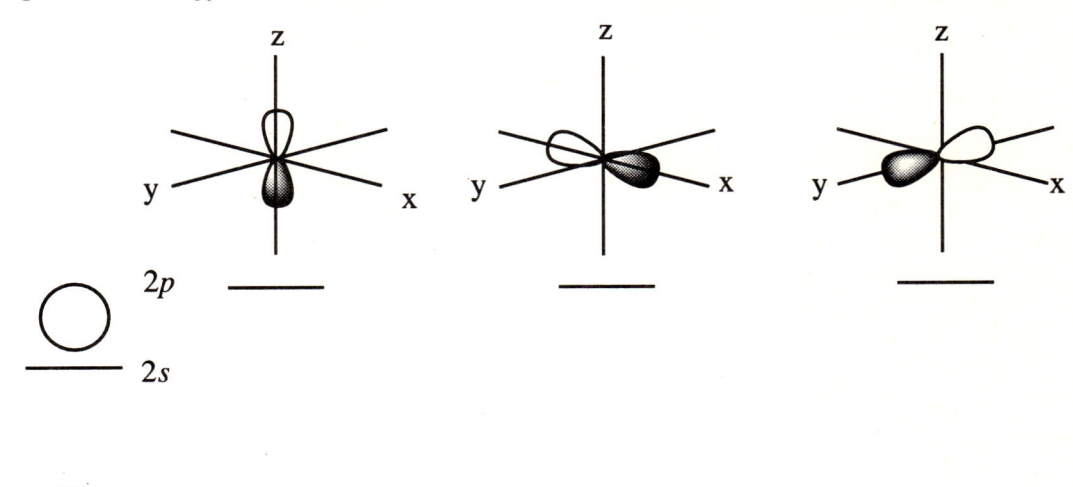

**Figure 9.5** Energy Levels of $n = 3$ Subshells (Including Shapes)

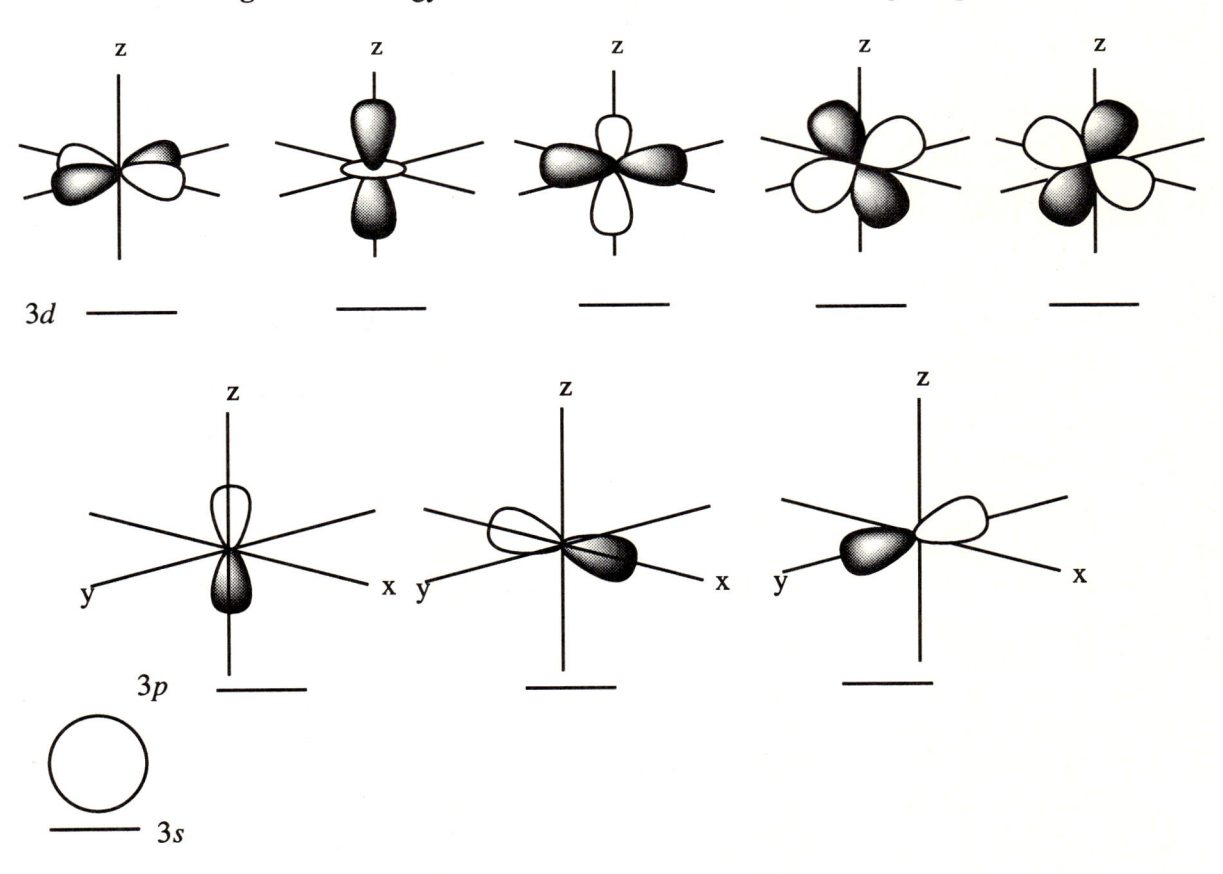

**Figure 9.6** Energy Levels Through 6p Orbital

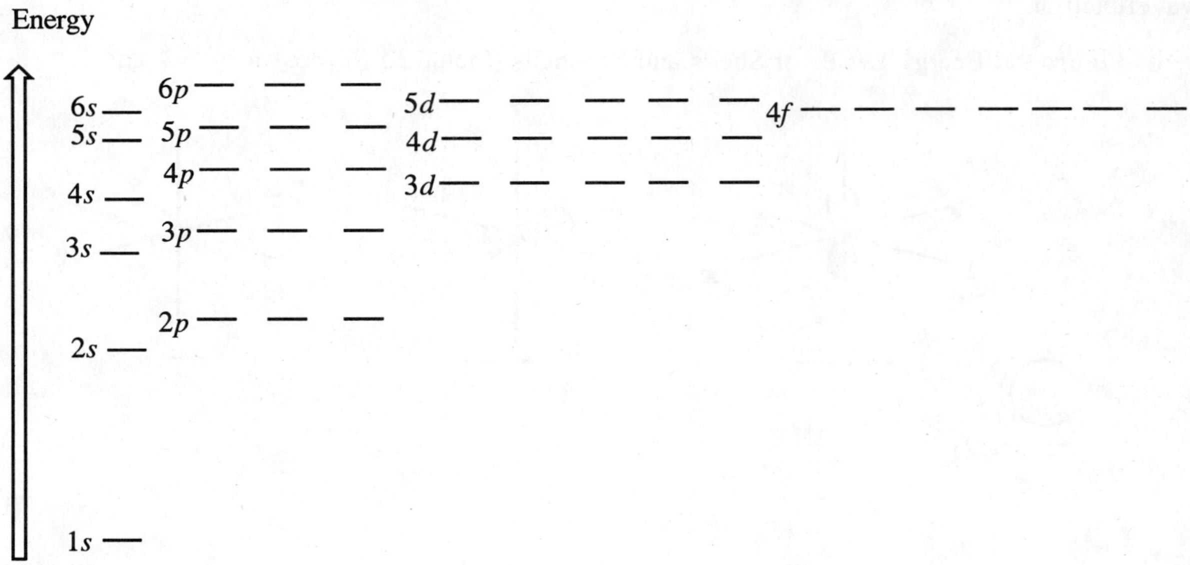

Energy

```
6s ___  6p — — — —        5d — — — — —                    4f — — — — — — —
5s ___  5p — — —          4d — — — —
4s ___  4p — — —          3d — — — — —
3s ___  3p — — —
2s ___  2p — — —
1s ___
```

Notice that the energy level of the 4s orbital is lower than the 3d orbitals. Similarly, the 5s orbital is lower than the 4d orbitals. Also notice that the following energy order for higher orbitals:

$$6s < 4f < 5d < 6p$$

They are close in energy, but there are subtle differences that can be detected by experiment. There is no need to memorize the orbital energy order. In Section 9.6 you will learn how to use the periodic table to predict the order of orbitals.

## Practice Problems

**9.11** How many total orbitals are in $n = 1$? $n = 2$? $n = 3$?

**9.12** How many orbitals are in the 2p subshell?

**9.13** How many orbitals are in any f subshell?

**9.14** Which of the following subshells **cannot** exist?
   1s, 7s, 1p, 2d, 3f, 7g

**9.15** Which orbital, 3d or 4s, has higher energy?

**9.16** Which orbital, 4d or 5p, has higher energy?

## 9.4 Electron Configurations of the First 18 Elements

Now that you understand that orbitals are mathematical (wave) functions that describe where an electron can be located, it is time to talk about where electrons are.

In the **ground** state, electrons seek the lowest energy levels possible.

The **Pauli exclusion principle** tells us that no orbital can hold more than two electrons.

Look at some examples, which show how electrons are described.

**Example 9.6:** Write the ground state electron configuration for hydrogen.

**Solution:** This atom has only one electron, and if it is in its lowest energy state, the electron will be in the 1s orbital (see Figure 9.3b).

$$H: 1s^1$$

The superscript 1 refers to the number of electrons in that particular subshell.

**Example 9.7:** Write the ground state electron configuration for fluorine. Place the electrons in orbitals like those in Figure 9.6.

**Solution:** This atom has 9 total electrons. When in its lowest energy state, the electrons will fill from the bottom up (see Figure 9.6).

$$F: 1s^2 2s^2 2p^5$$

The superscripts refer to the number of electrons in each particular subshell. On an energy scale, the electrons would look as follows. Notice that in each orbital there are a maximum of two electrons. When there are two, one is shown pointing up and one pointing down, which means they have opposite spins.

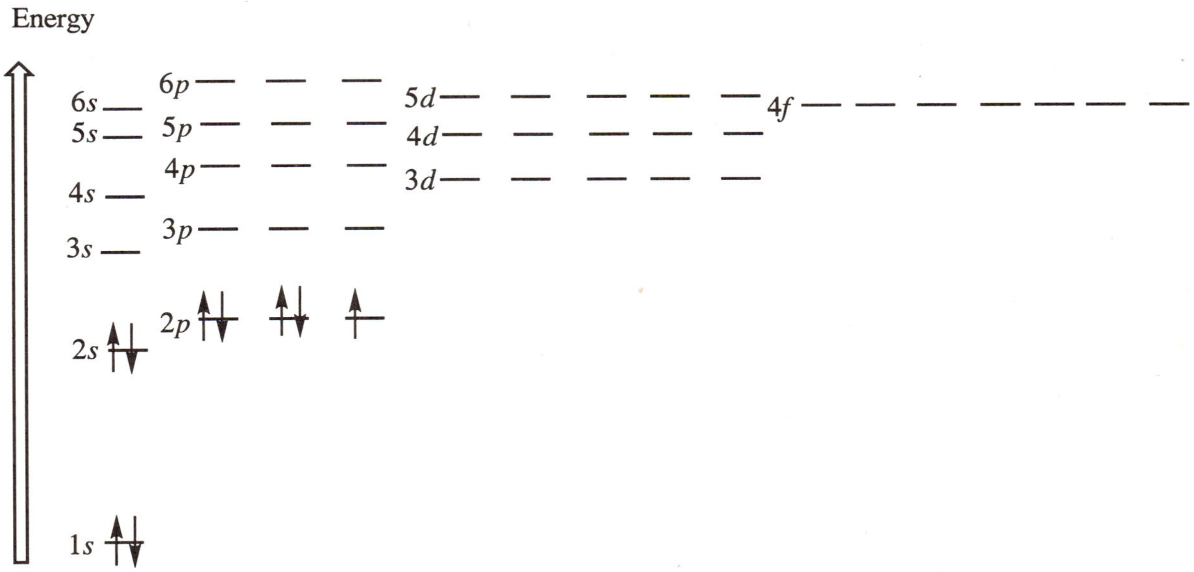

Energy

One final rule about electron structure is covered in this section. That is **Hund's rule**.

---

When electrons are filling orbitals of increasing energy, they pair up before filling the next higher level. However, when dealing with electrons filling degenerate orbitals within a subshell, they fill singly, before pairing up.

---

An example demonstrates this.

**Example 9.8:** Write the ground state electron configuration for phosphorus. Place the electrons in orbitals like those in Figure 9.6.

**Solution:** This atom has 15 total electrons, and if it is in its lowest energy state, the electrons will fill from the bottom up (see Figure 9.6).

$$P: 1s^2 2s^2 2p^6 3s^2 3p^3$$

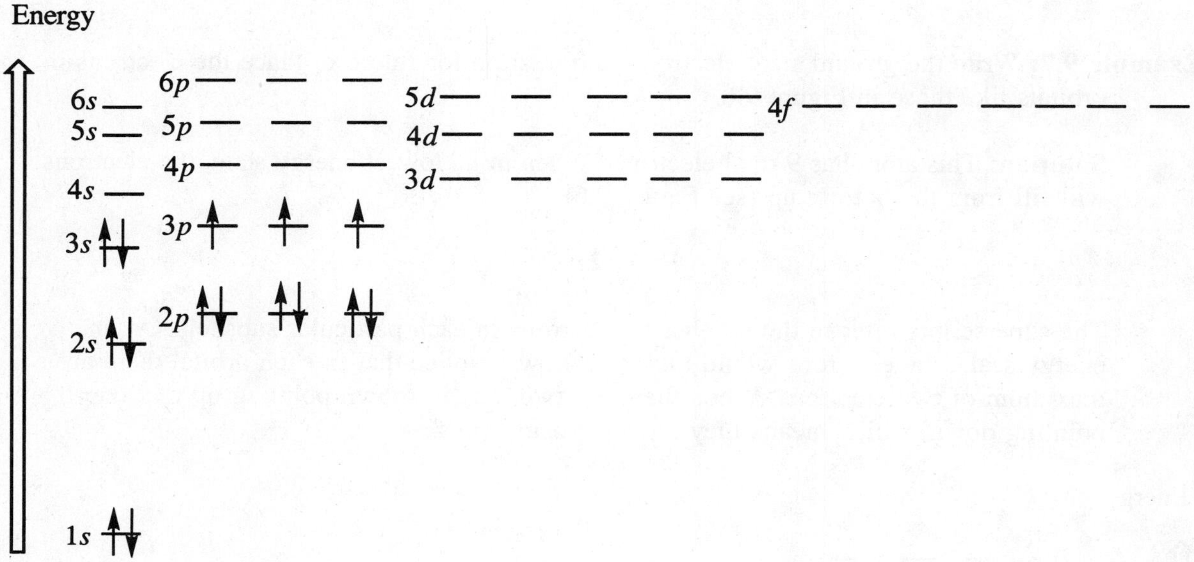

Energy

[Energy diagram showing orbitals 1s, 2s, 2p, 3s, 3p, 4s, 3d, 4p, 5s, 4d, 5p, 6s, 5d, 6p, 4f with electrons filled: 1s↑↓, 2s↑↓, 2p↑↓ ↑↓ ↑↓, 3s↑↓, 3p↑ ↑ ↑]

Notice the application of **Hund's rule**. One electron is placed in each 3p orbital before any orbital in a subshell is completely filled.

---

### Practice Problems

**9.17** Write the ground state electron configuration for He.

**9.18** Write the ground state electron configuration for Na.

**9.19** Write the ground state electron configuration for S. Put these electrons on an energy diagram.

**9.20** What element is represented by $1s^2 2s^2$?

**9.21** What element is represented by $1s^2 2s^2 2p^6 3s^2 3p^1$?

**9.22** What element is represented by $1s^2 2s^2 2p^6 3s^2 3p^6 4s^1$?

## 9.5 Periodicity of Electron Configurations

This section deals with the layout of the periodic table. It was originally laid out because of the similar characteristics of elements in groups (columns). The reason for this similar behavior is the similarity of the electrons in the outer shell.

**Example 9.9:** Compare the electron configurations of sodium and potassium. Note how they are alike.

**Solution:** Na: $1s^2 2s^2 2p^6 3s^1$ and K: $1s^2 2s^2 2p^6 3s^2 3p^6 4s^1$. Both have *one* electron in an outer *s* orbital.

**Example 9.10:** What electron configuration makes chlorine and bromine similar?

**Solution:** Cl: $1s^2 2s^2 2p^6 3s^2 3p^5$ and Br: $1s^2 2s^2 2p^6 3s^2 3p^6 4s^2 3d^{10} 4p^5$. Both have *five* electrons in an outer *p* orbital.

---

**Practice Problems**

**9.23** What are three elements that have 4 electrons in a *p* orbital as their highest energy electrons?

**9.24** What are three elements that have 5 electrons in a *d* orbital as their highest energy electrons?

**9.25** What are three elements that have 2 electrons in an *s* orbital as their highest energy electrons?

**9.26** What are three elements that have 1 electron in an *s* orbital as their highest energy electron?

**9.27** What are three elements that have 6 electrons in a *p* orbital as their highest energy electrons?

---

## 9.6 Using the Periodic Table to Obtain Electron Configurations

This section is very important, because it helps you extract information from the periodic table. Understanding this will decrease the need for memorization. To aid in this section, look at Figure 9.20 in *Introductory Chemistry*. It highlights:

- Blocks of elements in groups: *s* block, *p* block, *d* block, *f* block.
- Electron groups and row numbers.

Notice the size of the blocks. They match the number of electrons that can occupy each subshell.

**Example 9.11:** What are the number of electrons in the *s* and *p* subshells or any given *n*?

**Solution:** To solve this, just count the number of electrons in any row of the given block.

$$s = 2 \text{ electrons; the } p = 6 \text{ electrons}$$

Notice that the subshell containing the highest energy electron of an element in the ground state can be determined just by examining the periodic table. If this last electron is added to an *s* or *p* subshell, the number for *n* exactly matches the period number.

**Example 9.12:** What subshell contains the last electron added to oxygen?

**Solution:** To solve this, just locate oxygen. It is in the second row, in the *p* block; therefore, the last electron is added to a **2p** subshell. In fact, you can count from the beginning of the row in block *p* and determine that oxygen has 4 electrons in *p* orbitals.

Notice that the *n* number of the *d* subshell is one behind the period number. The *n* number for the *f* subshell is two behind the period number.

**Example 9.13:** What subshell contains the last electron added to iron? to uranium?

**Solution:** To solve for iron just locate it in the *d* block. It is in the fourth row, but the *d* block is one behind; therefore, the last electron is added to a **3d** subshell.

For U, it is located in the *f* block. Notice that the actinides are removed from period 7. Since the *f* block elements have *n* that is 2 numbers behind, the last electron added to uranium is in a **5f** subshell.

---

## Practice Problems

**9.28** How many electrons can be in the *d* and *f* subshells?

**9.29** What subshell has the last electron added to bromine?

**9.30** What subshell has the last electron added to lead?

**9.31** Name two elements which have final electrons in the 5*p* subshell?

**9.32** Name two elements which have final electrons in the 5*d* subshell?

**9.33** Name two elements which have final electrons in the 1*s* subshell?

---

### 9.7 Valence-Shell Configuration of a Main-Group Element

Valence electrons are those in the outermost *n* of the atom.

> In the main group (A) elements, the number of valence electrons is the same as the group number.

**Example 9.14:** How many valence electrons are present in lithium? in chlorine?

**Solution:** Lithium is in Group IA, so **Li has 1 valence** electron.

Chorine is in Group VIIIA, so **Cl has 7 valence** electrons.

> **Practice Problems**
>
> **9.34** Name two elements with five valence electrons.
>
> **9.35** Name two elements with four valence electrons.
>
> **9.36** Name two elements with two valence electrons.
>
> **9.37** How many valence electrons does barium have?
>
> **9.38** Note the number of valence electrons in K, Ca, and Al and compare to the charge they develop in ionic compounds.
>
> **9.39** Note the number of valence electrons in N, O, and F and compare to the charge they develop in ionic compounds.
>
> **9.40** What do each of the charged ions have in common with a noble gas?

## 9.8 Periodicity of Atomic and Ionic Radii

This section encourages use of the periodic table to recognize trends in atomic and ionic sizes.
   You might think that atomic size (distance from center of the atom to the outermost electrons) would be directly related to atomic number, but this is only partially true.
   Two trends should be kept in mind (look at Figure 9.23 in *Introductory Chemistry*):

- The size of the atomic radius increases as you go down a group.
- The size of the atomic radius increases as you go from right to left in a row.

This could be easily cued as: atomic radius ↑ as you go ↓ ← in the periodic table.

**Example 9.15:** Which of each pair has a larger atomic radius?
   Ca or K
   K or Na

   **Solution:**   For the first pair, potassium is to the right of calcium; therefore, **K > Ca.**
      For the second pair, potassium is below sodium; therefore, **K > Na.**

   Comparisons can also be made with **ionic** size. Ions have the same trend as you go down a column. However, for isoelectronic ions, the more negatively charged have larger ionic radii.

> Isoelectronic ions (atoms) are those which have the same number of electrons.

**Example 9.16:** Compare the magnesium and calcium ionic sizes. Compare the chlorine and sulfur ionic sizes.

   **Solution:**   $Ca^{2+}$ is larger than $Mg^{2+}$ because it is lower in the column.
      $S^{2-}$ is larger than $Cl^-$. Both have 18 electrons (they are isoelectronic).
      The sulfide ion is more negatively charged, and therefore, larger.

## Practice Problems

**9.41** Which has the larger atomic radius, bromine or iodine?

**9.42** Which has the larger atomic radius, bromine or selenium?

**9.43** Which has the larger ionic radius, bromine or iodine?

**9.44** Which has the larger ionic radius, bromine or selenium?

### 9.9 Periodicity of Ionization Energies

Remember that a photon of energy might excite an electron from the ground state to a higher orbital. If this photon has high enough energy, it can excite the electron to infinity. This is called the **ionization energy**.

> **Ionization energy** is that needed to remove an electron from an atom. It could also be the amount of energy to remove electrons from a mol of a given element.

The ionization is summarized in EQ 9.4:

$$X + \text{energy} \rightarrow X^+ + e^- \quad \text{(EQ 9.4)}$$

Two trends about ionization energy should be noted. Remember, a **high** ionization energy means that it is **difficult** to remove the electron.

- The ionization energy increases as you go up a column (group) of elements.
  Elements with small $n$ have the electron held more tightly by the positive nucleus. Elements, such as cesium, with a $6s$ valence electron take very little energy to remove the electron.
- The ionization energy increases as you go from left to right in a row of elements.
  As you look at elements like nonmetals, remember that they tend to gain electrons, not lose them; thus they have a very high ionization energy. Their electrons are held more tightly by the nucleus.

**Example 9.17:** Which of the following pairs has a higher ionization energy?
  Na or K
  S or Cl

**Solution:** Potassium is below sodium in Group IA, thus ionization energy for **Na > K**.
Sulfur is to the left of chlorine in row 3, therefore the ionization energy for **Cl > S**.

**Electron affinity** is an atom's ability to accept an extra electron. Elements that tend to form anions are going to have strong electron affinities. Electron affinities increase as you go left to right across a row and up a column.

> A **strong electron** affinity means a willingness to accept an extra electron. A weak electron affinity means an unwillingness to accept an extra electron.

**Example 9.18:** Which of the following pairs has a stronger electron affinity?
Cl or S
Br or Cl

**Solution:** Chlorine is to the right of sulfur, therefore **Cl > S**.
Chlorine is above bromine; therefore, **Cl > Br**.

> Noble gases (Group VIIIA) do not abide by periodic trends (except those about columns). They tend to have very high ionization energies and very weak electron affinities.
> **Noble gases do not want to lose or gain electrons. They have completed** $n$.

Metallic character is proportional to an atom's ability to lose an electron. Conversely, nonmetallic character is proportional to an atom's ability to gain an electron.

Cesium (Cs) is very metallic; fluorine (F) is very nonmetallic. (Remember: omit VIIIA when looking at trends.)

**Example 9.19:** Which is more metallic, aluminum or sodium?

**Solution:** Na > Al for metallic character. The trend is the same as for ionization energy. A low ionization energy means a high metallic character.

## Practice Problems

**9.45** Which has a higher ionization energy, Cs or K?

**9.46** Which has a stronger electron affinity, O or F?

**9.47** Which is more metallic, Ca or K?

**9.48** Which has a higher ionization energy, Cl or Ar?

**9.49** Prepare a table summarizing all the trends covered in this chapter.

## Integration of Multiple Skills

This section is a good chance to blend skills obtained in this chapter with earlier ones. Some examples show how to do this.

**Example 9.20:** If the energy of one photon of blue light is $3.68 \times 10^{-19}$ J, what energy would a mole of these photons have?

**Solution:** $\dfrac{3.68 \times 10^{-19} \text{ J}}{\text{photon}} \times \dfrac{6.02 \times 10^{23} \text{ photons}}{\text{mol}} = 2.22 \times 10^5 \text{ J/mol}$

**Example 9.21:** Compare the size of the atomic radii of arsenic, antimony, phosphorus, and sulfur.

**Solution:** Use the trends from the table you prepared to determine

$$S < P < As < Sb$$

**Example 9.22:** Compare the ionic radii of ions of chlorine, fluorine, selenium, and sulfur.

**Solution:** Remember that the ionic radii increase going down a column and to the left for isoelectronic elements.

$$F^- < Cl^- < S^{2-} < Se^{2-}$$

---

## Practice Problems

**9.50** How much energy is contained in a mole of UV photons if one photon has $7.11 \times 10^{-19}$ J?

**9.51** Place in order of increasing electron affinity: O, P, S, Si.

**9.52** Place in order of increasing ionization energy: F, Cl, Ne, S.

**9.53** Place in order of increasing metallic character: Ba, Ca, Cs, Mg.

**9.54** Write four ions which are isoelectronic with krypton. Which ion is largest?

---

## Self-Exam

**I. (4 ea) Write T (true) or F (false) in each corresponding blank.**

_____ 1. Bohr's theory of the atom applies only to hydrogen.

_____ 2. An orbital is a mathematical description of an electron's probable location.

_____ 3. An orbital can have 0, 1, or 2 electrons.

_____ 4. An *s* orbital can be spherical or dumbbell in shape.

_____ 5. The $n = 7$ shell has 7 subshells, *s, p, d, f, g, h,* and *i*.

_____ 6. The $n = 7$ shell has 49 total orbitals.

_____ 7. A photon can be described as a particle or a wave.

_____ 8. The ionization energy for K is less than that for Ca.

_____ 9. The ionization energy for K is less than that for Na.

_____ 10. Nobel gases have very high ionization energies and very weak electron affinities.

**II. (4 ea) Multiple Choice.** Place the letter of the BEST answer in the corresponding blank.

_____1. Blue light of 430 nm is equal to how many meters?

    A. 4.30 m      B. 43000 m      C. $4.30 \times 10^{-7}$ m

    D. $2.32 \times 10^{-3}$ m      E. 2.32 m

_____2. Consider three photons: blue light (540 nm), red light (710 nm), and ultraviolet light (220 nm). Which statement(s) is(are) true?
    I. The UV light has the highest energy of the three.
    II. The red light has the highest frequency of the three.
    III. All three are types of electromagnetic energy.

    A. Only I     B. Only II     C. Only III     D. Only I & II     E. Only I & III

_____3. Which is isoelectronic with $Ag^+$?

    A. $Pd^+$     B. $Cd^{2+}$     C. $Au^{3+}$     D. $Cu^{2+}$     E. $Cu^+$

_____4. Which subshell has the highest energy level?

    A. 2s     B. 2p     C. 3p     D. 3d     E. 4s

_____5. Which of the following is(are) a p orbital(s)?

    A. ◯     B. (three-lobed shape)     C. (two-lobed shape)     D. (four-lobed shape)     E. Both A and C

_____6. How many electrons can be in an $n = 3$ shell?

    A. 2     B. 3     C. 8     D. 10     E. 18

_____7. How many electrons can be in an $n = 3d$ shell?

    A. 2     B. 3     C. 8     D. 10     E. 18

_____8. What type of electron is the last electron added to chromium?

    A. 3s     B. 4s     C. 3d     D. 3p     E. 4d

_____9. Which has the most metallic character?

    A. N     B. C     C. Cl     D. Ca     E. Ba

_____10. Which of the following has the strongest electron affinity?

    A. N     B. C     C. Cl     D. Ca     E. Ba

**III. Provide Complete Answers**

1. (5 pts) Place the following in order of size (smallest to largest) of atomic radii.
    Al, K, Mg, Na

2. (5 pts) Place the following in order of size (smallest to largest) of ionic radii.
    $Cl^-$, $F^-$, $N^{3-}$, $O^{2-}$

3. (5 pts) Write the electron configuration for the ground state of tin.

4. (5 pts) Draw the energy levels for the ground state electrons in phosphorus.

## Answers to Practice Problems

| | | | | | |
|---|---|---|---|---|---|
| 9.1 | 1 | 9.2 | 36 | 9.3 | $7.2 \times 10^{-7}$ m |
| 9.4 | 710 nm | 9.5 | X-rays | 9.6 | Violet light |
| 9.7 | $n = 5$ | 9.8 | Takes energy | 9.9 | Releases energy |

9.10 Four lines: $n = 5 \rightarrow n = 4$; $n = 5 \rightarrow n = 3$; $n = 5 \rightarrow n = 2$; $n = 5 \rightarrow n = 1$

9.11 $n = 1$ (1); $n = 2$ (4); $n = 2$ (9)

| | | | | | |
|---|---|---|---|---|---|
| 9.12 | 3 | 9.13 | 7 | 9.14 | $1p$, $2d$, and $3f$ can't exist |
| 9.15 | $3d$ is higher. | 9.16 | $5p$ is higher. | 9.17 | He: $1s^2$ |
| 9.18 | Na: $1s^22s^22p^63s^1$ | 9.19 | S: $1s^22s^22p^63s^23p^4$ | | |

Energy

[Energy level diagram showing orbitals: 1s (filled), 2s (filled), 2p (filled), 3s (filled), 3p (2 paired, 1 unpaired), 4s, 3d, 4p, 5s, 4d, 5p, 6s, 5d, 6p, 4f]

| | | | | | |
|---|---|---|---|---|---|
| 9.20 | Be | 9.21 | Al | 9.22 | K |
| 9.23 | O, S, Se, et. al. | 9.24 | Mn, Tc, Re | 9.25 | Be, Mg, Ca, et. al. |
| 9.26 | H, Li, Cs, et. al. | 9.27 | Ne, Ar, Kr, et. al. | 9.28 | $d$: 10, $f$: 14 |
| 9.29 | $4p$ | 9.30 | $6p$ | 9.31 | Sn, I, et. al. |
| 9.32 | Au, Hg, et. al. | 9.33 | H, He | 9.34 | N, P, et. al. |
| 9.35 | C, Si, et. al. | 9.36 | Ca, Mg, et. al. | 9.37 | 2 |
| 9.38 | K: 1 valence electron, +1<br>Ca: 2 valence electrons, +2<br>Al: 2 valence electrons, +3 | | | 9.39 | N: 5 valence electrons, -3<br>O: 6 valence electrons, -2<br>F: 7 valence electrons, -1 |
| 9.40 | $K^+$ (18 e⁻) same as Ar<br>$Ca^{2+}$ (18 e⁻) same as Ar<br>$Al^{3+}$ (10 e⁻) same as Ne | | $N^{3-}$ (18 e⁻) same as Ne<br>$O^{2-}$ (18 e⁻) same as Ne<br>$F^-$ (10 e⁻) same as Ne | | |
| 9.41 | I | 9.42 | Se | 9.43 | $I^-$ |

9.44  Se²⁻       9.45  K       9.46  F
9.47  K          9.48  Ar

9.49

| Parameter | Periodic Trends that Increase |
|---|---|
| Size of atomic radius | ← ↓ |
| Size of ionic radius | ← (for isoelectronic) ↓ |
| Ionization energy | ↑ → |
| Electron affinity | ↑ → |
| Metallic character | ← ↓ |

9.50  $4.28 \times 10^5$ J/mol      9.51  Si < P < S < O

9.52  S < Cl < F < Ne      9.53  Mg < Ca < Ba < Cs

9.54  Br⁻, Se²⁻, Rb⁺, Sr²⁺, selenium ion is largest of the four.

## Answers to Self-Exam

**I. T/F**
1. T   2. T   3. T   4. F   5. T   6. T   7. T   8. F   9. T   10. T

**II. MC**
1. C   2. E   3. B   4. D   5. C   6. E   7. D   8. C   9. E   10. C

**III. Complete Answers**

1. Al < Mg < Na < K

2. F⁻ < Cl⁻ < O²⁻ < N³⁻

3. Sn: $1s^22s^22p^63s^23p^64s^23d^{10}4p^65s^24d^{10}5p^2$

4. See the solution for **Example 9.8**.

# 10
# Chemical Bonding

This chapter, like Chapter 9, is very heavy on new conceptual ideas. Very few computational skills are employed in the mastery of this material.

Keys for success in this chapter:

- Read the chapter to understand the new concepts.
- Learn the definitions of the new terms.
- Practice writing Lewis dot structures.
- Use the periodic table knowledge from Chapter 9 to apply ideas about valence electrons.
- Try to draw as many structures as possible.

## Summary of Verbal Knowledge

**ionic bond:** This is the strong attractive force that exists between a positive ion and a negative ion in an ionic compound.
**octet rule:** This says that atoms tend to lose or gain electrons when bonding to give eight electrons in their valence shells. (Exceptions are Li and Be, which tend to be isoelectronic with He when forming ions.)
**electron-dot symbol (Lewis symbol):** A representation of an atom or ion in which the valence-shell electrons are represented by dots placed around the letter symbol of the element.
**covalent bond:** This is a chemical bond formed by the sharing of electrons between two atoms.
**electron-dot formula (Lewis formula):** A representation of the electrons in covalent molecules using dots or lines (meaning two electrons).
**single bond:** A covalent bond formed by the sharing of a single pair (2 e$^-$) of electrons between two atoms.
**double bond:** A covalent bond formed by the sharing of two pairs (4 e$^-$) of electrons between two atoms.
**triple bond:** A covalent bond formed by the sharing of three pairs (6 e$^-$) of electrons between two atoms.
**bonding pair:** Two electrons shared between two atoms.
**lone pair (nonbonding pair):** Two electrons which belong to one atom in a molecule.
**electronegativity:** A measure of the ability of an atom in a covalent bond to draw bonding electrons to itself.
**polar covalent bond (polar bond):** This is a bond between two atoms with different electronegativities.
**polar molecule (dipole):** This is a covalent molecule in which the centers of partially positive charge and partially negative charge are separated.
**bond length:** This is the distance between atoms in a molecule (usually in picometers).

**valence-shell electron-pair repulsion (VSEPR) model:** A model for predicting the shapes of molecules and ions, in which the valence-shell electron pairs are arranged about each atom in such a way as to keep electron pairs as far away from one another as possible.

## Review of Mathematical and Calculator Skills

While there are no complex mathematical calculations required for this chapter, an ability to count *valence* electrons is very important. Recall how you did this in Chapter 9. When looking at valence electrons in a molecule or an ion, you simply sum all the valence electrons (and **add** or **subtract** those from any charge on an ion).

**Example 10.1:** How many valence electrons are in O? How many are in $H_2O$?

> **Solution:** Remember for main group elements that the number of valence electrons matches the Roman numeral of the group. Oxygen is in Group VIA.
> Thus **O has 6 valence electrons.**
>
> The number of valence electrons in a molecule such as water is simply the number of atoms of each element times the number of valence electrons in that atom. Then sum all the electrons.
>
> | | |
> |---|---|
> | 2 H × 1 valence electron  =  | 2 electrons |
> | 1 O × 6 valence electrons =  | 6 electrons |
> | Total | 8 electrons |

---

**Practice Problems**

**10.1** How many valence electrons are in N?

**10.2** How many valence electrons are in $N_2$?

**10.3** How many valence electrons are in Ca?

**10.4** How many valence electrons are in Cl?

---

## Application of Skills and Concepts

### 10.1 Forming an Ionic Bond from Atoms

The ideas to comprehend from this section are **how** and **why** ions (and ionic compounds) form.

The overall reason for the formation of ionic (and covalent) bonds is the lowering of overall energy. Compare to a rock on a slope. It seeks the lowest possible energy level. So does matter.

**Example 10.2:** What ions exist in KBr? What noble gas do they each resemble (isoelectronic with what Group VIIIA element)? Which process(es) in ion formation is(are) energy-

## Chapter 10

consuming (i.e., need energy to occur)? Which is(are) energy-releasing (i.e., give off energy)?

**Solution:** K is **K⁺** in ionic compounds. This ion has **18 electrons**, which is isoelectronic with argon.

Br is **Br⁻** in ionic compounds. This ion has **36 electrons**, which is isoelectronic with krypton.

K + energy → K⁺ + e⁻            (consumes energy)
Br + e⁻ → Br⁻ + energy            (releases energy)

Formation of ionic array of K⁺ Br⁻ **releases energy**.

---

## Practice Problems

**10.5** What ions exist in LiF? What noble gas do they resemble (isoelectronic)?

**10.6** What ions exist in CaI₂? What noble gas do they resemble (isoelectronic)?

**10.7** What ions exist in Na₂S? What noble gas do they resemble (isoelectronic)?

**10.8** Write the ion formation equations for Na₂S.

---

### 10.2 Describing Ionic Bond Formation with Electron-Dot Symbols

This section deals with representing atoms and ions with *dots*. Each dot represents a valence electron. Table 10.1 in your textbook, *Introductory Chemistry*, gives some examples of how dots are used to represent **valence** electrons.

When writing Lewis dot symbols for atoms (also known as **electron-dot symbols**), do the following:

- Write the chemical symbol for the element.
- Use the periodic table to determine the number of valence electrons.
- Put one dot around the symbol for each valence electron.

**Example 10.3:** How many valence electrons are in O? in Mg? Draw the neutral Lewis dot representation.

**Solution:**

O has 6 valence electrons      :Ö·

Mg has 2 valence electrons      ·Mg·

When writing Lewis dot symbols for ions (also known as **electron-dot symbols**) do the following:

- Write the chemical symbol for the element.
- Use the periodic table to determine the number of valence electrons.
- Add electron(s) for each negative charge or subtract electron(s) for each positive charge.
- Put one dot around the symbol for each valence electron.
  This will equal **eight** for anions (complete s and p subshells for largest n).
  This will equal **zero** for cations (complete s and p subshells for inner value of n).

**Example 10.4:** Draw the ionic Lewis dot representations for MgO.

**Solution:**

$$Mg^{2+} \; [:\ddot{\underset{..}{O}}:]^{2-}$$

## Practice Problems

**10.9** Draw Lewis dots for neutral calcium and neutral fluorine.

**10.10** Draw Lewis dots for neutral aluminum and neutral sulfur.

**10.11** Draw Lewis dots for ionic calcium and fluorine in calcium fluoride.

**10.12** Draw Lewis dots for ionic aluminum and sulfur in aluminum sulfide.

### 10.3 Covalent Bonding as a Sharing of Electron Pairs

The main concept to understand in this section is that you will be dealing with **nonmetals** bonding with **nonmetals**. For the valence electrons in these types of bonds:

- Electrons are shared between two atoms.
- Electrons try to form complete octets around all atoms (except H, which has 2 electrons).
- Two **atomic orbitals** overlap to form two **molecular orbitals**, which overlap.
- Two (maximum) electrons forming the bond can be in a **molecular orbital**.
- Electrons in covalent molecules (ions) can be of two types.
  **Bonding electrons**
  **Lone pairs (l.p.)**

Look at Table 10.1, for a summary of some of the bonds you might expect in neutral molecules.

**Table 10.1** Common Bonds and Lone Pairs in Neutral Molecules

| Type of Element | No. of Bonds | No. of Lone Pairs |
|---|---|---|
| H | 1 | 0 |
| Group IVA* | 4 | 0 |
| Group VA* | 3 | 1 |
| Group VIA* | 2 | 2 |
| Group VIIA* | 1 | 3 |

* Atoms in row $n = 3$ or larger can form more bonds than listed, because they can use $d$ orbitals for bonding.

**154  Chapter 10**

**Example 10.5:** Examine electron-dot symbols for H$_2$ and Cl$_2$, and note how to draw these so that the octet rule is obeyed (except hydrogen).

**Solution:** For each of these molecules, start by determining the total number of valence electrons in each diatomic molecule.

H$_2$ has **2** valence electrons

Cl$_2$ has **14** valence electrons

They can be distributed as follows.

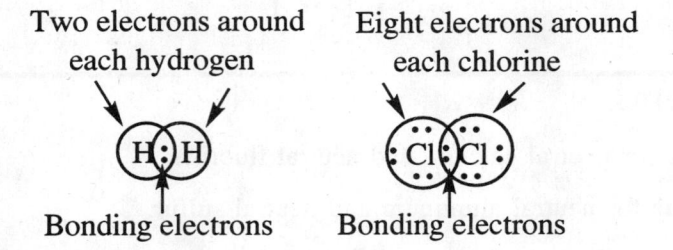

Two electrons around each hydrogen

Eight electrons around each chlorine

H:H  :Cl:Cl:

Bonding electrons    Bonding electrons

**Example 10.6:** Draw the correct electron dot symbol for the nitrogen molecule.

**Solution:** Counting valence electrons, you can determine that a total of **10** valence electrons must be distributed about the two atoms in N$_2$. Below are two examples of how those electrons might be distributed. Symbol **A** is **incorrect** because the octet rule is not obeyed. In symbol **A,** each atom has only 6 electrons around it. Symbol **B** is correct because all 10 electrons are arranged so that there are **8** electrons around each atom, thus obeying the **octet rule.**

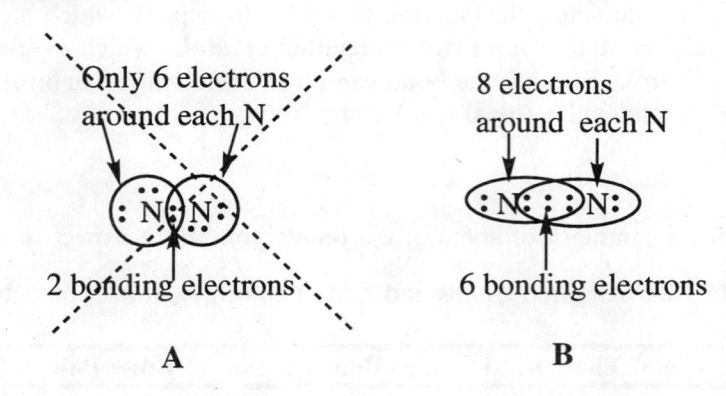

Only 6 electrons around each N

:N:N:

2 bonding electrons

**A**

8 electrons around each N

:N:::N:

6 bonding electrons

**B**

---

## Practice Problems

**10.13** Which pairs will form covalent bonds?
  Ca, F;  F, O;  C, Br;  C, O;  Al, Br

**10.14** Draw the electron-dot structure for O, for O$_2$.

**10.15** Draw the electron-dot structure of I, for I$_2$.

> **10.16** Draw the electron-dot structures for H, Cl, and HCl.
>
> **10.17** How many bonds will O form in neutral compounds?
>
> **10.18** How many lone pairs will phosphorus have in $PH_3$?

## 10.4 Electronegativity and Polar Covalent Bonds

Remember that **electronegativity** is the ability of an atom to draw electrons in a covalent bond toward itself.

Look at Figure 10.5 in your textbook, *Introductory Chemistry*. You will notice some numerical values given for electronegativities. The highest value is for fluorine, which is 4.0. Notice that Group VIIIA (noble gases) are omitted, since they don't tend to form compounds.

In general, you should note trends about electronegativities. They increase as you go up a given column and as you go from left to right across a row. Remembering this trend will allow you to make many predictions about polarity of covalent bonds without exact values.

> Electronegativity **increases** with these trends in the periodic table: ↑ →

Before looking at some consequences of electronegativity in covalent bonds, look at the summary of bond types in Table 10.2.

**Table 10.2** Bond Types

| Type of Bond | Reason Bond Forms | Type of Elements | Examples |
|---|---|---|---|
| Ionic | Attraction of opposite charges | Metallic and nonmetallic Large difference in electronegativities | NaCl $AlF_3$ CaS |
| Polar covalent | Sharing of electrons to obey the octet rule | Two nonmetallic Small difference in electronegativities | C—S S—O N—F |
| Pure covalent | Sharing of electrons to obey the octet rule | Two nonmetallic Zero difference in electronegativities | $H_2$ $N_2$ C—C |

There are several things you should have noticed in Table 10.2.

- Pure covalent bonds are between two nonmetal atoms of the **same** element.
- A line between two atoms can be used to represent **two** electrons in a covalent bond.

Even though bonds are placed in the three categories in Table 10.2, in reality they are on a continuum from **pure covalent** to very **ionic**. The degree of **covalent** character or **ionic** character is related to the difference between the absolute values of the electronegativities.

- A small difference in electronegativity increases the **covalent** character.
- A large difference in electronegativity increases the **ionic** character.

Look at Figure 10.1, which illustrates some bonds along this continuum.

## 156   Chapter 10

**Figure 10.1** Some Typical Bonds and Their Electronegativity Differences

Pure covalent　　　　　　　　Covalent　　　　　　　　　　Pure ionic
→

| Cl—Cl | H—C | H—Cl | Si—Cl | Al—Cl | Ga—F | Ba—O | Cs—F |
|-------|-----|------|-------|-------|------|------|------|
| 0     | 0.4 | 0.9  | 1.2   | 1.5   | 2.4  | 2.6  | 3.3  |

Absolute difference in electronegativities

When bonds are **polar covalent** they are sometimes denoted with the Greek letter delta ($\delta$) and a sign to show which atom in the bond is more electronegative. This is demonstrated in Figure 10.2.

**Figure 10.2** Some Typical Dipoles in Polar Covalent Bonds

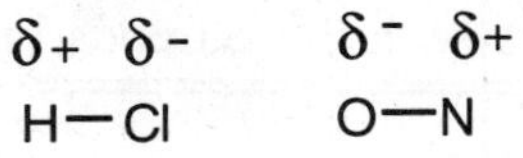

$$\overset{\delta+\ \ \delta-}{\text{H—Cl}} \qquad \overset{\delta-\ \ \delta+}{\text{O—N}}$$

The negative sign indicates the more electronegative atom in the bond. Many times you will be able to determine this just by looking at periodic trends. Sometimes you may need to refer to Pauling's electronegativity values.

**Example 10.7:** Draw each of the following bonds in such a way as to indicate if they are **ionic**, **polar covalent**, or **pure covalent**.

$$\text{NaI}, \qquad \text{HBr}, \qquad \text{NBr}_3, \qquad \text{I}_2$$

**Solution:** The bond in sodium iodide is obviously **ionic** because it is made up of sodium metal and iodine nonmetal.

The bond in hydrogen bromide is made up of two nonmetals, so it is **polar covalent**.

The bonds between nitrogen and each bromine in nitrogen tribromide are each **polar covalent**.

The bond between the two iodine atoms, which are equivalent nonmetals, is **pure covalent**.

The four molecules are drawn below (note that lone pairs on bromine and iodine are omitted from the covalent molecules for ease of viewing).

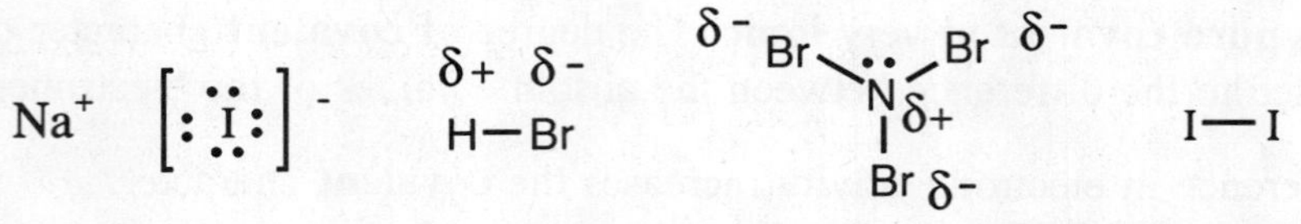

$$\text{Na}^+ \quad [:\!\ddot{\underset{..}{\text{I}}}\!:]^- \qquad \overset{\delta+\ \ \delta-}{\text{H—Br}} \qquad \overset{\delta-}{\text{Br}}\diagdown\overset{..}{\underset{|}{\text{N}}}\diagup\overset{\delta-}{\text{Br}} \qquad \text{I—I}$$
$$\qquad\qquad\qquad\qquad\qquad\qquad\qquad\qquad\ \ \underset{\text{Br }\delta-}{}$$

**Example 10.8:** Place the following bonds on a continuum from pure covalent to most ionic.

<pre>           F—N      O—F      CaO      MgS</pre>

**Solution:** The rules tell us that the fluorine nitrogen bond and the oxygen fluorine bond are **polar covalent**. Since both calcium and magnesium are metals and oxygen and sulfur are nonmetals, these two compounds have **ionic** bonds.

When trying to place the compounds on a continuum, do the following:

- Look at periodic table to try and determine which of the polar covalent bonds has a larger difference in electronegativities.
    F is the most electronegative, and from observing the periodic table, it is easy to see that O is more electronegative than N.

Therefore, the F—N bond is **more polar.**

- Look at the periodic table to compare the two metals and two nonmetals in the ionic compounds.
    Since Mg is higher in Group IIA than Ca, it is **more electronegative.**
    Since S is below O in Group VIA, it is **less electronegative**.

Calcium oxide will have a **greater** difference in the electronegativities, It's **more** ionic.

The final continuum looks like that in Figure 10.3.

**Figure 10.3** Continuum from Most Covalent to Most Ionic

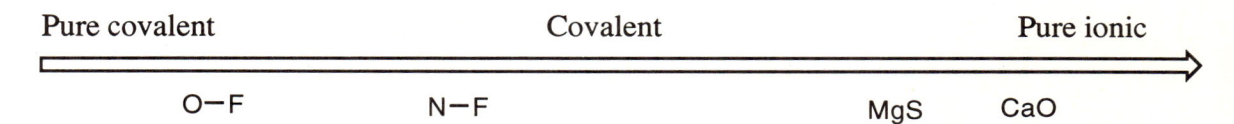

Pure covalent                Covalent                     Pure ionic
──────────────────────────────────────────────────────────────────▶
      O—F              N—F                     MgS      CaO

Notice that the problems below require you to apply your knowledge of *nomenclature* and *electronegativity* to arrive at solutions. When attempting to arrange bonds along a continuum, try to do this without the actual values. When there is an apparent tie (e.g., with O—F versus Cl—F), refer to Pauling's electronegativity values. The more you can use your knowledge of trends in the periodic table, the higher your skill will be.

---

### Practice Problems

**10.19** Which of the following have *pure covalent* bonds.
   a. hydrogen molecule     b. hydrogen chloride     c. oxygen molecule
   d. nitrogen molecule     e. carbon monoxide       f. nitrogen triiodide

**10.20** Which of the following have *polar covalent* bonds.
   a. hydrogen molecule     b. hydrogen chloride     c. oxygen molecule
   d. nitrogen molecule     e. carbon monoxide       f. nitrogen triiodide

**10.21** Arrange the following bonds on a continuum from *most covalent* to *most ionic*.
   a. oxygen-oxygen         b. carbon-oxygen         c. silicon-oxygen
   d. carbon-fluorine       e. aluminum-fluorine     f. aluminum-oxygen

## 158 Chapter 10

> **10.22** Arrange the following bonds on a continuum from *most covalent* to *most ionic*.
> a. hydrogen-nitrogen   b. hydrogen-carbon   c. carbon-carbon
> d. gallium-oxygen   e. carbon-silicon   e. aluminum-oxygen

## 10.5 A General Method for Writing Electron-Dot Formulas

This section expands on electron-dot formulas. You learned how to write these formulas for chlorine and nitrogen (Examples 10.4 and 10.5). Molecules that are more complicated than these diatomic examples can also be written with electron dots.

There are several types of covalent molecules you will learn to depict in this section.

- One central atom surrounded by more electronegative atoms
- One central atom surrounded by hydrogens
- One central atom surrounded by oxygen and hydrogen (oxyacids)

Some simple rules can be used to draw these simple covalent molecules:

- Count the total valence electrons present (sum all atomic valence electrons).
- Connect all atoms by single bond (one dashed line = 2 bonding electrons).
- Satisfy the octet rule with lone pairs on outer atoms.
- If valence electrons remain, put them on the central atom.
- If the octet rule is not obeyed for central atom, move one or more lone pairs to make double or triple bonds to central atoms.

Some examples illustrate how this is done.

**Example 10.9:** Draw the electron-dot formulas for carbon tetrachloride.

**Solution:** Carbon tetrachloride = $CCl_4$. The total valence electrons can be calculated:

$1\ C \times 4$ valence electron = 4 electrons
$4\ Cl \times 7$ valence electrons = 28 electrons
Total = 32 electrons

| Connect outside atoms to the central atom. | Satisfy octet rule with lone pairs on outside atoms. | Put any extra electrons on central atom to satisfy octet rule. |
|---|---|---|

$$\begin{array}{c} Cl \\ | \\ Cl-C-Cl \\ | \\ Cl \end{array} \qquad \begin{array}{c} :\!\ddot{C}l\!: \\ | \\ :\!\ddot{C}l-C-\ddot{C}l\!: \\ | \\ :\!\ddot{C}l\!: \end{array}$$

At this point, 8 electrons are used.    At this piont, all 32 electrons are used.    In this case, the final step is not necessary.

# Chemical Bonding 159

**Example 10.10:** Draw the electron dot structure for carbon dioxide.

   **Solution:** The first step of the solution involves writing the formula, **CO₂**.

   The number of valence electrons = **16**.

   The rest of the stepwise solution follows. Notice that lone pairs (l.p.) are drawn as dots, and bonds with 2 electrons are drawn as a line.

| Connect outside atoms to the central atom. | Satisfy octet rule with lone pairs on outside atoms. | Move lone pairs to central atom to satisfy octet rule |
|---|---|---|
| O—C—O | :Ö—C—Ö: | Ö=C=Ö |
| At this point, 4 electrons are used. | At this point, all 16 electrons are used, but central atom does not obey octet rule. | By moving a lone pair to a double bond, all atoms obey the octet rule. |

**Example 10.11:** Draw the structure for sulfuric acid.

   **Solution:** Sulfuric acid is **H₂SO₄**.

   The number of valence electrons = **32**.

   The rest of the solution follows.

| Connect outside atoms to the central atom. | Satisfy octet rule with lone pairs on outside atoms. | Put any extra electrons on central atom to satisfy octet rule. |
|---|---|---|
| H—O—S(=O)(=O)—O—H | H—Ö—S(:Ö:)(:Ö:)—Ö—H | No need, for final step, because all atoms (except H) obey octet rule |
| At this point 12 electrons are used | At this point all 32 electrons are used | |

**Example 10.12:** Draw the electron dot structure for the carbonate ion.

**Solution:** The carbonate ion is $CO_3^{2-}$.

The number of valence electrons = **22.** This polyatomic ion also has 2 extra electrons.

The total number of electrons in the electron-dot formula = 22 + 2 = **24.**

| Connect outside atoms to the central atom. | Satisfy octet rule with lone pairs on outside atoms. | Move a lone pair to central atom to satisfy octet rule |
|---|---|---|

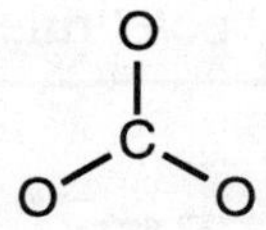

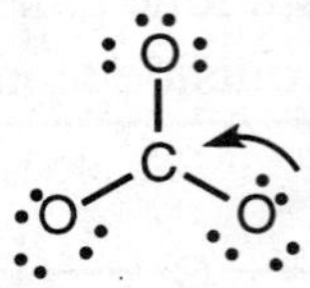

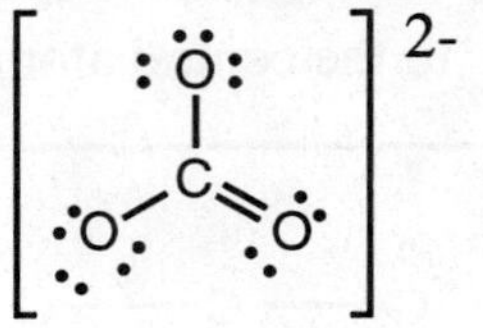

At this point, 6 electrons are used.

At this point, all 24 electrons are used, but central atom does not obey octet rule.

By moving a lone pair to a double bond, all atoms obey the octet rule.

Note that a charge of -2 must be added to the final structure. Also note that the double bond is equally present on all three oxygen atoms. These are known as **resonance contributions** and are shown completely drawn below.

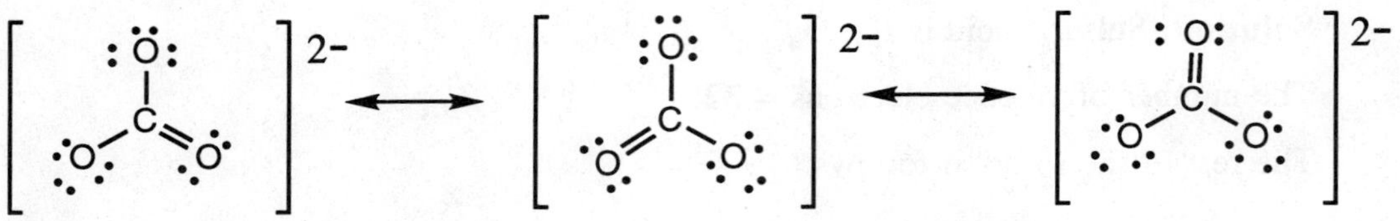

## Practice Problems

**10.23** Write an electron-dot formula for $H_2S$.

**10.24** Write an electron-dot formula for carbon monoxide.

**10.25** Write an electron-dot formula for dinitrogen monoxide.

**10.26** Write an electron-dot formula for sulfur dioxide.

**10.27** Write an electron-dot formula for $C_2H_6$.

**10.28** Write an electron-dot formula for $C_2H_2$.

**10.29** Write an electron-dot formula for the ammonium ion.

**10.30** Write an electron-dot formula for the cyanide ion.

**10.31** Write an electron-dot formula for hydroxide ion.

## 10.6 Molecular Structure

The structures using electron-dots depict molecules as being flat (planar). In actuality they often have three-dimensional shapes. Sometimes they are linear, sometimes planar, and sometimes other shapes.

**Example 10.13:** Look at the shape of water, which is known to have an angle between the two hydrogens of 105°, not 180°, as the Lewis dot structure indicates. Why is the shape the way it is?

**Solution:** Lewis dot drawings depict water to look like **A**, which has the two O—H bonds 180° apart from each other but only 90° from each lone pair. Experiments show that water really has the shape shown in **B**. Here the two O—H bonds are 105° apart, but they are also further apart from the lone pairs (approximately 110°). This structure puts mutually repulsive electrons the furthest away from each other.

In structure **B** the two lone pairs are shown occupying orbitals which come out and go into the plane of the page. The three atoms, H, O, and H, are all in the plane of the paper.

The information in Table 10.3 can be helpful in trying to predict what shape a given molecule will have. After examining this information, try to understand how you can quickly predict shapes.

**Table 10.3 Predicting Molecular Shapes**

| Shape | Atoms About Central Atom | Lone Pairs on Central Atom |
|---|---|---|
| Linear* | 2 | 0 |
| Bent planar | 2 | 1 or 2 |
| Trigonal planar | 3 | 0 |
| Trigonal pyramidal | 3 | 1 |
| Tetrahedral | 4 | 0 |

\* All diatomic molecules are linear.

**Example 10.14:** What shapes do $NH_4^+$, $CO_3^{2-}$, and HCN have?

**Solution:** The ammonium ion has **8** total electrons.
>It has one central atom, N, surrounded by four hydrogens.
>The central atom has **four** atoms attached and **zero** lone pairs.
>Therefore, it is **tetrahedral** in shape.

>The carbonate ion has **24** electrons (see Example **10.11**).
>It has a central atom, C, surrounded by three oxygens.
>The central atom has **three** atoms attached and **zero** lone pairs.
>Therefore, it is **trigonal planar** in shape.

>The hydrogen cyanide molecule has **10** total electrons.
>It has one central atom, C, attached to two other atoms.
>The central atom has **zero** lone pairs (when octet rule is obeyed for C and N).
>Therefore, it is **linear** in shape. (H–C≡N:)

---

**Practice Problems**

**10.32** Categorize the following molecules as linear, planar, or tetrahedral.
CO, $N_2$, $CCl_4$, $H_2S$

---

### 10.7 The VSEPR Model of Molecular Shape

The definition of VSEPR is listed at the beginning of the chapter. Some simple rules will help you to predict (and draw) molecules in the correct shapes.

- Draw the electron-dot (Lewis) formula.
- Look at the lone pairs (l.p.) on the central atom.
- Consult Table 10.3 to predict shape.

There are some common shapes that you will encounter in this course. Table 10.4 summarizes some of these. (There are other shapes that are known to exist but will not be covered at this level of study).

**Example 10.15:** Why is sulfur dioxide bent?

>**Solution:** It has 2 atoms (O) attached to a central atom (S) and a lone pair on the sulfur. There are a total of 18 electrons in this molecule. It is shown below (with all atoms obeying the octet rule).

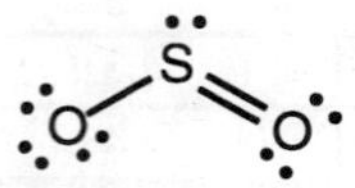

Table 10.4 Some Molecular Shapes and Their Approximate Angles Between Bonds

| Shape | Approximate Angles | Examples |
|---|---|---|
| Linear | 180° | $CO_2$<br>$N_2$<br>HCN |
| Bent planar | 105° | $H_2O$<br>$SO_2$ |
| Trigonal planar | 120° | $CO_3^{2-}$<br>$BF_3$<br>$NO_3^-$ |
| Trigonal pyramidal | 107° | $NH_3$<br>$PCl_3$ |
| Tetrahedral | 109.5° | $CH_4$<br>$H_2SO_4$<br>$NH_4^+$ |

**Example 10.16:** Why is any molecule with carbon at the center of four single bonds tetrahedral?

**Solution:** The drawing shown below has carbon at the center surrounded by four groups. The groups are called $R_1$, $R_2$, $R_3$, and $R_4$. These groups can represent any atom which might be covalently attached to the central carbon.
The furthest four bonding electron pairs can get from each other is **109.5°** apart. This represents the four corners of a **tetrahedron.**

Notice in the drawing that the **3-D** aspect of the molecule is depicted by:

- Using **straight** bond lines to represent an electron pair within the **plane** of the paper.
- Using **wedged** bond lines to represent an electron pair **coming out** of the paper.
- Using **dashed** bond lines to represent an electron pair **going behind** the paper.

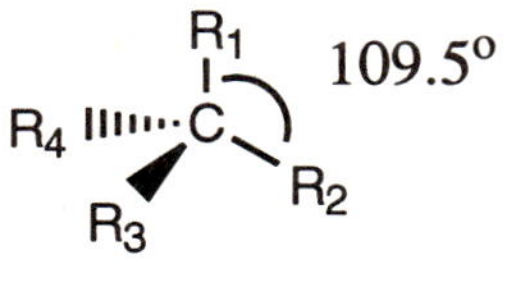

**Example 10.17:** Why is a molecule with N at the center of three bonds trigonal planar?

**Solution:** The nitrogen is in Group VA, therefore it will have **three** bonds and **one** l.p. in a neutral molecule (see Table 10.1). The lone pair takes up space, so it pushes the three bonds into a **tripod** type of configuration. This is called **trigonal pyramidal** for obvious reasons.

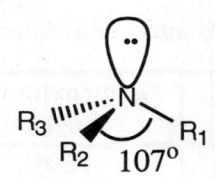

**Example 10.18:** What is the polarity of each bond in water? in carbon dioxide? What is the overall polarity of each molecule?

**Solution:** The water has a **bent** formation and the $CO_2$ has a **linear** formation as shown in the structures below. The polarity of each bond is shown also.

For the water, the two bonds are both polarized up at an angle toward the oxygen. This gives the **molecules a net overall polarity**.

For the carbon dioxide molecule the two bonds are polarized away from the carbon toward the more electronegative oxygens. However, in this case the two polarities cancel each other, because they are equal and opposite along a line in this linear molecule. The **molecule has an overall zero polarity.**

Many molecules have polar bonds, but are not polar if the polarities all cancel. Another example of a nonpolar molecule with polar bonds is $CCl_4$. If the polarities do not cancel, the molecule will be overall polar, like water.

Notice that the lone pairs are omitted in the structural drawings above. This does not mean that their presence is not recognized. It is assumed that readers know they are there, but they have been omitted to make the drawing easier to read.

---

## Practice Problems

**10.33** What is the shape of hydrogen bromide?

**10.34** What is the shape of carbon disulfide?

**10.35** What is the shape of dihydrogen monosulfide?

**10.36** What is the shape of carbon monoxide?

**10.37** What is the shape of chloroform ($CHCl_3$)?

**10.38** For each of the molecules above, which have polar bonds?

**10.39** For each of the molecules above, which are overall polar?

## Integration of Multiple Skills

This section involves some *integration* of current and previous observations. Use your powers of observation to follow the examples of comparisons in periodic trends. Try to make some additional observations of your own.

**Example 10.19:** How do periodic table trends for electronegativity compare with those for metallic character?

**Solution:** They are just the opposite. Electronegativity **increases** with ↑ → trends in the periodic table. Metallic character **increases** with ↓ ← trends in the periodic table.

**Example 10.20:** What type of compound is sodium cyanide?

**Solution:** It is an **ionic** compound which has an array of Na⁺ ions and CN⁻ ions. The polyatomic cyanide anion is **linear**.

---

## Practice Problems

**10.40** Compare trends in electronegativity to trends in ionization energy and atomic radius.

**10.41** What type of compound is ammonium nitrate?

**10.42** What type of compound is calcium sulfate?

---

## Self-Exam

**I. (4 ea) Write T (true) or F (false) in each corresponding blank.**

_____ 1. The two types of bonds are *covalent* and *molecular*.

_____ 2. Potassium can form ions of K⁺ or K²⁺.

_____ 3. A Lewis dot representation of an ion can have zero to eight dots about an atom.

_____ 4. Electrons are shared in covalent bonds.

_____ 5. A molecular orbital can hold four to eight electrons.

_____ 6. A triple bond has three lone pairs.

_____ 7. It takes more energy to break a double bond than a single bond.

_____ 8. Neon is more electronegative than fluorine.

_____ 9. Sulfur dioxide has 18 valence electrons.

_____ 10. The shape of water is linear.

## 166   Chapter 10

**II. (4 ea) Multiple Choice. Place the letter of the BEST answer in the corresponding blank.**

_____ 1. Which forms +2 ions?
   A. Na      B. Sr      C. O      D. Al      E. Cs

_____ 2. Which is the correct Lewis dot symbol for $O^{2-}$?
   A. :Ö·      B. :Ö:      C. :O:      D. :O      E. O :

_____ 3. Which of the following compounds has a polar covalent bond?
   A. $N_2$      B. $N_2O$      C. $Na_3N$      D. CsI      E. MgO

_____ 4. How many electrons are around carbon in $CCl_4$?
   A. 0      B. 2      C. 4      D. 8      E. 28

_____ 5. Which of the following molecules has(have) a triple bond?
   A. $N_2$      B. $O_2$      C. $C_2H_4$      D. HCN      E. A & D

_____ 6. How many valence electrons are in $CO_2$?
   A. 8      B. 4      C. 16      D. 6      E. 12

_____ 7. How many total electrons are in the Lewis dot formula for $CO_3^{2-}$?
   A. 4      B. 6      C. 10      D. 22      E. 24

_____ 8. Arrange the following bonds from most *ionic* to purest *covalent*.
   I. Na—Cl      II. N—Cl      III. Cs—Cl      IV. Cl—Cl

   A. I, II, III, IV      B. II, III, I, IV      C. IV, III, II, I
   D. I, III, II, IV      E. III, II, I, IV

_____ 9. Which of the following is linear?
   A. HCN      B. $CO_3^{2-}$      C. $H_2O$      D. $SO_2$      E. $BF_3$

_____ 10. Which of the following is tetrahedral?
   A. $CH_2Cl_2$      B. $NO_3^-$      C. $H_2S$      D. $CO_2$      E. $NH_3$

**III. Provide Complete Answers**

1. (5 pts) How do *electronegativity* trends in the periodic table compare to *metallic character* trends?

2. (5 pts) Draw the Lewis dot (electron-dot) symbols for the nitrogen molecule and hydrogen cyanide.

3. (5 pts) Draw the Lewis dot (electron-dot) symbol for the nitrate ion.

4. (5 pts) Draw the shapes of carbon tetrabromide and carbon disulfide.

# Answers to Practice Problems

10.1   5                10.2   10
10.3   2                10.4   7

**10.5** Li⁺ (isoelectronic with He)
F⁻ (isoelectronic with Ne)

**10.6** Ca²⁺ (isoelectronic with Ar)
I⁻ (isoelectronic with Xe)

**10.7** Na⁺ (isoelectronic with Ne)
S²⁻ (isoelectronic with Ar)

**10.8** $2Na + energy \rightarrow 2Na^+ + 2e^-$
$S + 2e^- \rightarrow S^{2-} + energy$
Formation of array of ions is energy releasing

**10.9** ·Ca·   :F̈·

**10.10** ·Al·   :S̈·

**10.11** [:F̈:]⁻ Ca²⁺ [:F̈:]⁻

**10.12** [:S̈:]²⁻ Al³⁺ [:S̈:]²⁻ Al³⁺ [:S̈:]²⁻

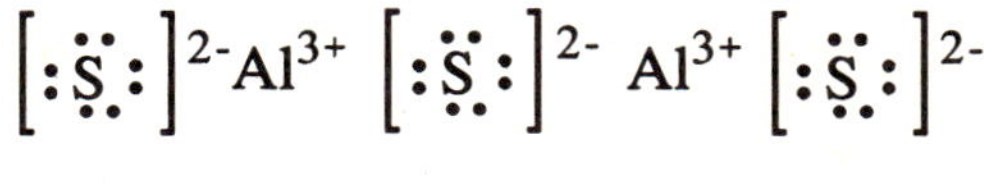

**10.13** F, O;  C, Br;  C, O

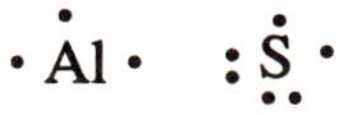

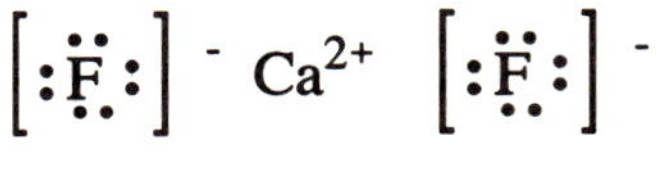

**10.14** :Ö·    Ö::Ö

**10.15** :Ï·    :Ï:Ï:

**10.16** H·   :C̈l·   H:C̈l:

**10.17** Two bonds

**10.18** One lone pair

**10.19** H₂, N₂, O₂

**10.20** HCl, CO, NI₃

**10.21** O₂, C—O, Si—O, C—F, Al—O, Al—F

**10.22** C—C, H—C, C—Si, H—N, Al—O, Ga—O

**10.23** H—S̈—H

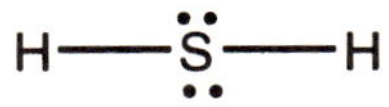

**10.24** :C≡O:

**10.25** Ö=N=N̈

**10.26** :Ö—S̈=Ö

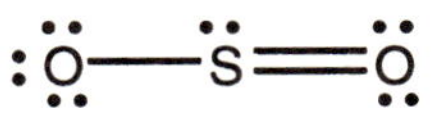

**10.27** 
```
    H H
    | |
H—C—C—H
    | |
    H H
```

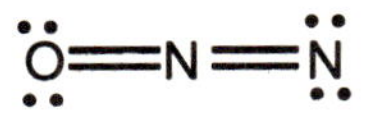

**10.28** H—C≡C—H

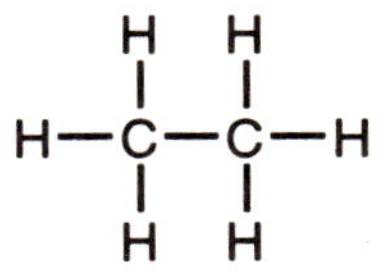

**10.29** 
```
   ⎡  H    ⎤⁺
   ⎢  |    ⎥
   ⎢H—N—H⎥
   ⎢  |    ⎥
   ⎣  H    ⎦
```

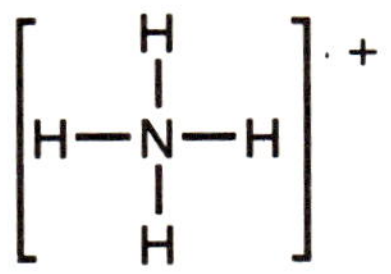

**10.30** [:C≡N:]⁻

**10.31** [:Ö—H]⁻

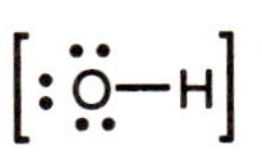

**10.32** CO and N₂ are linear; CCl₄ is tetrahedral; H₂S is bent planar.

168    Chapter 10

**10.33**  HBr is linear.  **10.34**  $CS_2$ is linear.

**10.35**  $H_2S$ is bent planar  **10.36**  CO is linear.

**10.37**  $CHCl_3$ is tetrahedral.  **10.38**  All have polar bonds.

**10.39**  All but $CS_2$ are polar molecules.

**10.40**  Electronegativity increases ↑ →, ionization energy increases ↑ →, atomic radius increase ↓ ←.

**10.41**  It is an ionic compound with an array of alternating $NH_4^+$ and $NO_3^-$ ions. The $NH_4^+$ ion is tetrahedral, and the $NO_3^-$ ion is trigonal planar.

**10.42**  It is an ionic compound with an array of alternating $Ca^{2+}$ and $SO_4^{2-}$ ions. The $SO_4^{2-}$ ion is tetrahedral.

## Answers to Self-Exam

### I. T/F
1. F   2. F   3. T   4. T   5. F   6. F   7. T   8. F   9. T   10. F

### II. MC
1. B   2. B   3. B   4. D   5. E   6. C   7. E   8. E   9. A   10. A

### III. Complete Answers

1. They are the opposite. Electronegativity increases ↑ → and metallic ↓ ←

2. :N≡N:   and   H–C≡N:

3. The ion contains 24 electrons in the structure. It is correctly drawn with all three resonance contributors.

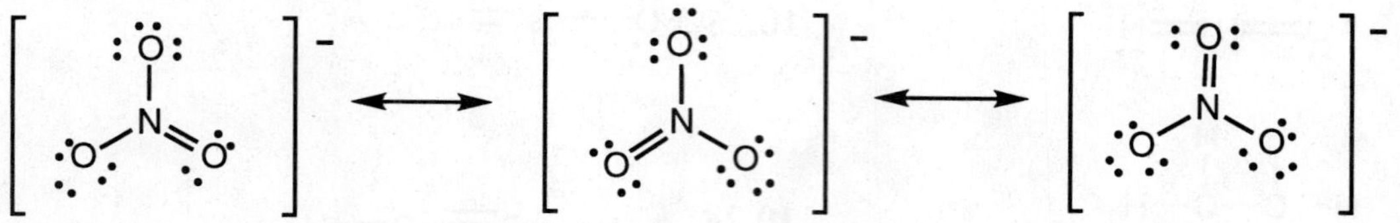

4. The two are drawn below. The lone pairs on the bromine are not shown to make the 3-D drawing clear.

# 11
# The Gaseous State

This chapter brings a return to calculations in the study of chemistry. There are some new terms and concepts to master also. This chapter is quite rigorous compared to the last two, so plan on scheduling more time than usual for studying this material.
   Keys for success in this chapter:

- Read the chapter to understand the new concepts.
- Learn the definitions of the new terms.
- Review mathematical concepts, including handling exponents and dimensional analysis.
- Review the basics of stoichiometry (see Chapter 8).
- Learn the key units for pressure, temperature, and volume.
- Learn how to convert from one unit to another.

## Summary of Verbal Knowledge

**pressure:** The force exerted on a unit area.
**barometer:** A device used to measure atmospheric pressure.
**millimeters of mercury:** This is the unit for measuring gas pressure.
**torr:** This is equal to 1 mm of mercury. It is named for Evangelista Torricelli, who invented the barometer.
**atmosphere:** This is a unit of pressure measurement. It is equal to 760 mm Hg or 760 torr.
**pascal:** This is a unit of pressure measurement in the SI system. It is abbreviated **Pa**. 760 mm Hg is equal to $1.01325 \times 10^5$ Pa.
**manometer:** This is a device that measures a gas's pressure in a container.
**kinetic molecular theory of gases:** This applies to a gas that obeys a simple mathematical relationship between its pressure, volume, temperature, and amount.
**ideal gas:** A substance in the vapor state that obeys the kinetic molecular theory mathematical relationships.
**Boyle's law:** This states that the *volume* of a fixed molar amount of a gas at a given temperature is *inversely* proportional to the applied *pressure*.
**Charles's law:** This states that the *volume* of a fixed amount of gas is *directly* proportional to its *Kelvin temperature* at constant pressure.
**combined gas law:** This combines Charles's law and Boyle's law.
**Avogadro's law:** This states that equal volumes of any two gases at the same temperature and pressure contain equal moles of molecules. It could also be interpreted in another way.
   At constant T and P, the volume is *directly* proportional to the number of moles.
**standard temperature and pressure (STP):** These are arbitrarily chosen parameters of 0°C (273 K) and 1 atm (760 mm Hg). At STP, 1 mol of *any ideal gas* occupies 22.4 L.

**ideal gas law:** This relates the parameters of pressure (atm), volume (L), amount (no. of moles), and temperature (K) to each other. It can be summarized as PV = nRT (where R is a constant).
**partial pressure:** This is the pressure exerted by one component of a mixture of gases.
**Dalton's law of partial pressures:** This states that the total pressure of a mixture of gases is equal to the sum of the partial pressures of all the components in the mixture.

## Review of Mathematical and Calculator Skills

There are some algebraic concepts which need to be reviewed before you begin the problems in this chapter. You may need to go back and look at some skills covered in Chapters 1 and 2.

**Example 11.1:** Given the equation PV = nRT, solve for R.

    **Solution:** Divide both sides of the equation by nT.

$$\frac{PV}{nT} = R$$

**Example 11.2:** Determine the units for R in the above equation, given the following:
    P (atm)
    V (L)
    n (mol)
    T (K)

    **Solution:** Simply insert the units into the solution for Example 11.1.

$$\frac{atm \cdot L}{mol \cdot K} = \text{units of R}$$

**Example 11.3:** Look at the given the algebraic equalities below:
    If: a = c and b = c
      Then: a = b

    If $P_1V_1 = a$ and $P_2V_2 = a$, how are these two related to each other?

    **Solution:** $P_1V_1 = P_2V_2$

---

### Practice Problems

**11.1** Solve PV = nRT for n.

**11.2** Solve PV = nRT for P.

**11.3** Solve PV = nRT for V.

**11.4** Solve PV = nRT for T.

11.5 Does $\dfrac{PV}{nRT} = 1$?

11.6 If $\dfrac{V_1}{T_1} = b$ and $\dfrac{V_2}{T_2} = b$, how are these two related to each other?

11.7 If $\dfrac{V_1 P_1}{T_1} = c$ and $\dfrac{V_2 P_2}{T_2} = c$, how are these two related to each other?

11.8 If $\dfrac{V_1}{n_1} = d$ and $\dfrac{V_2}{n_2} = d$, how are these two related to each other?

## Application of Skills and Concepts

### 11.1 The Nature of Gases

There are many different types of gases. Some are elements, and others are compounds. Table 11.1 lists some common gases, and some of their physical properties.

Table 11.1 Some Common Gases and Their Properties at STP

| Name | Mol. Form. | Color | Odor | Molar Mass (g) | Density (g/L) | Boiling Point (°C) |
|---|---|---|---|---|---|---|
| Hydrogen | H$_2$ | None | None | 2.0 | 0.090 | -253 |
| Nitrogen | N$_2$ | None | None | 28.0 | 1.251 | -196 |
| Oxygen | O$_2$ | None | None | 32.0 | 1.529 | -183 |
| Fluorine | F$_2$ | Pale Yellow | None | 38.0 | 1.696 | -188 |
| Chlorine | Cl$_2$ | Green-Yellow | Pungent | 70.9 | 3.214 | -35 |
| Helium | He | None | None | 4.0 | 0.179 | -269 |
| Neon | Ne | None | None | 20.2 | 0.900 | -246 |
| Argon | Ar | None | None | 39.9 | 1.784 | -186 |
| Krypton | Kr | None | None | 83.8 | 3.733 | -152 |
| Carbon monoxide | CO | None | None | 28.0 | 1.977 | -192 |
| Carbon dioxide | CO$_2$ | None | None | 44.0 | 1.250 | -79 (sub) |
| Nitrogen dioxide | NO$_2$ | Yellow-Brown | Irritating | 46.0 | 1.449 | Dimerize |
| Dinitrogen monoxide | N$_2$O | None | None | 44.0 | 1.977 | -89 |
| Sulfur hexafluoride | SF$_6$ | None | None | 146.1 | 6.602 | -65 (sub) |
| Methane | CH$_4$ | None | None | 16.0 | 0.555 | -164 |
| Ammonia | NH$_3$ | None | Pungent | 17.0 | 0.771 | -33 |
| Methanethiol | CH$_3$SH | None | Foul | 62.1 | 0.867 | 6.0 |
| Hydrogen chloride | HCl | None | Pungent | 36.5 | 1.000 | -85 |
| Hydrogen cyanide | HCN | None | Almond | 27.0 | 0.901 | 26 |

**Example 11.4:** Is there any relationship between molar mass and the density of a gas?

## 172    Chapter 11

**Solution:** They are directly proportional. $H_2$ has the lowest molar mass, 2.0 g, and the lowest density. $SF_6$ has the highest molar mass and the highest density. (See Table 11.1.)

---

**Practice Problem**

**11.9** Are there any other trends you can find in Table 11.1?

---

### 11.2 Gas Pressure

There are two types of problems you need to be able to solve in this section.

- Conversion of one pressure unit to another.
- Calculation of pressure from information about barometric pressure and a manometer reading.

The pressure of a gas is measured for a unit area. There are some standard pressures that all others are compared to. The **reference pressure** (or **standard pressure**) refers to that exerted by the earth's atmosphere at sea level. The following units are all different, but they are all equivalent.

$$1 \text{ atm} = 1.01325 \times 10^5 \text{ Pa} = 760 \text{ mm Hg} = 760 \text{ torr}$$

Some examples will illustrate how to convert measured pressures from one unit to another. Notice that the units of *mm Hg* are exactly equivalent to *torr*.

**Example 11.5:** How many torr are equal to 1.25 atmospheres? How many pascals?

**Solution:** Notice that you are given the measurement in atmospheres (atm) and need to convert it to torr and pascals. These are two separate dimensional analysis problems.

$$1.25 \text{ atm} \times \frac{760 \text{ torr}}{1 \text{ atm}} = 950 \text{ torr}$$

$$1.25 \text{ atm} \times \frac{1.01325 \times 10^5 \text{ Pa}}{1 \text{ atm}} = 1.27 \times 10^5 \text{ Pa}$$

**Example 11.6:** What is the pressure of a gas in a rigid container (connected to a manometer) if the height of the mercury in the open arm is 152 mm higher than that in the container? (Assume that barometric pressure is 760 mm Hg)

**Solution:** The solution to this problem is partially mathematical and partially logical. The latter part can only be accomplished by understanding the principles of the barometer and the manometer. The barometer measures the ambient atmospheric pressure. In this case it is 760 mm Hg. This is the pressure of the atmosphere pushing down on the open end of the manometer (see Figure 11.5 in *Introductory Chemistry*). The gas inside the container pushes on the closed end of the manometer.

In the case above, the pressure of the gas in the container must be *greater* than 760 mm Hg because it is pushing the mercury in the open arm up higher. In fact, it pushes it up 152 mm higher, and therefore the pressure of the gas in the container is

$$P_{gas} = 760 \text{ mm Hg} + 152 \text{ mm Hg} = \textbf{912 mm Hg} = \textbf{912 torr}$$

When trying to solve manometer problems, think about what type of answer makes sense, logically. In all cases you will either add or subtract the height difference with the barometric pressure.

## Practice Problems

**11.10** A pressure of 642 mm Hg is equal to how many atm?

**11.11** A pressure of 642 mm Hg is equal to how many Pa?

**11.12** A pressure of $8.452 \times 10^4$ Pa is equal to how many atm?

**11.13** A pressure of $8.452 \times 10^4$ Pa is equal to how many torr?

**11.14** A pressure of 1.867 atm is equal to how many torr?

**11.15** If the arm of a manometer connected to a container of gas is 85 mm Hg lower than the arm open to the atmosphere (760 torr), what is the pressure of the gas in the container?

**11.16** If the arm of a manometer connected to a container of gas is 105 mm Hg higher than the arm open to the atmosphere (772 torr), what is the pressure of the gas in the container?

### 11.3 The Kinetic Molecular Theory of Gases

This theory can be summarized as follows:

- The particles composing a gas are very small compared to the distances between them.
- The particles are in constant random motion. They move in straight lines unless they collide with one another or the wall of the container.
- The attractions and/or repulsions between particles are small enough to be negligible.
- The average kinetic energy (K.E.) of these particles is directly proportional to the temperature in kelvins.

**Example 11.7:** If the molecules in the rigid flask containing nitrogen are at 250 K, how does their kinetic energy change when the temperature is raised to 500 K? What if the temperature is raised to 300 K?

**Solution:** The kinetic energy is directly proportional to kelvins. Thus, if the Kelvin temperature is doubled, the kinetic energy is doubled.

$$\text{K.E.} \times 500 \text{ K}/250 \text{ K} = 2 \text{ K.E. (K.E. is doubled)} \quad \textbf{(EQ 11.1)}$$

## 174 Chapter 11

For the second case, the K.E. is raised proportionally as shown in EQ 11.2.

$$\text{K.E.} \times 300 \text{ K}/250 \text{ K} = 1.2 \text{ K.E.} \text{ (K.E. is 1.2 times as large)} \quad \text{(EQ 11.2)}$$

---

**Practice Problems**

**11.17** If the temperature of oxygen in a rigid container is decreased from 752 K to 376 K how is the K.E. changed?

**11.18** If the temperature of oxygen in a rigid container is decreased from 840 K to 560 K how is the K.E. changed?

---

### 11.4 Boyle's Law (Pressure and Volume)

This law relates pressure and volume. For a gas, if P or V increases, the other must decrease. In mathematical language this can be summarized in equations (EQ 11.3 and EQ 11.4).

$$PV = a \quad \text{(EQ 11.3)}$$

In EQ 11.3, the general statement that pressure times volume is a constant (a) is demonstrated.

$$P_1V_1 = P_2V_2 \quad \text{(EQ 11.4)}$$

The most important skill you need to attain (after reading and understanding the background of Boyle's work) in this section is how to apply EQ 11.4. Some examples show how to do this.

A secondary skill you want to master is the ability to make predictions about pressure and volume for real-life situations by applying your understanding of Boyle's law.

---

One of the most important things you can do in determining solutions for these types of problems is to read the problem carefully enough to correctly identify $P_1$, $P_2$, $V_2$, and $V_2$.

---

**Example 11.8:** A gas is at a pressure of 1 atm and a volume of 2 L in an elastic container. If the container size is reduced to 1 L, what will the pressure exerted by the gas be? (Assume the temperature is constant.)

**Solution:** The volume has been reduced (think of squeezing a balloon), so the pressure exerted by the gas inside the container must be increased. You can rearrange EQ 11.4 as shown below to solve for the new pressure ($P_2$).

$$P_2 = \frac{P_1V_1}{V_2} = \frac{(1 \text{ atm})(2 \text{ L})}{1 \text{ L}} = 2 \text{ atm}$$

Notice that if you *think* about the magnitude of the correct answer, you will decrease mistakes made from misassigning the values for $P_1$, $P_2$, $V_2$, and $V_2$. In the case above, the predicted new pressure is *increased,* which is consistent with the calculated value.

**Example 11.9:** A gas is at a pressure of 1.00 atm and a volume of 2.00 L in an elastic container. If the pressure is reduced to 0.854 atm, what will the new volume of the gas be? (Assume the temperature is constant.)

**Solution:** The pressure has been reduced, so the volume must be increased. Notice that EQ 11.4 can be rearranged to solve for the new volume ($V_2$).

$$V_2 = \frac{P_1 V_1}{P_2} = \frac{(1.00 \text{ atm})(2.00 \text{ L})}{0.854 \text{ atm}} = 2.34 \text{ L}$$

You should have noticed that in the cases solved above, all the units canceled, except for the one needed on the solution.

When solving the following problems, always apply good *number sense*. If a volume is *decreased,* the new pressure **must** *increase,* etc.

## Practice Problems

**11.19** What is the new volume if a gas which occupies 2.50 L at a pressure of 0.896 atm has its pressure increased to 1.25 atm?

**11.20** What is the new pressure if a gas which occupies 1.04 L at a pressure of 1.22 atm has its volume increased to 2.56 L?

**11.21** What is the new pressure if a gas which occupies 250 mL at a pressure of 710 torr has its volume decreased to 150 mL?

**11.22** What is the new volume if a gas which occupies 250 mL at a pressure of 710 torr has its pressure decreased to 120 torr?

**11.23** You are traveling from sea level to the Rocky Mountains (atmospheric pressure will decrease). At sea level you buy a back of potato chips. What will happen to this package as you ascend?

## 11.5 Charles's Law (Volume and Temperature)

This law relates temperature and volume. It can also be applied to problems relating temperature and pressure. For a gas, if

- The temperature increases (at constant P and molar amount), the volume must increase.
- The temperature decreases (at constant P and molar amount), the volume must decrease.

In mathematical language this law can be summarized in equations (EQ 11.5 to EQ 11.6):

$$\frac{V}{T} = b \qquad \textbf{(EQ 11.5)}$$

The fact that volume divided by temperature is a constant is demonstrated in EQ 11.5. The working equation for a system with constant pressure is EQ 11.6.

$$\frac{V_1}{T_1} = \frac{V_2}{T_2} \qquad \text{(EQ 11.6)}$$

The most important skill you need to attain (after reading and understanding the background of Charles's work) in this section is how to apply EQ 11.6. Some examples show how to do this.

A secondary skill you want to master is the ability to make predictions about temperature and volume for real-life situations by applying your understanding of Charles's law.

**Example 11.10:** A gas is at a temperature of 295 K and a volume 2 L in an elastic container. If the container size is reduced to 1.7 L, what will the new temperature of the gas be? (Assume the pressure is constant.)

**Solution:** The solution to this problem involves the rearrangement of EQ 11.6 as shown below. Before actually making the calculation you should be able to predict that since the new volume is lower than the initial volume, then the new temperature should also be *lower* than 295 K.

$$T_2 = \frac{V_2 T_1}{V_1} = \frac{(1.7 \text{ L})(295 \text{ K})}{2.0 \text{ L}} = 251 \text{ K}$$

**Example 11.11:** A gas is at a temperature of 95°C and a volume of 24 L in an elastic container. If the temperature is reduced to 24°C, what will the new volume of the gas be? (Assume the pressure is constant.)

**Solution:** Before you begin this problem remember:

| Temperature must be in kelvins for all gas law calculations. |
|---|

95°C = 368 K and 24°C = 297 K

The solution can be obtained by rearranging EQ 11.6.

$$V_2 = \frac{V_1 T_2}{T_1} = \frac{(24 \text{ L})(297 \text{ K})}{368 \text{ K}} = 19 \text{ L}$$

## Practice Problems

**11.24** A gas is at a temperature of 265 K and a volume of 24 L in an elastic container. If the temperature is increased to 642 K, what will the new volume of the gas be? (Assume the pressure is constant.)

> **11.25** A gas is at a temperature of 305°C and a volume of 236 mL in an elastic container. If the temperature is reduced to 37.0°C, what will the new volume of the gas be? (Assume the pressure is constant.)
>
> **11.26** If a balloon at room temperature has liquid nitrogen (-180°C) poured over it, what will happen to it?

## 11.6 The Combined Gas Law (Pressure, Volume, and Temperature)

This section simply does what the title implies. The working equation is shown in EQ 11.7. You simply need to rearrange this equation to isolate the unknown of any given problem.

The most challenging part of solving these problems entails correctly identifying all the given measurements in a story problem.

$$\frac{P_1 V_1}{T_1} = \frac{P_2 V_2}{T_2} \qquad \text{(EQ 11.7)}$$

**Example 11.12:** A gas at 1.00 atm and 0°C has a volume of 11.2 L. If the temperature is raised to 25.0°C and the pressure is reduced to 715 mm Hg, what is the new volume?

**Solution:** Rearrange EQ 11.7 to isolate $V_2$. Before inserting given measurements, make sure that the temperatures are written in kelvins.

$$V_2 = \frac{P_1 V_1 T_2}{T_1 P_2} = \frac{(760 \text{ mm Hg})(11.2 \text{ L})(298 \text{ K})}{(273 \text{ K})(715 \text{ mm Hg})} = 13.0 \text{ L}$$

Notice that all units canceled except liters. Also reason that since the temperature *increased* and the pressure *decreased*, you can expect the volume to *increase* (see Table 11.2 in Section 11.8).

> ## Practice Problems
>
> **11.27** 250 mL of gas is in an elastic container at 0.50 atm and 50°C. What will the new volume be if the temperature is lowered to 0°C and the pressure is increase to 1.00 atm?
>
> **11.28** A 2.52-L sample of oxygen is at 37°C and 450 mm Hg pressure. If the pressure is increased to 600 mm Hg and the volume is decreased to 1.57 L, what is the new temperature in Celsius?
>
> **11.29** A 450-mL sample of nitrogen is at 25.0°C and 0.472 atm pressure. If the volume is increased to 825 mL and the temperature is increased to 52.5°C, what is the new pressure?
>
> **11.30** 1.00 L of gas is collected in a sealed elastic container in outer space where the pressure is $1.54 \times 10^{-4}$ mm Hg and the temperature is -185°C. What will the volume be if the container is moved to sea level (760 mm Hg) and room temperature (24.0°C)?

## 11.7 Avogadro's Law (Volume and Moles)

This law relates the amount of gas (in moles) and volume. It can also be applied to problems relating temperature and pressure. For a gas, if

- The number of moles increases (at constant P and T), the volume must increase.
- The number of moles decreases (at constant P and T), the volume must decrease.

In mathematical language this law can be summarized in equations (EQ 11.8 and EQ 11.9):

$$\frac{V}{n} = d \qquad \text{(EQ 11.8)}$$

The fact that volume divided by the number of moles (n) is a constant (d) is demonstrated in EQ 11.8. The working equation for Avogadro's law is EQ 11.9.

$$\frac{V_1}{n_1} = \frac{V_2}{n_2} \qquad \text{(EQ 11.9)}$$

In this section you should learn how to solve two types of problems.

- Solve for an unknown using EQ 11.9.
- Determine the molar mass of an unknown gas given the volume, number of moles of the gas, and the mass of the gas.

Examples of each follow.

**Example 11.13:** If an elastic container has 2.00 mol of helium gas and occupies 40.0 L, what would the new volume be if the amount of gas was doubled?

**Solution:** To solve this rearrange EQ 11.9 so that $V_2$ is isolated. Plug in the given values to arrive at the solution. (Double 2.00 mol to 4.00 mol to find $n_2$.)

$$V_2 = \frac{V_1 n_2}{n_1} = \frac{(40.0 \text{ L})(4.00 \text{ mol})}{2.00 \text{ mol}} = 80.0 \text{ L}$$

---
**STP is standard temperature and pressure.**
These values are arbitrarily assigned.
Standard temperature is exactly 0°C.
Standard pressure is exactly 1 atmosphere.

---

You can also use Avogadro's law to determine mass.

**Example 11.14:** An unknown gas has a mass of 11.3 g and occupies 5.72 L of space at STP. What is the molar mass of the gas?

**Solution:** Before this problem can be solved, some additional information needs to be obtained. Experimental results give us this needed information.

> At STP, 1 mole of an **ideal gas** occupies 22.4 L.

An ideal gas can be any gas (element or compound) that is behaving under the rules of kinetic molecular theory.

With the information given in the problem and in the box, the solution can now be obtained using dimensional analysis. Remember you are trying to determine **molar mass**. This has units of g/mol, so this should be kept in mind when setting up the solution.

$$\frac{11.3 \text{ g gas}}{5.72 \text{ L gas}} \times \frac{22.4 \text{ L gas at STP}}{1 \text{ mol gas at STP}} = 44.3 \text{ g/mol}$$

## Practice Problems

**11.31** An elastic container has a volume of 32.7 L when occupied by 1.57 mol of $O_2$. Some gas is removed, leaving a new volume of 28.3 L. How many moles of gas are left?

**11.32** A rigid 5.0-L container has 0.245 mol of neon gas at a pressure of 760 torr. If 0.736 additional moles of gas are added, what is the new pressure?

**11.33** A balloon occupies 2.5 L with 0.305 mol of carbon dioxide. How much $CO_2$ must be added to the balloon to triple its size?

**11.34** An unknown gas has a mass of 23.4 g and occupies 3.59 L at STP. What is the molar mass of the gas?

**11.35** An unknown gas has a mass of 0.764 g and occupies 819 mL at 25.0°C and 642 mm Hg. What is the molar mass of the gas? Use Table 11.14 to determine what the identity of the gas might be. (*Hint:* First find the volume at STP.)

### 11.8 The Ideal Gas Law

This law combines Boyle's law, Charles's law, and Avogadro's law. One constant for all four relationships can be arrived at. This constant is called R. This is shown in EQ 11.10.

$$\frac{PV}{nT} = R \quad \text{(EQ 11.10)}$$

**Example 11.15:** Experiments show that at STP, 1 mol of an ideal gas occupies 22.4 L. Calculate the value of R from this information.

**Solution:** If experimental values are substituted into EQ 11.10, a value for R can be obtained.

$$\frac{(22.4 \text{ L})(1 \text{ atm})}{(273 \text{ K})(1 \text{ mol})} = 0.08206 \frac{\text{L} \cdot \text{atm}}{\text{K} \cdot \text{mol}} = R$$

## 180    Chapter 11

Notice that the units on R make sense. Now the **ideal gas law** can be shown in its usual form (see EQ 11.11).

$$PV = nRT \qquad \text{(EQ 11.11)}$$

Some of the predictable changes from all the gas laws are summarized in Table 11.2. You should not make an attempt to memorize this table. These predictions can be made just from looking at the ideal gas law.

**Table 11.2** Measured Parameter for Ideal Gas and Its Affect on Others

| Parameter that Increases | Affect on Parameter (Assuming Others Are Constant) | Affect on Parameter (Assuming Others Are Constant) | Affect on Parameter (Assuming Others Are Constant) |
|---|---|---|---|
| P ↑ | V ↓ | T ↑ | n ↑ |
| V ↑ | P ↓ | T ↑ | n ↑ |
| T ↑ | V ↑ | P ↑ | n ↓ |
| n ↑ | V ↑ | P ↑ | T ↓ |

Try to use the ideal gas law to solve some problems. For each problem, try to isolate the unknown, and then substitute given values.

> Remember to use correct units for ideal gas law problems (see Example 11.13).

**Example 11.16:** What volume does 1.00 mol Kr gas occupy at room temperature and 760 mm Hg?

**Solution:** First convert all units, then solve EQ 11.11 for V and substitute the given values into the equation.

$$V = \frac{nRT}{P} = \frac{(1 \text{ mol})(0.08206 \text{ L} \cdot \text{atm/K} \cdot \text{mol})(298 \text{ K})}{1.00 \text{ atm}} = 24.5 \text{ L}$$

## Practice Problems

**11.36** How many moles of $N_2O$ are in a 2.57-L container where the pressure is 0.786 atm and the temperature is 22.5°C?

**11.37** What is the pressure of 12.0 mol of oxygen in 57.6 L at 24.0°C?

**11.38** What volume will 15.0 mol of hydrogen at 24.5°C and 642 mm Hg have?

**11.39** How many moles of nitrogen are present in a 2.00-L tank at 3.0 atm and 18.0°C?

**11.40** What is the mass of the $N_2$ in problem 11.39? What is the density?

> **11.41** Acetylene ($C_2H_2$) has a mass of 3.95 kg in a 65.0 L tank at 25.2°C. What is the pressure of the gas in the tank? (*Hint*: First determine how many moles of gas are present and then use the ideal gas law.)

## 11.9 Dalton's Law of Partial Pressures

This simply states that for a mixture of gases, the total pressure of the gases is equal to the sum of the individual pressures. It is summarized in EQ 11.12:

$$P_T = P_A + P_B + \cdots + P_N \qquad \text{(EQ 11.12)}$$

**Example 11.17:** A rigid 25.0-L container at 22.5°C has 0.525 mol of oxygen and 2.10 mol of nitrogen. What is the partial pressure of each? What is the total pressure in the container?

**Solution:** Use the ideal gas law to calculate the pressure of each of the gases separately.

$$P_{oxygen} = 0.509 \text{ atm and } P_{nitrogen} = 2.03 \text{ atm}$$

The total pressure can be calculated by applying Dalton's law.

$$P_T = 0.509 \text{ atm} + 2.03 \text{ atm} = \textbf{2.54 atm}$$

> ## Practice Problems
> (Remember to make all pressure units the same for Dalton's Law Problems)
>
> **11.42** If the total pressure in a mixture of oxygen, nitrogen, and argon is 810 mm Hg, and the partial pressure of oxygen is 160 mm Hg, and nitrogen is 638 mm Hg, what is the partial pressure of argon?
>
> **11.43** A mixture of oxygen and water vapor has a total pressure of 0.875 atm at 22.0°C. The pressure from water vapor at this temperature is 19.8 mm Hg. What is the pressure of oxygen?
>
> **11.44** A mixture of oxygen gas ($P_{ox}$ = 342 torr) and hydrogen gas ($P_{hyd}$ = 684 mm Hg) has what total pressure (in mm Hg and in atm)?
>
> **11.45** A 2.56-L container with 0.0526 mol of hydrogen and 0.0782 mol helium at 27.0°C has what partial pressures and what total pressure?
>
> **11.46** A 49.0-L container at 25.0°C contains 71.0 g of $Cl_2$ and 2.0 g of $H_2$. What is the partial pressure of each gas and the total pressure in the container?

## 11.10 Stoichiometry of Reactions Involving Gases

This section uses the equalities about ideal gases in combination with stoichiometry from Chapter 8 to solve problems involving volumes of substances in a balanced equation.

**Example 11.18:** The reaction of hydrogen with nitrogen produces $NH_3$ gas. If 4.00 g of hydrogen is used with an excess of nitrogen, how many liters of $NH_3$ will be produced at STP?

**Solution:** To solve Example 11.18, refer to Figure 11.1. Notice that this expands on the ideas from Chapter 8, where a balanced equation is used to convert from one substance (**A**) in a balanced equation to another (**B**).

The first thing that needs to be done is to write a balanced equation.

$$N_2 + 3H_2 \rightarrow 2NH_3$$

Now you can read the problem and identify that **A** is hydrogen and **B** is $NH_3$. In this case you know the mass of $H_2$ and want to find liters of $NH_3$ at STP. Follow the scheme in Figure 11.1 and note the transformations that are needed. The complete solution follows.

$$4.00 \text{ g } H_2 \times \frac{1 \text{ mol } H_2}{2.00 \text{ g } H_2} \times \frac{2 \text{ mol } NH_3}{3 \text{ mol } H_2} \times \frac{22.4 \text{ L } NH_3 \text{ at STP}}{1 \text{ mol } NH_3 \text{ at STP}} = 29.9 \text{ L at STP}$$

**Example 11.19:** If 50.0 L of each reactant in the equation above are combined at STP, how many grams of $NH_3$ will be produced?

**Solution:** You should recognize that this is a limiting reactant problem. Simply convert 50.0 L of each starting reagent to grams of product, and note that the smaller amount will be the correct answer.

$$50.0 \text{ L } H_2 \times \frac{1 \text{ mol } H_2}{22.4 \text{ L } H_2 \text{ at STP}} \times \frac{2 \text{ mol } NH_3}{3 \text{ mol } H_2} \times \frac{17.0 \text{ g } NH_3}{1 \text{ mol } NH_3} = 25.3 \text{ g } NH_3$$

$$50.0 \text{ L } N_2 \times \frac{1 \text{ mol } N_2}{22.4 \text{ L } N_2 \text{ at STP}} \times \frac{2 \text{ mol } NH_3}{1 \text{ mol } N_2} \times \frac{17.0 \text{ g } NH_3}{1 \text{ mol } NH_3} = 75.9 \text{ g } NH_3$$

The actual amount of ammonia that can be produced is **25.3 g**.

Use the equation for the production of ammonia ($NH_3$) to solve some stoichiometry problems. Refer to Figure 11.1 to help design solutions.

## Practice Problems

**11.47** How many liters of $NH_3$ can be produced from 36.5 L of $H_2$ and an excess of $N_2$ at STP?

**11.48** How many grams of $NH_3$ can be produced from 98.5 L of each reagent at STP?

**11.49** How many liters (at STP) of each reagent are necessary to produce 549 g of $NH_3$?

# The Gaseous State

**Figure 11.1 Scheme for Conversions Within a Balanced Equation**

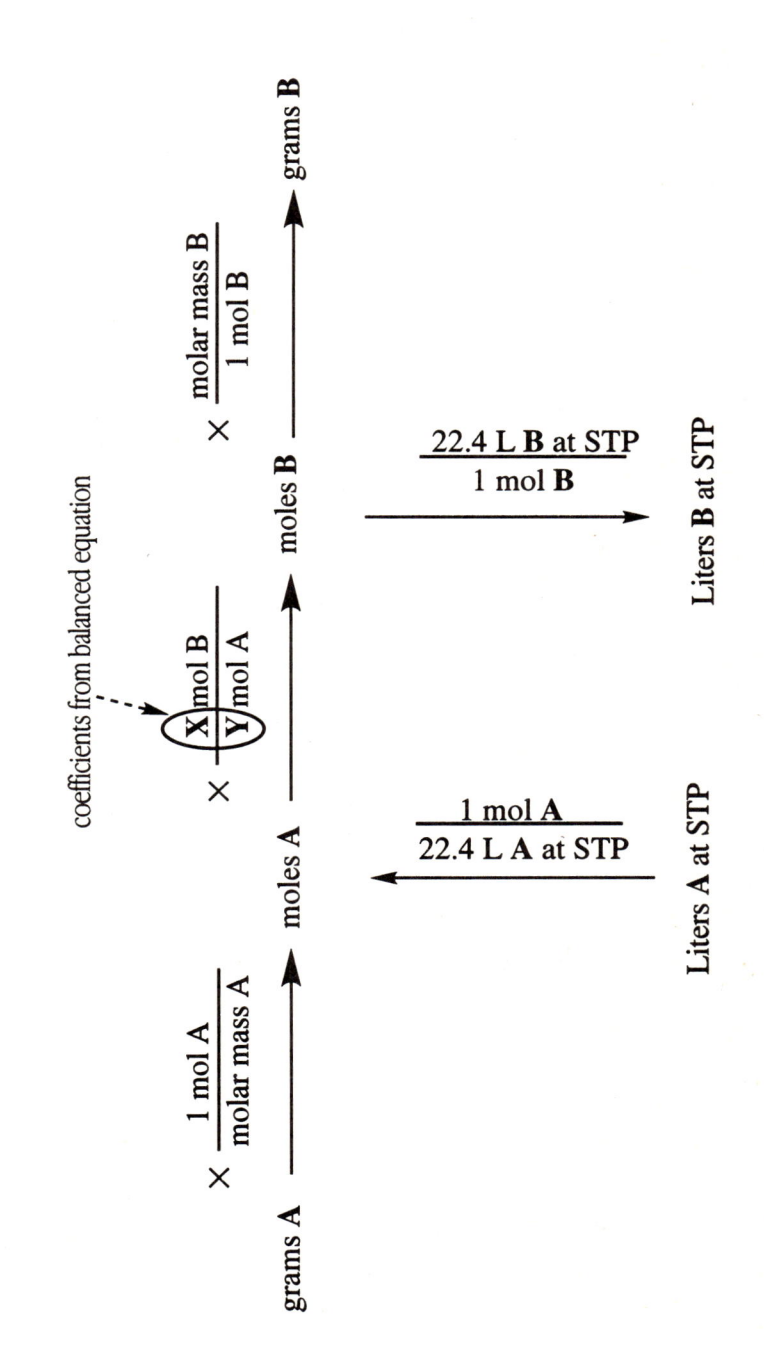

$$\text{grams A} \xrightarrow{\times \dfrac{1 \text{ mol A}}{\text{molar mass A}}} \text{moles A} \xrightarrow{\times \dfrac{\mathbf{X} \text{ mol B}}{\mathbf{Y} \text{ mol A}}} \text{moles B} \xrightarrow{\times \dfrac{\text{molar mass B}}{1 \text{ mol B}}} \text{grams B}$$

coefficients from balanced equation

$$\text{Liters A at STP} \xrightarrow{\dfrac{1 \text{ mol A}}{22.4 \text{ L A at STP}}} \text{moles A}$$

$$\text{moles B} \xrightarrow{\dfrac{22.4 \text{ L B at STP}}{1 \text{ mol B}}} \text{Liters B at STP}$$

# 184   Chapter 11

## Integration of Multiple Skills

The rigid boxes shown below contain $N_2$ and $O_2$ at the specified conditions.

| $N_2$ 2 L, 27°C | $O_2$ 2 L, 27°C | Both boxes are at 700 torr |

**Example 11.20:** Answer the following true/false questions about the gases above.
- A. Each box contains the same number of molecules.
- B. Each box contains the same number of moles.
- C. Each box has the same mass.
- D. Each box has the same density.
- E. The molecules in the nitrogen box are moving at the same speed as the molecules in the oxygen box. (*Hint:* K.E. $\propto$ T, and K.E. = 1/2 mv²)

**Solution:** The solutions with brief explanations follow. None of these require that calculations be made but do require application of current and past principles.

- **A. True** — Since both boxes are under the same T, P (remember that 700 torr = 700 mm Hg), and V, they must have the same number of particles.
- **B. True** — Since both boxes have the same number of molecules, they must have the same number of moles, since 1 mol of anything is $6.02 \times 10^{23}$ molecules.
- **C. False** — Since oxygen has a higher molar mass (32.0 g) than nitrogen (28.0 g) and both boxes have the same number of moles, then the oxygen has the greater mass.
- **D. False** — Since density is mass/volume, the oxygen has the higher density.
- **E. False** — Since both boxes are at the same temperature, they must have the same kinetic energy (K.E.). However, since K.E. is related to the mass times the velocity squared, the smaller molecules ($N_2$) must be moving faster.

**Example 11.20:** For the combustion of gasoline shown below, how many liters of carbon dioxide gas will be produced from the burning of 2.50 kg of $C_8H_{18}$ at 648 mm Hg and 28.0°C?

$$2C_8H_{18} + 25O_2 \rightarrow 16CO_2 + 18H_2O$$

**Solution:** You need to apply stoichiometry to solve this problem, in combination with several other techniques. You will need to convert kilograms to grams and convert the volume determined from stoichiometry to volume of gas at the conditions of the reaction. The solution follows on the next two lines. The last two factors convert from STP to experimental conditions.

$$2.50 \text{ kg } C_8H_{18} \times \frac{1000 \text{ g } C_8H_{18}}{1 \text{ kg } C_8H_{18}} \times \frac{1 \text{ mol } C_8H_{18}}{114 \text{ g } C_8H_{18}} \times \frac{16 \text{ mol } CO_2}{2 \text{ mol } C_8H_{18}} \times$$

$$\frac{22.4 \text{ L } CO_2 \text{ STP}}{1 \text{ mol } CO_2} \times \frac{760 \text{ mm Hg}}{648 \text{ mm Hg}} \times \frac{301 \text{ K}}{273 \text{ K}} = 5.08 \times 10^3 \text{ L of } CO_2$$

## Practice Problems

**11.50** Use the equation for the combustion of gasoline to determine how many liters of water vapor are produced from burning 2.50 kg of $C_8H_{18}$ (at 700 mm Hg, 25°C)?

**11.51** How much (in grams) $C_8H_{18}$ must be burned to produce 1 million liters of $CO_2$ (at 700 mm Hg, 25°C)?

**11.52** If the density of $C_8H_{18}$ is 0.703 g/mL, how many gallons of $C_8H_8$ are required to produce 1 million liters of $CO_2$ (at 700 mm Hg, 25°C)?

**11.53** If the temperature of oxygen in a rigid container is decreased from 327°C to 27°C how is the K.E. changed?

**11.54** What is the new volume if a gas which occupies 250 mL at a pressure of 710 mm Hg has its pressure changed to 1.15 atm?

## Self-Exam

**I. (4 ea)** Write T (true) or F (false) in each corresponding blank.

_____ 1. A barometer is identical to a manometer.

_____ 2. Boyle's law relates the temperature of ideal gases to their pressures.

_____ 3. The temperature in gas law calculations must be in kelvins.

_____ 4. 760 mm Hg is equal to 760 atm.

_____ 5. A temperature of -269°C is equal to 4 K.

_____ 6. The kinetic energy of gas molecules at 60°C is double that of the gas at 30°C.

_____ 7. Kinetic molecular theory says that attractions and repulsions between particles of an ideal gas are negligible.

_____ 8. 1 mol of an ideal gas *always* occupies 22.4 L.

_____ 9. The partial pressures of a mixture of gases are proportional to the number of moles of each type of gas.

_____ 10. A gas that has its pressure tripled and its temperature doubled, will have a volume 6 times as great.

**II. (4 ea)** Multiple Choice. Place the letter of the BEST answer in the corresponding blank.

_____ 1. Which condition(s) favor ideal gas behavior?
    A. Low T    B. High T    C. Low P    D. High P    E. Both B and C

_____ 2. What is STP?
    A. 0°C and 1 atm    B. 25°C and 1 atm    C. 0 K, 1 atm
    D. 0°C and 0 mm Hg    E. 0°C, 760 atm

# 186   Chapter 11

_____ 3. Whose law deals with the number of moles of an ideal gas?
   A. Boyle's
   B. Charles's
   C. Avogadro's
   D. Dalton's
   E. Pascal's

_____ 4. If 2.56 L of carbon monoxide has a pressure of 0.526 atm, what will its new volume be if the pressure is increased to 0.965 atm?
   A. 4.70 L    B. 1.35 L    C. 2.47 L    D. 4.39 L    E. 1.40 L

_____ 5. A gas occupies 1.56 L in an elastic container at 27.0°C. What is its new volume if the temperature is raised to 54.0°C?
   A. 3.12 L    B. 0.780 L    C. 1.70 L    D. 1.43 L    E. 42.1 L

_____ 6. An elastic container (at constant T and P) has 5.0 mol of $H_2$ and occupies 115 L. If 5.0 mol of He is added, what is the new volume?
   A. 115 L    B. 57.5 L    C. 153 L    D. 230 L    E. 575 L

_____ 7. What is the volume of 1.0 mol of carbon monoxide at 25.0°C and 1 atm?
   A. 20.6 L    B. 22.4 L    C. 24.5 L    D. 230 L    E. 46.4 L

_____ 8. A mixture of Ne and Ar has a total pressure of 0.942 atm. If the partial pressure of Ne is 358 mm Hg, what is the partial pressure of Ar?
   A. 584 mm Hg
   B. 0.584 atm
   C. 337 mm Hg
   D. 471 mm Hg
   E. 0.471 atm

_____ 9. An unknown gas has a mass of 0.854 g and occupies 4.78 L at STP. What is the most likely identity of the gas?
   A. $H_2$    B. He    C. Ne    D. $O_2$    E. CO

_____ 10. A diver at a depth of 90 ft is breathing air at approximately 4 atm pressure. Which of the following statements is true about this diver?
   A. The density of the air in his lungs is 4 times what it was at sea level.
   B. The volume of his lungs is 4 times what it was at sea level.
   C. The volume of his lungs is 1/4 what it was at sea level.
   D. The pressure in his lungs is 1/4 what it was at sea level.
   E. The temperature of his lungs is 4 times what it was at sea level.

## III. Provide Complete Answers

1. (8 pts) Answer the following about the ideal gases in the rigid containers shown.

   | $N_2$ 2 L, 27°C | $O_2$ 2 L, 27°C | Both boxes are at 700 torr |

   A. Which box (if any) has a larger mass?
   B. Which box (if any) has more molecules?
   C. Which box (if any) has a greater kinetic energy?
   D. How many moles of gas are in each box?

2. (6 pts) Calculate the molar mass of an unknown gas if 9.53 g of the gas occupies 5.62 L at STP. What is the density of the gas?

3. (6 pts) For the reaction where water is decomposed to hydrogen and oxygen gas, how many liters of each product will be produced from 54.0 g of water at STP?
How many liters of each will be produced at 24.0°C and 658 mm Hg?

2 H$_2$O → 2 H$_2$(g) + O$_2$(g)

# Answers to Practice Problems

**11.1** $\dfrac{PV}{RT} = n$      **11.2** $\dfrac{nRT}{V} = P$      **11.3** $\dfrac{nRT}{P} = V$

**11.4** $\dfrac{PV}{nR} = T$      **11.5** Yes, $\dfrac{PV}{nRT} = 1$. It is also equal to $\dfrac{nRT}{PV}$.

**11.6** $\dfrac{V_1}{T_1} = \dfrac{V_2}{T_2}$      **11.7** $\dfrac{P_1V_1}{T_1} = \dfrac{P_2V_2}{T_2}$      **11.8** $\dfrac{V_1}{n_1} = \dfrac{V_2}{n_2}$

**11.9** Some other trends include noting that the gaseous elements are all nonmetals. The gaseous compounds also are covalent molecules made up of two or more nonmetal atoms. You should note a general trend between molar mass and boiling point. There are some exceptions. Helium (M.M. = 4.0 g) has a lower boiling point than hydrogen (M.M. = 2.0 g). In fact, helium has the lowest temperature for liquification (-269°C or 4 K), which is nearly at absolute zero. You will notice some other small molecules with high boiling points, such as NH$_3$. This you will learn later is due to stronger than usual *attractive* forces between molecules of this gas. While one of the postulates of kinetic molecular theory says that these are *nearly* negligible, for some compounds (those that are polar) this is less true.

| | | | |
|---|---|---|---|
| **11.10** | 0.845 atm | **11.11** | 8.56 × 10$^4$ Pa |
| **11.12** | 0.834 atm | **11.13** | 634 mm Hg |
| **11.14** | 1419 mm Hg | **11.15** | 845 mm Hg |
| **11.16** | 667 mm Hg | **11.17** | The K.E. is 1/2 what it was initially. |
| **11.18** | The K.E. is 2/3 what it was initially. | **11.19** | 1.79 L |
| **11.20** | 0.500 atm | **11.21** | 1200 mm Hg    **11.22**   1500 mL |

**11.23** As you ascend the volume of the bag would increase (because the atmospheric pressure is decreasing). Eventually, this volume could become so large that the bag would burst.

| | | | |
|---|---|---|---|
| **11.24** | 58 L | **11.25** | 127 mL |
| **11.26** | Volume will decrease. | **11.27** | 106 mL |
| **11.28** | -16°C (257 K) | **11.29** | 0.281 atm |
| **11.30** | 6.84 × 10$^{-7}$ L | **11.31** | 1.36 mol O$_2$ |
| **11.32** | 3040 mm Hg | | |
| **11.33** | 0.610 mol must be added to give a total of 0.915 mol (triple the original amount). | | |
| **11.34** | 146.1 g/mol | **11.35** | 27.0 g/mol could be HCN |

| | | | |
|---|---|---|---|
| **11.36** | $8.33 \times 10^{-2}$ mol | **11.37** | 5.08 atm |
| **11.38** | 433 L $H_2$ | **11.39** | 0.252 mol $N_2$ |
| **11.40** | 7.04g $N_2$, 3.53 g/L = density | | |
| **11.41** | 57.0 atm | **11.42** | $P_{ar}$ = 12 mm Hg |
| **11.43** | $P_{ox}$ = 645 mm Hg | **11.44** | 1030 mm Hg (1.35 atm) |
| **11.45** | 0.506 atm $H_2$, 0.752 atm He, 1.26 atm total | | |
| **11.46** | 0.499 atm $Cl_2$, 0.499 atm $H_2$, 0.998 atm total | | |
| **11.47** | 24.3 L $NH_3$ | **11.48** | 49.8 g $NH_3$ |
| **11.49** | 1090 L $H_2$, 362 L $N_2$ | **11.50** | $5.24 \times 10^3$ L $H_2O$ gas |
| **11.51** | $5.37 \times 10^5$g $C_8H_{18}$ | **11.52** | 203 gal $C_8H_{18}$ |
| **11.53** | K.E. is 1/2 initial K.E. (T decreases 600K to 300K). | | |
| **11.54** | 203 mL | | |

## Answers to Self-Exam

**I. T/F**
1. F   2. F   3. T   4. F   5. T   6. F   7. T   8. F   9. T   10. F

**II. MC**
1. E   2. A   3. C   4. E   5. C   6. D   7. C   8. E   9. B   10. A

**III. Complete Answers**
1. A. $O_2$ box has greater mass.
   B. Both boxes have the same number of molecules.
   C. Both boxes have the same K.E.
   D. 0.07 mol of each.

2. Molar mass of the gas is 38.0 g/mol.
   Density is 1.696 g/L.

3. At STP: 67.2 L $H_2$, and 33.6 L $O_2$
   At 24.0°C and 658 mm Hg: 84.4 L $H_2$, and 42.2 L $O_2$

# 12
# Liquids, Solids, and Attractions Between Molecules

Chapter 11 dealt with the gaseous state of matter. This chapter will look at liquids and solids from both a microscopic (molecular) and macroscopic (visible properties) approach.

This chapter has both calculational and conceptual topics to master. Compared to Chapter 11, you will spend much less time doing calculations in Chapter 12. You will need to spend time trying to understand new ideas and finding patterns in molecular behavior.

Keys for success in this chapter:

- Review definitions about states of matter from Chapter 3.
- Learn the definitions of the new terms.
- Understand how energy is involved in changes of state and changes of temperature.
- Comprehend the attractions of molecules/atoms to each other which bring about liquid and solid formations.
- Examine and compare the different types of solids.
- Practice energy problems and Dalton's law problems.

## Summary of Verbal Knowledge

**gas:** A state of matter that has neither a fixed volume nor a fixed shape.
**liquid:** A state of matter that has a fixed volume but does not have specific shape.
**solid:** A state of matter that has both a fixed volume and a specific shape.
**melting:** This is also called *fusion*. It is the change of solid into a liquid.
**freezing:** This is the change of a liquid into a solid.
**vaporization (evaporation):** This is the change of a liquid into a gas.
**condensation:** This is the change of a gas to a liquid.
**sublimation:** This is the change of a solid directly to a gas.
**deposition:** This is the change of a gas directly to a solid.
**normal melting point:** This is the temperature where a solid changes into a liquid at 1 atm pressure.
**surface tension:** Energy required to stretch a unit area.
**normal boiling point:** This is the constant temperature at 1 atm pressure at which a liquid changes into a gas.
**molar heat of fusion:** The energy required to melt (fuse) 1 mol of a solid.
**molar heat of vaporization:** The energy required to vaporize 1 mol of a liquid.
**vapor pressure:** This is the pressure exerted by molecules of a substance in the vapor phase which are in equilibrium with molecules of the same substance in the liquid phase.

**boiling point:** The temperature at which the vapor pressure of a liquid equals the atmospheric pressure.
**dispersion forces:** These are weak attractive forces between temporarily polarized atoms (or molecules) caused by the varying positions of the electrons during their motion about their nuclei.
**dipole-dipole force:** This is a permanent attractive intermolecular force resulting from the interaction of the positive end of one molecule with the negative end of another.
**hydrogen bond:** This is a dipole-dipole attractive force that exists between two polar molecules containing a hydrogen atom covalently bonded to an atom of nitrogen, oxygen, or fluorine.
**crystalline solid:** A solid that is composed of one or more crystals with each crystal having a well-defined, ordered arrangement of structural units.
**amorphous solid:** A solid that lacks order and, as a result, does not have a well-defined arrangement of structural units.
**atomic solid:** A solid containing atoms from a nonmetallic element.
**metallic solid:** A solid containing atoms from a metallic element.
**covalent network solid:** A solid containing atoms held in large networks or chains by covalent bonds.
**molecular solid:** A solid consisting of molecules.
**ionic solid:** A solid consisting of an array of ions (attractive forces of opposite charges holding them together.)

## Review of Mathematical and Calculator Skills

You should review the process of converting moles to grams and grams to moles. If you have forgotten how to do this, please refer to Chapter 7. Also review the problems in Chapter 3 dealing with energy changes.

**Example 12.1:** If you have 54.0 g of $H_2O$, how many moles are present?

$$\text{Solution: } 54.0 \text{ g } H_2O \times \frac{1 \text{ mol } H_2O}{18.0 \text{ g } H_2O} = 3.0 \text{ mol } H_2O$$

### Practice Problems

**12.1** How many grams are in 2.54 mol $H_2O$?

**12.2** How many moles are in $1.86 \times 10^3$ g of $H_2O$?

**12.3** How many moles are in 98.7 g of $C_6H_6$?

# Liquids, Solids, and Attractions Between Molecules

## Application of Skills and Concepts

### 12.1 The States of Matter and Changes in These States

This section describes macroscopic (large enough to see) changes of states of matter. Before trying some sample problems, do the following:

- Review definitions at the beginning of the chapter.
- Learn the names of changes of state.
- Learn about concepts of energy changes connected with changes of state.

Table 12.1 Summary of Changes of State

| Name of Process | Change | Example | Energy |
|---|---|---|---|
| Freezing | l → s | Water becomes ice | − |
| Melting (fusion) | s → l | Solid gallium becomes liquid | + |
| Condensation | v → l | HCl gas becomes liquid HCl | − |
| Vaporization (evaporation) | l → v | Liquid alcohol becomes a vapor | + |
| Deposition | v → s | Water vapor becomes frost | − |
| Sublimation | s → v | Dry ice (solid $CO_2$) becomes a gas | + |

g = gas; l = liquid; s = solid
+ means energy needs to be added for the change of state to occur.
− means energy is released during the change of state.

**Example 12.2:** A bottle of rubbing alcohol is left out on a counter with the cap off for three weeks. The volume of liquid decreases. What process occurred? Was energy needed for this to happen?

**Solution:** The alcohol evaporated, which required heat.

---

### Practice Problems

**12.4** What is the process where solid $SF_6$ becomes a vapor?

**12.5** When steam condenses on a surface, what energy change occurs?

**12.6** Frost forms on the outside of a glass of ice-cold lemonade, what process occurs? What energy change occured?

## 12.2 The Energy for a Change of State

Changes of states either require or release energy. This might be easy to grasp by examining the heating curve in Figure 12.1. This curve is *heating* when going from left to right. When going from right to left it is a *cooling curve*.

**Figure 12.1** A Heating Curve

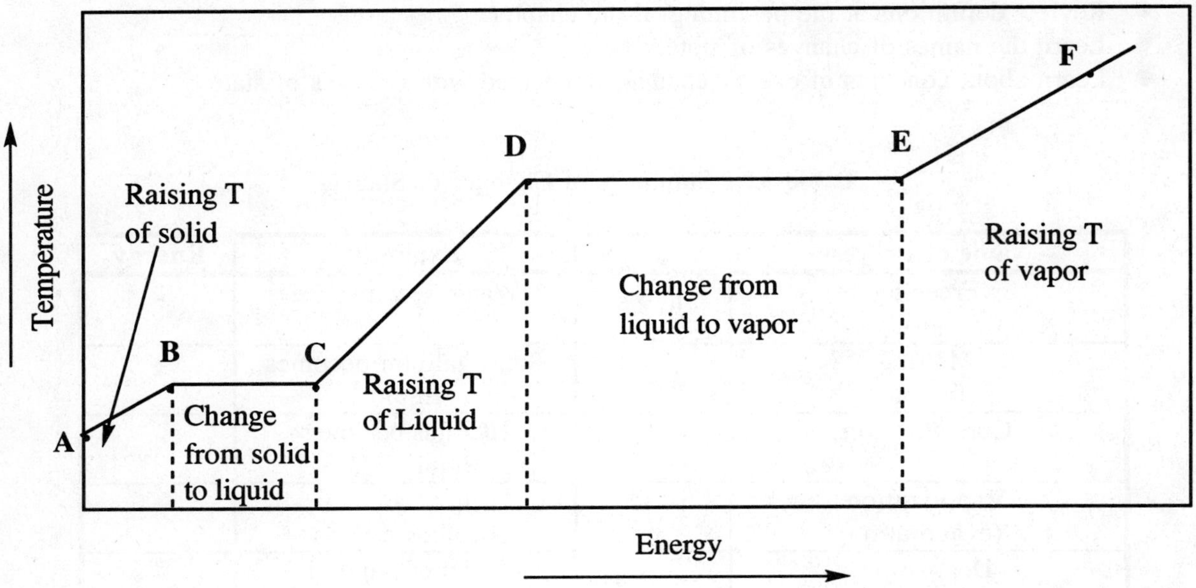

Look at all the points on the curve.

**A.** The substance is a solid at a temperature below the freezing (melting) point.
   When heat is added, the temperature rises.
**B.** The substance is a solid at the freezing point.
    When heat is added, the phase changes, but the temperature does not.
**C.** The substance is a liquid at the freezing point.
   When heat is added, the temperature rises.
**D.** The substance is a liquid at the boiling point.
   When heat is added the phase changes, but the temperature does not.
**E.** The substance is a vapor at the boiling point.
   When heat is added the temperature rises.
**F.** The substance is a vapor at a temperature above the boiling point.

Each substance has a unique energy associated with a change of state. Some of these are summarized in Table 12.2.

Also included in this table are the normal melting points and boiling points. The molar energies for changes of state are determined for these points.

Use the information in Figure 12.1 and Table 12.2 to solve the practice problems.

**Liquids, Solids, and Attractions Between Molecules** 193

Table 12.2 Some Examples of Molar Heats of Fusion and Vaporization

| Substance | Normal m.p. (°C) | Molar Heat Fusion (kJ/mol) | Normal b.p. (°C) | Molar Heat Vaporization (kJ/mol) |
|---|---|---|---|---|
| Water, $H_2O$ | 0 | 6.01 | 100 | 40.7 |
| Methane, $CH_4$ | -182 | 0.94 | -164 | 10.4 |
| Hydrogen sulfide, $H_2S$ | -86 | 2.39 | -61 | N/A |
| Mercury, Hg | -39 | 2.3 | 357 | 59.3 |
| Argon, Ar | -189.2 | 1.3 | -186.7 | 6.3 |
| Ethyl ether, $C_2H_5OC_2H_5$ | -116 | 6.9 | 34.6 | 26.0 |
| Ethanol, $CH_3CH_2OH$ | -117 | 7.61 | 78.5 | 39.3 |
| Benzene, $C_6H_6$ | 5.5 | 10.9 | 80.1 | 31.0 |

**Example 12.3:** If 25.5 g of ice is melted at 0°C to 25.5 g of liquid water, then heated to 100°C, and finally changed to steam at 100°C, how much energy is required for each step? How much total energy is required?

**Solution:** The scheme shown below summarizes the energy needed for solving this problem. Try to follow what energy is needed by examining the heating curve in Figure 12.1, and use any necessary data from Table 12.2.

$$\boxed{\begin{array}{c}\text{Total}\\\text{energy}\\\text{needed}\end{array}} = \boxed{\begin{array}{c}\text{Heat for}\\\text{melting at}\\\text{normal m.p.}\end{array}} + \boxed{\begin{array}{c}\text{Heat for}\\\text{changing T}\\\text{from m.p. to b.p.}\end{array}} + \boxed{\begin{array}{c}\text{Heat for}\\\text{vaporizing}\\\text{at normal b.p.}\end{array}}$$

Take each step separately. The first step involves the energy of a change of state.

$$25.5 \text{ g } H_2O(s) \times \frac{1 \text{ mol } H_2O}{18.0 \text{ g } H_2O} \times \frac{6.01 \text{ kJ}}{\text{mol } H_2O} = 8.51 \text{ kJ}$$

The second step involves the heat of changing the temperature of the water from 0°C to 100°C. This is a 100°C change. Also needed in this step is the specific heat for water (See Table 12.3). Note that this energy is in units of joules, while the changes of state are in units of kilojoules. In order to make all the units the same, the last factor in the second step involves changing J to kJ.

$$25.5 \text{ g } H_2O \times \frac{4.184 \text{ J}}{\text{g °C}} \times 100° \text{C} \times \frac{1 \text{ kJ}}{1000 \text{ J}} = 10.7 \text{ kJ}$$

Table 12.3 Some Specific Heats of Substances

| Substance | Specific Heat (J/g °C) |
|---|---|
| H₂O(l) | 4.184 |
| H₂O(s) | 1.99 |
| H₂O(g) | 2.03 |
| Hg(l) | 0.139 |
| Ethanol(l) | 2.46 |
| Al(s) | 0.900 |
| Au(s) | 0.129 |
| Cu s) | 0.385 |
| C (graphite) | 0.720 |
| C (diamond) | 0.502 |

The third step is the change of state from liquid to vapor. (See Table 12.2.)

$$25.5 \text{ g } H_2O(l) \times \frac{1 \text{ mol } H_2O}{18.0 \text{ g } H_2O} \times \frac{40.7 \text{ kJ}}{\text{mol } H_2O} = 57.7 \text{ kJ}$$

The total energy is just the sum of all the energies above, which equals **76.9 kJ**.

## Practice Problems

**12.7** How much heat is needed to melt 12.6 g of ice at 0°C?

**12.8** How much heat is require to raise the temperature of 12.6 g H₂O at 0°C to 100°C?

**12.9** How much heat is required to evaporate 12.6 g H₂O?

**12.10** How much total heat is required for all three steps (problems 12.7 to 12.9)?

**12.11** How much heat is required to vaporize 28.7 g Hg?

**12.12** How much heat is released when 458 g of water freezes at 0°C?

## 12.3 Vapor Pressure and Evaporation

This section examines the pressure exerted by molecules in the vapor phase over the same substance in the liquid phase. This is known as **vapor pressure**. For any given liquid, some molecules will develop enough energy to vaporize. This number increases as the temperature increases. Note the following:

- Molecules in the vapor phase exert some pressure at any temperature.
- The molecules in the vapor phase are in equilibrium with those in liquid phase.
- As temperature increases, the vapor pressure increases, but not linearly.
- At the boiling point, the **vapor pressure** equals the **atmospheric pressure**.
- Each substance has a unique vapor pressure curve.

# Liquids, Solids, and Attractions Between Molecules

Look at some data taken from two clear, colorless liquids, water and diethyl ether. Note the differences in vapor pressures at various temperatures. Examine Table 12.3 and Figure 12.2, which is a graphical representation of the data in Table 12.3. Then try some problems to test your unerstanding of vapor pressure.

**Example 12.4:** What is the vapor pressure of oxygen collected over water at 28°C if the atmospheric pressure is 715 mm Hg and the volume of gas collected is 132 mL? How many moles of oxygen are in the vessel?

**Solution:** The solution to this problem involves several steps. The first is recognizing that the total pressure (Dalton's law) is equal to the partial pressure of the water vapor and the partial pressure of the oxygen. ($P_T = P_{oxygen} + P_{water\ vapor}$)

The partial pressure of the water vapor can be obtained from empirical data. Using Table 12.4, note that at 28°C, the $P_{water\ vapor}$ = 28.3 mm Hg. Solve Dalton's law for $P_{oxygen}$.

$P_{oxygen}$ = 715 mm Hg - 28.3 mm Hg = 687 mm Hg

The moles of oxygen can be determined using the ideal gas law. Note that units have been changed for use in ideal gas law.

$$n = \frac{PV}{RT} = \frac{(0.904\ atm)(0.132\ L)}{(0.08206\ L \cdot atm/mol \cdot K)(301\ K)} = 4.83 \times 10^{-3}\ mol\ O_2$$

**Table 12.4** Some Vapor Pressures of Water and Ether ($C_4H_{10}O$) at Various Temperatures

| Temperature (°C) | Vapor Pressure of Water (mm Hg) | Vapor Pressure of Ether (mm Hg) | Temperature (°C) | Vapor Pressure of Water (mm Hg) | Vapor Pressure of Ether (mm Hg) |
|---|---|---|---|---|---|
| -74.3 |  | 1 | 24 | 22.4 |  |
| -48.1 |  | 10 | 25 | 23.8 |  |
| -27.7 |  | 40 | 28 | 28.3 |  |
| -11.5 |  | 100 | 30 | 31.8 |  |
| 0 | 4.6 |  | 34.6 |  | 760 |
| 5 | 6.5 |  | 40 | 55.3 |  |
| 10 | 9.2 |  | 50 | 92.5 |  |
| 15 | 12.8 |  | 60 | 149.4 |  |
| 17.9 |  | 400 | 70 | 233.7 |  |
| 20 | 17.5 |  | 80 | 355.1 |  |
| 21 | 18.7 |  | 90 | 525.8 |  |
| 22 | 19.8 |  | 100 | 760 |  |

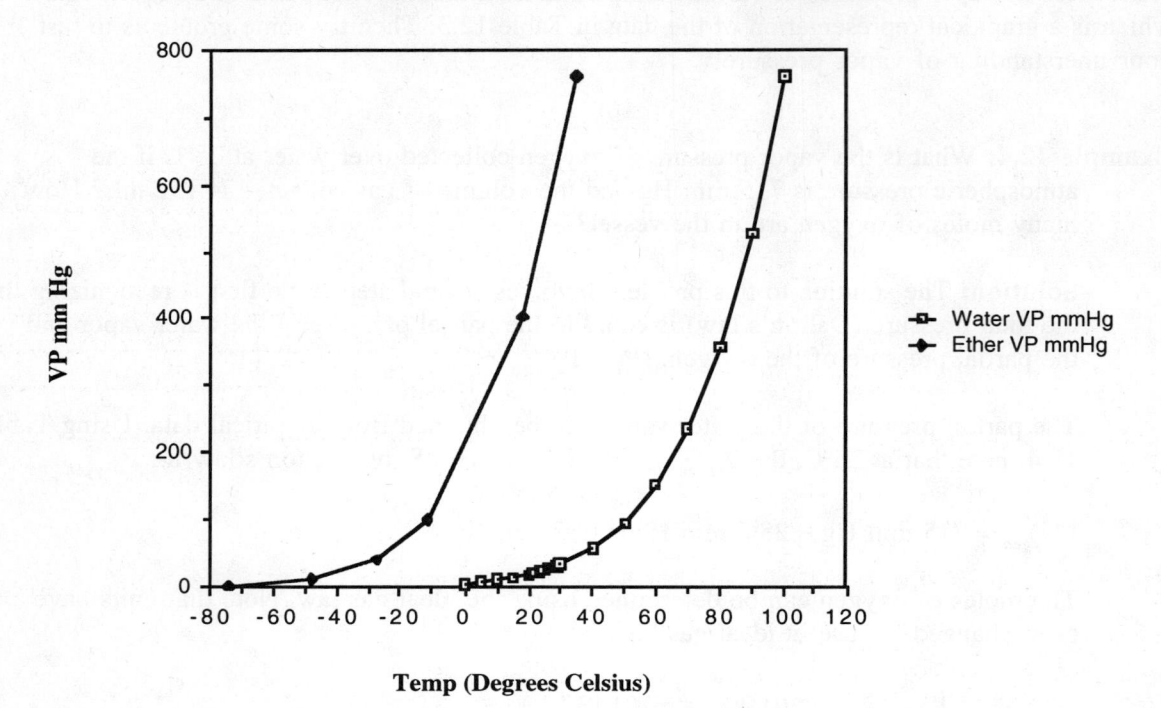

### Figure 12.2 Graph of Vapor Pressure as Function of Temperature

### Practice Problems

**12.13** What is the vapor pressure of water at 30°C?

**12.14** What is the vapor pressure of ether at 17.9°C?

**12.15** If hydrogen is collected over water at 22°C where the atmospheric pressure is 738 mm Hg and the volume is 78.5 mL, how many moles of $H_2$ are present? What is partial pressure of $H_2$?

**12.16** If helium is collected (in a 1.28-L vessel) over water at 25°C (atmospheric pressure = 692 mm-Hg), what is the partial pressure of He? How many moles of He are present? How many grams of He?

**12.17** If the atmospheric pressure is 526 mm Hg, at what temperature will water boil (see definition of vapor pressure)?

## 12.4 A Kinetic Molecular Description of Gases, Liquids, and Solids

This section examines the three different phases of matter at the microscopic level. The ideas are summarized in Table 12.5 and applied in Example 12.5.

**Example 12.5:** Kinetic molecular theory says that in the gas phase, the molecules are very far apart. What changes in condition might bring these particles close together?

  **Solution:** If the *temperature* decreases, the molecules will move slower (have lower kinetic energy) and thus become closer together (interact).

# Liquids, Solids, and Attractions Between Molecules

If the *pressure* increases, the molecules will be crowded closer together. If the closeness of the molecules is great enough, they will have attractions and repulsions which are characteristic of the liquid phase.

Table 12.5 Summary of Microscopic Properties of States of Matter

|  | Gas | Liquid | Solid |
|---|---|---|---|
| **Molecules (particles)** | Very far apart | Close enough to interact, but they still can move. | Close enough to be held in definite shape |
| **Attractions/repulsions** | Negligible | Important | Important |
| **Kinetic energy** | Very high | Moderate | Low |

## Practice Problems

**12.18** What happens to molecules as the temperature increases?

**12.19** What happens to molecules as the temperature decreases?

**12.20** What happens to molecules as the pressure increases?

**12.21** What happens to molecules as the pressure decreases?

## 12.5 The Liquid State

There are three types of forces *between* molecules which can bring them into the liquid phase. These are

- Dispersion
- Dipole interactions
- Hydrogen bonding

These are summarized in Table 12.6.

Table 12.6 Some Interactive Forces Between Molecules in the Liquid/Solid Phases

| Type of Force | Types of Molecules Where It Is Primary | Strength | Examples of Molecules Where It Is Primary |
|---|---|---|---|
| Dispersion | Nonpolar | Weak | $Br_2$, Ne, $CCl_4$ |
| Dipole interaction | Polar | Moderate | HBr, $CH_2Cl_2$, $H_2S$ |
| Hydrogen bonding | Molecules with H-F, H-N, H-O bonds | Very Strong | $H_2O$, HF, $NH_3$, $CH_3OH$ |

The boiling point is the temperature where the molecules in a liquid are in equilibrium with the molecules in the vapor phase to such a high extent that the vapor pressure of those molecules in the vapor phase equals the atmospheric pressure. Each liquid has a characteristic boiling point (b.p.). This point is dependent on how much energy it takes to break the forces binding the molecules to one another.

Some aids in predicting boiling points:

- For molecules with similar forces, b.p. increases with molar mass.
- For molecules with differing forces, H-bonding will increase b.p. even for small molecules.

Look at Figures 12.3-12.5, which are pictorial representations of each of the types of intermolecular forces which are in operation for the liquid or solid phases of matter.

**Figure 12.3** Weak Dispersion Forces

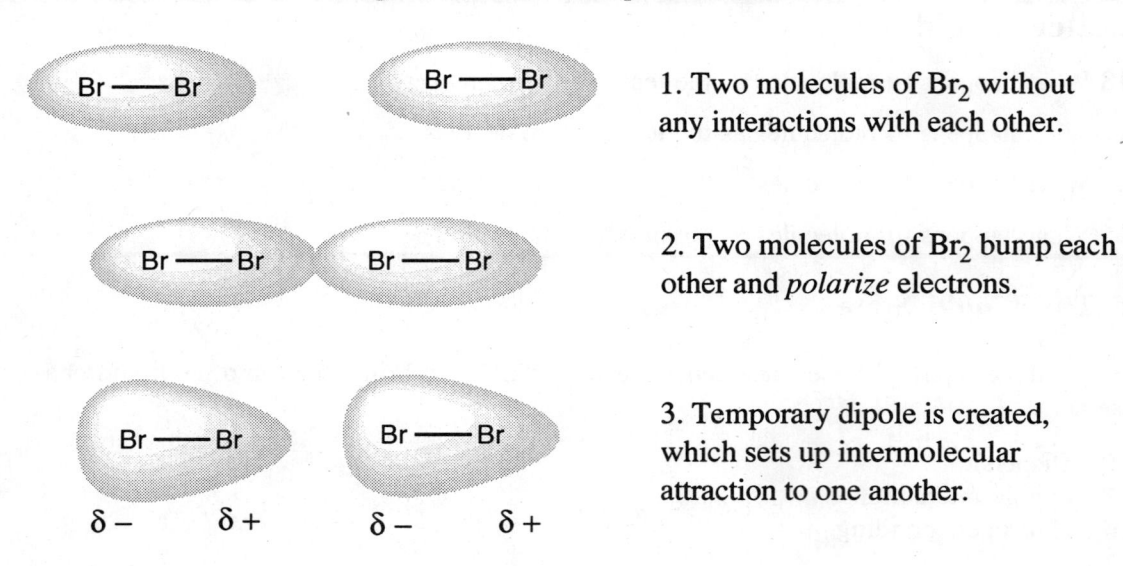

1. Two molecules of $Br_2$ without any interactions with each other.

2. Two molecules of $Br_2$ bump each other and *polarize* electrons.

3. Temporary dipole is created, which sets up intermolecular attraction to one another.

The intermolecular forces in nonpolar molecules are created when the molecules get close enough to each other to interact (attractions and repulsions). This means that conditions must be such that *kinetic molecular theory* is no longer obeyed.

The larger the molecules, the easier it is to create a temporary dipole. This is because the electrons are easier to move such that a temporary $\delta+$ and $\delta-$ area can be created.

This property of atoms is called **polarizability.** Bromine is larger than chlorine; therefore, it is more *polarizable*.

**Figure 12.4** Dipole-Dipole Interactions

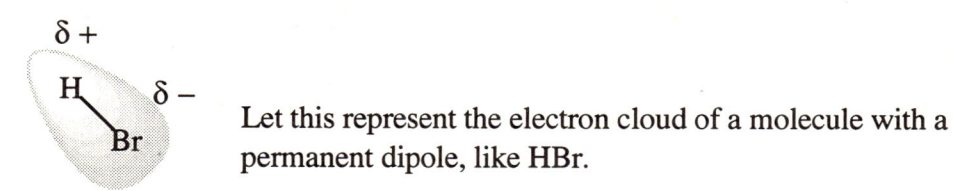

Let this represent the electron cloud of a molecule with a permanent dipole, like HBr.

In a liquid state, these molecules are close enough to interact so that they align the partial charges in an attractive manner. These molecules can still move in a liquid, but are attracted to one another.

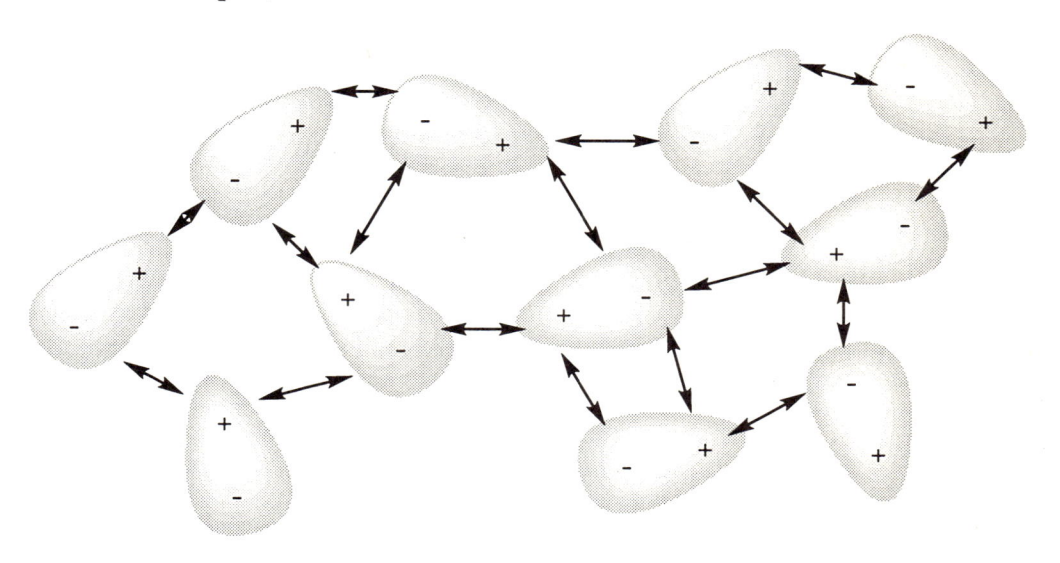

The arrows represent attractive interactions, but not bonds.

In general the dipole-dipole forces are stronger than the weak dispersion forces found in nonpolar molecules. There can be exceptions, however. The dispersion forces in iodine are strong enough for this nonpolar molecule to be a solid at room temperature. $I_2$ is very polarizable, and the dispersion forces are strong in this molecule.

The final type of force is the strongest intermolecular force. The very polar *covalent* bond between hydrogen and very small, electronegative atoms of fluorine, oxygen, and nitrogen allow the formation of a *hydrogen* bond. In Figure 12.5, these are depicted as dotted lines, while the covalent bond is depicted as a solid line.

The strength of the hydrogen bond is smaller than a covalent bond (which needs a chemical reaction to break), but it is strong enough to require a lot of energy to break. This explains why a small molecule like water (molar mass = 18.0 g) has such a high boiling point.

**Figure 12.5** Hydrogen Bonding

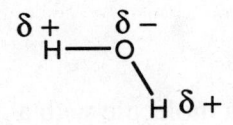

δ+ δ−
H—O
   H δ+

The highly polarized bonds of water are able to form strong intermolecular attractions.

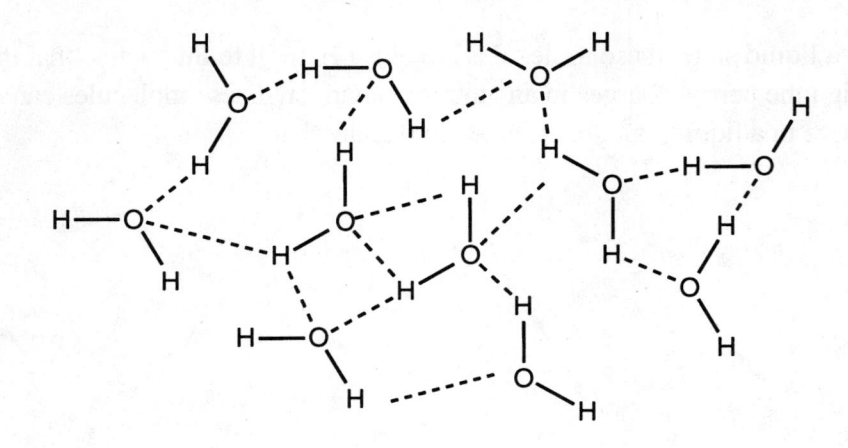

Some normal boiling points of hydrides are summarized in Table 12.7. These boiling points are depicted, graphically in Figure 2.6. Use Table 12.7 and Figure 12.6 to solve Example 12.6 and Problems 12.22 to 12.27.

**Table 12.7** Some Boiling Points of Hydride Molecules

| Group IVA | Boiling Point (°C) | Group VA | Boiling Point (°C) | Group VIA | Boiling Point (°C) | Group VIIA | Boiling Point (°C) |
|---|---|---|---|---|---|---|---|
| $CH_4$ | −164 | $NH_3$ | −33 | $H_2O$ | 100 | HF | 20 |
| $SiH_4$ | −112 | $PH_3$ | −87 | $H_2S$ | −61 | HCl | −85 |
| $GeH_4$ | −89 | $AsH_3$ | −55 | $H_2Se$ | −42 | HBr | −67 |
| $SnH_4$ | −52 | $SbH_3$ | −17 | $H_2Te$ | −2 | HI | −35 |

**Figure 12.6** Graph of Boiling Points of Compounds from Various Groups with Hydrogen

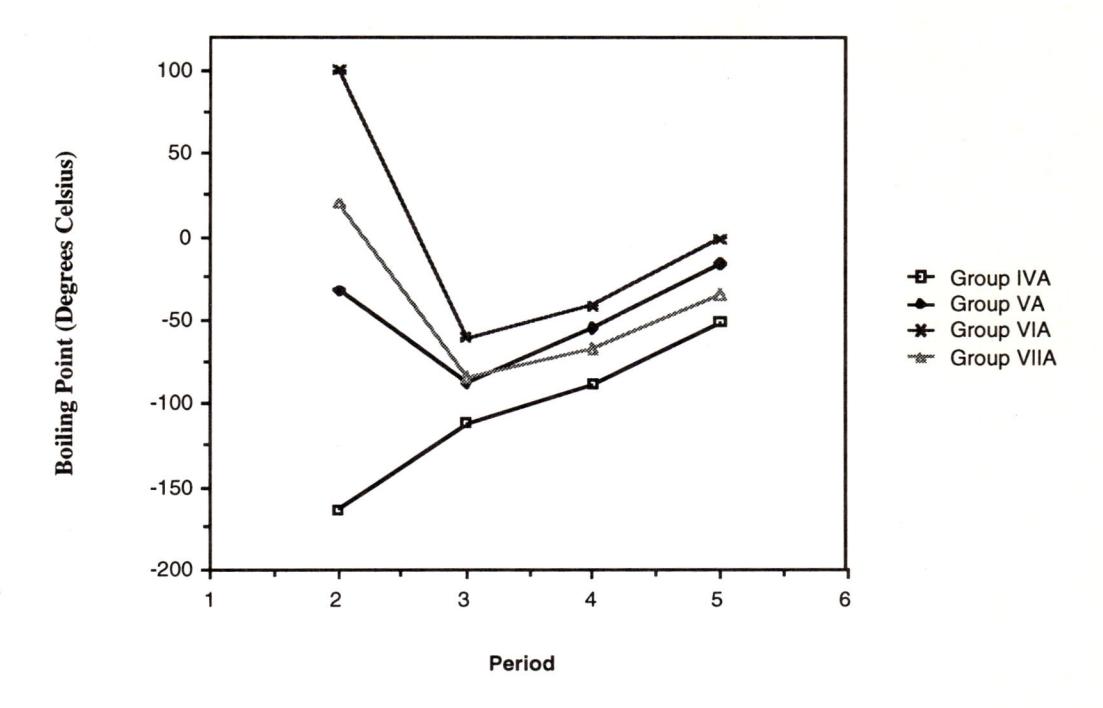

**Example 12.6:** Look at Figure 12.6. For Group IVA molecules, rank the four compounds for increasing boiling point, and state the reason for these properties.

Why do the lines for Groups VA-VIIA have patterns where the molecules in period 2 have such high boiling points?

**Solution:** For Group IVA, the boiling points increase in the following order:
$$CH_4 < SiH_4 < GeH_4 < SnH_4$$

These molecules are all nonpolar with only weak dispersion forces holding them together. As the molar mass increases, the boiling point increases, because it takes a higher temperature to provide enough energy to get a heavier molecule into the vapor phase.

For the other groups, the period 2 molecules for Group V ($NH_3$), Group VI ($H_2O$), and Group VII (HF) all have very strong hydrogen bonds between them which take more energy to break than the heavier molecules in the group, which only have dipole interactions. Other forces for Group VA to VIIA are dipole-dipole.

---

**Practice Problems**

**12.22** What forces cause $Br_2$ to be a liquid at room temperature (ca. 23°C)?

**12.23** What forces cause $NH_3$ to have a higher boiling point than the heavier HCl?

**12.24** Place the following in order of increasing boiling point: $H_2O$, $H_2S$, $H_2Se$, $H_2Te$.

> **12.25** Place the following in order of increasing vapor pressure:
> $H_2O$ (b.p. = 100°C), $C_2H_5OC_2H_5$ (b.p. = 34.6°C), $CH_3CH_2OH$ (b.p. = 78.5°C).
>
> **12.26** Place the following in order of increasing boiling point: $CH_4$, $C_2H_6$, $C_3H_8$, $C_4H_{10}$.
>
> **12.27** Place the following in order of increasing boiling point: $F_2$, HCl, Ne.

## 12.6 The Solid State

The attractive forces in the solid state become so strong that the molecules no longer flow. The types of solids are summarized in Table 12.8.

**Table 12.8** Types of Solids

| Type | Subtype | Description | Structure | Melting Point | Examples |
|---|---|---|---|---|---|
| Crystalline | Atomic | Nonmetallic atoms held together by dispersion forces | Well-defined array. Close-packed structure | Low | He Ne Ar |
|  | Metallic | Metallic atoms held together in a sea of e⁻. | Well-defined array. Close-packed structure | Usually high, but varies. | Cu Hg Au |
|  | Molecular | Consists of molecules held together by dispersion forces, dipole and H-bonding | Varies | Low to high | $F_2$ $P_4$ $H_2O$ $H_2S$ |
|  | Ionic | Ions held together by attraction of opposite charges | Well-defined array. Close-packed structure | High | NaCl CaO $Al(OH)_3$ |
|  | Covalent Network | Nonmetals held together by covalent bonds | Tetrahedral or planar | High | C (diamond) C (graphite) Si |
| Amorphous |  | Non-defined solid. | No well-defined array | Medium | Glass |

**Example 12.7:** Classify an describe the solid of each of the following using Table 12.8.
Ar, $C_6H_{12}O_6$ (glucose), NaCl

**Solution:** First try to categorize the substance, and then use the table to predict the type of solid it will form.

Argon is a noble gas and thus will be an atomic solid, held together by only weak dispersion forces. It will have a low melting point.

Glucose is a molecule that would have a moderate melting point and would be categorized as a molecular crystalline solid.

Sodium chloride is an ionic compound, would have a high melting point, and would be categorized as an ionic crystalline solid.

## Practice Problems

**12.28** Which has a lower melting point, Ar or Xe? Why?

**12.29** Which has a lower melting point, $Br_2$ or Rn? Why?

**12.30** What type of solid is each of the following?
Gold, Krypton, Glass, Sodium chloride, carbon dioxide, silicon dioxide, sulfur ($S_8$)

**12.31** Which has the lower melting point, KBr or $Br_2$? Why?

**12.32** Which has the lower melting point, $CO_2$ or diamond (C)? Why?

**12.33** Arrange the following in order of increasing melting point.
Kr, KI, $C_{10}H_{22}$

## Integration of Multiple Skills

One type of problem you might encounter is an energy problem where you need to calculate the energy needed to convert a substance in the solid phase (below the melting point) to one in the vapor phase (above the boiling point). This involves the integration of skills you learned in Chapter 3, for calculating energy needed to change the temperature of a substance, with those you learned in this chapter.

**Example 12.8:** How much heat is necessary to convert 5.78 g of ice at -38°C to 5.78 g of steam at 115°C?

**Solution:** This solution involves five energy calculations and then the addition of all the individual results.

Step 1: Calculate the energy to raise the temperature of ice from -38°C to 0°C. (Use Table 12.3 for this step, Step 3, and Step 5.)

$$5.78 \text{ g } H_2O(s) \times \frac{1.99 \text{ J}}{\text{g °C}} \times 38°C \times \frac{1 \text{ kJ}}{1000 \text{ J}} = 0.437 \text{ kJ}$$

Step 2: Calculate the energy to change from solid to liquid at the m.p (Use Table 12.2).

$$5.78 \text{ g } H_2O \times \frac{1 \text{ mol } H_2O}{18.0 \text{ g } H_2O} \times \frac{6.01 \text{ kJ}}{\text{mol } H_2O} = 1.93 \text{ kJ}$$

Step 3: Calculate the energy to raise the temperature of water from 0°C to 100°C.

$$5.78 \text{ g } H_2O(s) \times \frac{4.184 \text{ J}}{\text{g °C}} \times 100°C \times \frac{1 \text{ kJ}}{1000 \text{ J}} = 2.42 \text{ kJ}$$

Step 4: Calculate the energy to change water to vapor at the b.p (Use Table 12.2).

$$5.78 \text{ g H}_2\text{O} \times \frac{1 \text{ mol H}_2\text{O}}{18.0 \text{ g H}_2\text{O}} \times \frac{40.7 \text{ kJ}}{\text{mol H}_2\text{O}} = 13.1 \text{ kJ}$$

Step 5: Calculate the energy to raise the temperature of vapor from 100°C to 115°C.

$$5.78 \text{ g H}_2\text{O(s)} \times \frac{2.03 \text{ J}}{\text{g °C}} \times 15°\text{C} \times \frac{1 \text{ kJ}}{1000 \text{ J}} = 0.176 \text{ kJ}$$

Add the individual amounts: 0.437 kJ + 1.93 kJ + 2.42 kJ + 13.1 kJ + 0.176 kJ = **18.1 kJ**

---

## Practice Problems

**12.34** How much heat is needed to convert 12.6 g of ice at −24°C to steam at 128°C?

**12.35** How much heat does it take to convert 15.2 g Hg at 100°C to 15.2 g Hg vapor at 357°C? (*Hint:* Use information from Tables 12.2 and 12.3)

**12.36** Compare trends for electronegativity and polarizability for Group VIIA elements.

---

## Self-Exam

**I. (4 ea) Write T (true) or F (false) in each corresponding blank.**

_____ 1. The state of matter with fixed volume, but no specific shape is *liquid*.

_____ 2. All three states of matter obey *kinetic molecular theory*.

_____ 3. The change of SF$_6$ solid to SF$_6$ vapor at −65°C is called *sublimation*.

_____ 4. The change of water to ice requires the input of heat energy.

_____ 5. In a liquid, the more molecules are attracted to each other, the higher the boiling point will be (assuming similar molar masses).

_____ 6. Glass is an amorphous solid.

_____ 7. The boiling point of water is *always* 100°C.

_____ 8. For a given substance, the molar heat of vaporization is much higher than the molar heat of fusion.

_____ 9. 10.0 g of liquid water at 100°C will melt the same amount of ice as 10.0 g of steam at 100°C.

_____ 10. A hydrogen bond is a covalent bond.

**II. (4 ea) Multiple Choice. Place the letter of the BEST answer in the corresponding blank.**

_____ 1. Which of the following molecules will **not** hydrogen bond?
   A. CH$_4$     B. H$_2$O     C. CH$_3$OH     D. NH$_3$     E. HF

_____2. What is the process where liquid water becomes ice?
    A. Freezing        B. Melting        C. Deposition
    D. Condensation   E. Vaporization

_____3. A 545-mL sample of oxygen is collected over water at 28°C. The total pressure of the gas is 698 mm Hg. What is the partial pressure of oxygen?
    A. 153 mm Hg    B. 525 mm Hg    C. 670 mm Hg
    D. 698 mm Hg    E. 720 mm Hg

_____4. How much heat is given off when 39.7 g of water vapor at 100°C condenses?
    A. 1620 kJ    B. 733 kJ    C. 239 kJ    D. 89.8 kJ    E. 18.5 kJ

_____5. How much energy does it take to convert 32.6 g of solid $H_2O$ to liquid at 0°C?
    A. 196 kJ    B. 108 kJ    C. 10.9 kJ    D. 2.25 kJ    E. 1.81 kJ

_____6. Which of the following has the highest boiling point?
    A. $H_2$    B. $F_2$    C. $Cl_2$    D. $Br_2$    E. $I_2$

_____7. Which of the following has the highest vapor pressure?
    A. $H_2O$    B. HCl    C. HBr    D. $CH_2Cl_2$    E. $CHCl_3$

_____8. Which will have **only** dispersion forces in the liquid state?
    A. $Br_2$    B. HBr    C. $CBr_4$    D. $CHBr_3$    E. Both A and C

_____9. Which is going to have dipole forces as the strongest attractions in the liquid state?
    A. $Cl_2$    B. HCl    C. $H_2$    D. $CCl_4$    E. NaCl

_____10. Which process is sublimation?
    A. Frost forming from the air    B. Water forming from steam
    C. Ice melting to water    D. Solid moth balls vaporizing
    E. Liquid gold forming a solid

### III. Provide Complete Answers

1. (6 pts) How much heat is needed to convert 10.8g of ice at 0°C to 10.8g of vapor at 100°C?

2. (6 pts) Carbon dioxide (4.37 L) was collected over water at 50°C, where the total pressure was 0.975 atm. How many moles of carbon dioxide were in the container? How many grams?

3. (4 pts) Place the following in order of increasing boiling point.
    $NH_3$, HBr, Ne, NaBr

4. (4 pts) Put the following in order of increasing melting point.
    NaI, $CO_2$, $O_2$, $S_8$

## Answers to Practice Problems

| | | | |
|---|---|---|---|
| 12.1 | 45.7 g $H_2O$ | 12.2 | 103 mol $H_2O$ |
| 12.3 | 1.27 mol $C_6H_6$ | 12.4 | Sublimation |
| 12.5 | Heat given off | 12.6 | Deposition; heat given off |
| 12.7 | 4.21 kJ | 12.8 | 5.27 kJ |

| | | | |
|---|---|---|---|
| **12.9** | 28.5 kJ | **12.10** | 38.0 kJ |
| **12.11** | 8.48 kJ | **12.12** | 153 kJ |
| **12.13** | 31.8 mm Hg | **12.14** | 400 mm Hg |

**12.15** $P_{hydrogen}$ = 718 mm Hg; $3.06 \times 10^{-3}$ mol $H_2$

**12.16** $P_{helium}$ = 668 mmHg; $4.60 \times 10^{-2}$ mol He; 0.184g He

**12.17** 90°C

**12.18** They have more kinetic energy, and get further from each other.

**12.19** They have less kinetic energy, and get closer to each other.

**12.20** They get closer together; attractions and repulsions increase.

**12.21** They get further apart; attractions and repulsions decrease.

**12.22** Dispersion       **12.23** Hydrogen bonding

**12.24** $H_2S < H_2Se < H_2Te < H_2O$

**12.25** $H_2O < CH_3CH_2OH < C_2H_5OC_2H_5$

**12.26** $CH_4 < C_2H_6 < C_3H_8 < C_4H_{10}$     **12.27** $Ne < F_2 < HCl$

**12.28** Ar has lower m.p. because it is smaller.

**12.29** Rn has lower m.p. because it has weaker dispersion forces (atomic solid)

**12.30** Au (metallic); Kr (atomic); glass (amorphous); NaCl (ionic); $CO_2$ (molecular); $S_8$ (molecular)

**12.31** $Br_2$ has lower m.p. because it is held together by weak dispersion forces.

**12.32** $CO_2$ is molecular crystal held together by dispersion forces.

**12.33** $Kr < C_{10}H_{22} < KI$      **12.34** 39.3 kJ

**12.35** 0.543 kJ to change T of liquid Hg from 100°C to 357°C; 4.49 kJ to vaporize mercury at its boiling point: 5.04 kJ total energy needed

**12.36** The trends are just the opposite. Electronegativity increases as you go up the column, while the polarizability increases as you go down.

# Answers to Self-Exam

**I. T/F**
1. T   2. F   3. T   4. F   5. T   6. T   7. F   8. T   9. F   10. F

**II. MC**
1. A   2. A   3. C   4. D   5. C   6. E   7. B   8. E   9. B   10. D

**III. Complete Answers**
1. 32.6 kJ                     2. 0.141 mol $CO_2$, 6.19 g $CO_2$
3. $Ne < HBr < NH_3 < NaBr$    4. $O_2 < CO_2 < S_8 < NaI$

# 13
# Solutions

This chapter has a lot of new material and requires time to absorb concepts that haven't been encountered before. You'll also need time to learn how to handle new calculations.

The first sections (13.1-13.6) are mainly conceptual and explain empiricisms of solutions. The last sections (13.7-13.11) involve problem solving with solution chemistry.

Keys for success in this chapter:

- Learn new terms.
- Practice problems for concentration calculations.
- Review stoichiometry and practice problems using this in connection with solutions.
- Learn how to apply principles of colligative properties.

## Summary of Verbal Knowledge

**solution:** A homogeneous mixture of two or more substances.
**solvent:** The substance in a solution that dissolves another substance; if it is not clear which substance does the dissolving, it is the substance in the solution that is in the greater amount.
**solute:** The substance in a solution which is considered to be dissolved into another substance in the solution.
**miscible:** This describes two liquids which completely mix to form a solution.
**immiscible:** This describes two liquids which **do not** mix to any appreciable amount.
**solubility:** This is the maximum amount of substance that dissolves in a given volume of solvent at a specified temperature.
**saturated solution:** A solution where the maximum amount of substance is dissolved in a given amount of solvent at a given temperature.
**unsaturated solution:** A solution where less than the maximum amount of a substance is dissolved in a given amount of solvent at a given temperature.
**supersaturated solution:** A solution where the maximum amount of a substance was dissolved at a high temperature and then the solution was slowly cooled without disturbance. This would leave more solute dissolved than is *usually* possible at a given temperature.
**dilute solution:** One in which a small amount of solute is dissolved in the solvent. This is a *qualitative* term.
**concentrated solution:** One in which a large amount of solute is dissolved in the solvent. This is a *qualitative* term.
**mass percent of solute:** This is a *quantitative* term for an expression of concentration. It is defined as the mass of solute divided by the mass of total solution times 100%.
**molarity:** This is a *quantitative* term for an expression of concentration. It is defined as moles of solute per liters of solution.

**titration:** This is an experimental technique where the known concentration of one substance is added quantitatively to a given amount of a second substance. Knowledge about the nature of the two substances and how they react chemically can be used along with experimental values to determine the concentration of the second substance.

**equivalent of an acid:** This is the amount of an acid that yields 1 mol of hydrogen ions, $H^+$.

**equivalent of a base:** This is the amount of a base that yields 1 mol of hydroxide ions, $OH^-$.

**normality:** This is a *quantitative* term for an expression of concentration, which is closely related to molarity. It is equal to the number of equivalents of acid or base solute in one liter of solution.

**molality:** This is a *quantitative* term for an expression of concentration. It is defined as moles of solute in kilograms of solvent.

**colligative property:** A property of a solution that depends only on the number of solute particles (molecules and/or ions) in a given amount of solution and not on the particular characteristics of the solute particles.

**freezing-point depression:** This is the lowering of the freezing point of a solvent upon the addition of solute.

**boiling-point elevation:** This is the raising of the boiling point of a solvent upon the addition of solute.

**osmosis:** The process of solvent flow through a semipermeable membrane in order to equalize the concentrations of solutes on the two sides of the membrane.

**osmotic pressure:** The pressure that must be exerted on a solution to stop osmosis.

## Review of Mathematical and Calculator Skills

Two skills should be reviewed for this section. They are

- Working with percents
- Dimensional analysis

**Example 13.1:** 15 g is what percent of 75 g?

$$\text{Solution: } \frac{15 \text{ g}}{75 \text{ g}} \times 100\% = 20\%$$

**Example 13.2:** How many moles of $H^+$ ions are present in 37.6 g of $H_2SO_4$?

$$\text{Solution: } 37.6 \text{ g } H_2SO_4 \times \frac{1 \text{ mol } H_2SO_4}{98.1 \text{ g } H_2SO_4} \times \frac{2 \text{ mol } H^+}{1 \text{ mol } H_2SO_4} = 0.767 \text{ mol } H^+ \text{ ions}$$

Try some problems applying these skills.

---

**Practice Problems**

13.1 22 g is what percent of 96 g?

**13.2** What is 25% of 48 g?

**13.3** How many moles of NaOH are present in 12.6 g of NaOH?

**13.4** How many grams of HCl are present in 4.27 mol HCl?

**13.5** How many moles of H$^+$ ions are present in 5.62 g of H$_3$PO$_4$?

**13.6** How many moles of OH$^-$ ions are present in 22.4 g of KOH?

## Application of Skills and Concepts

### 13.1 Some Terms Used to Describe Solutions

Use the definitions at the beginning of this chapter to answer the practice problems.

**Practice Problems**

**13.7** A teaspoon of sugar is dissolved in a cup of water. What is the sugar called?

**13.8** For problem 13.7, what is the water called?

**13.9** For problem 13.7, what is the mixture called?

**13.10** When oil is mixed with water, the oil floats on top. What are these two liquids to each other?

### 13.2 Types of Solutions

Some of the types of solutions are summarized in Table 13.1.

**Table 13.1** Examples of Solutions

| Physical State, Solute | Physical State, Solvent | Physical State, Solution | Examples |
|---|---|---|---|
| Gas | Gas | Gas | Air |
| Gas | Liquid | Liquid | Oxygen in water |
| Gas | Solid | Solid | Hydrogen in palladium |
| Liquid | Liquid | Liquid | Ether in carbon tetrachloride |
| Liquid | Solid | Solid | Mercury in silver |
| Solid | Liquid | Liquid | Sugar in water |
| Solid | Solid | Liquid | Potassium-sodium alloy |
| Solid | Solid | Solid | Brass alloy (zinc in copper) |

**Example 13.3:** Using Table 13.1 as a guide, what type of solution is carbon dioxide dissolved in water?

**Solution:** Under normal conditions CO$_2$ is a gas and water is a liquid. Therefore, it is a liquid solution where a gas solute is dissolved in a liquid solvent.

## Practice Problems

What type of solution is each of the following?

**13.11** Oxygen in nitrogen

**13.12** Sodium chloride in water

**13.13** HCl gas in water

**13.14** Bromine in carbon tetrachloride

**13.15** $Cr_2O_3$ in $Al_2O_3$

## 13.3 General Properties of Solutions

Some of the properties of solutions are summarized below:

- Homogeneous (same throughout, macroscopically)
- They can be separated by physical means.
- The solute is uniformly distributed throughout the solvent.

The above properties can be used to determine which substances are solutions.

**Example 13.4:** Potassium nitrate is mixed into pure distilled water. In a second instance, silver chloride is stirred into pure distilled water.

What is each substance described? What type of substance is the mixture in each case?

**Solution:** In the first case, $KNO_3$ is an ionic solid (a pure substance). Pure distilled water is a molecular substance that is in the liquid phase. When $KNO_3$ is stirred into water, it will dissolve (all compounds with Group IA cations are soluble), forming a **solution.**

In the second case, AgCl is an ionic solid (a pure substance). When stirred into the pure distilled water, very little of the AgCl will dissolve (see Chapter 6). After stirring, the solid will precipitate out in the bottom of the container. The mixture is **not** a solution but a **heterogeneous** mixture.

## Practice Problems

For each of the following, label the substance as a solution, heterogeneous mixture, or pure substance.

**13.16** Salt water

**13.17** A balloon filled with 0.4 mol of $H_2$ and 0.2 mol of $O_2$

**13.18** A balloon filled with 0.5 mol of $SF_6$

**13.19** Carbon tetrachloride

13.20 Oxygen dissolved in water

13.21 Red blood cells in plasma

## 13.4 Saturated, Unsaturated, and Supersaturated Solutions

This section looks at the descriptive terms in the title. Review the definitions at the beginning of the chapter. For a given solute and solvent, there are a variety of amounts that will dissolve at a given temperature. Table 13.2 lists several substances and the number of grams that will dissolve in water at 20°C and at 100°C.

**Table 13.2** Some Common Solubilities in Water

| Name | Formula | Solubility in Cold, g in 100 cc at 20°C | Solubility in Hot, g in 100 cc at 100°C |
|---|---|---|---|
| Sodium chloride | NaCl | 35.7 | 39.1 |
| Magnesium iodate | $Mg(IO_3)_2$ | 10.2 | 19.3 |
| Sodium thiosulfate | $Na_2S_2O_3$ | 50 | 231 |
| Sodium acetate | $NaC_2H_3O_2$ | 119 | 170 |
| Potassium nitrate | $KNO_3$ | 13.3 | 247 |
| Calcium carbonate | $CaCO_3$ | 0.0014 | 0.0018 |

Notice that for some compounds, heating the solvent increases the solubility. For others it has little effect. There are some compounds where heating decreases the solubility, although this is rare.

**Example 13.5:** How many grams of sodium acetate will dissolve in 50.0 mL of water at 20°C?

**Solution:** Look at Table 13.2 and note that 119 g of $NaC_2H_3O_2$ will dissolve in 100 mL of water at this temperature. Using dimensional analysis, the solution follows.

$$50.0 \text{ ml } H_2O \times \frac{1 \text{ cc } H_2O}{1 \text{ mL } H_2O} \times \frac{119 \text{ g } NaC_2H_3O_2}{100 \text{ cc } H_2O} = 59.5 \text{ g } NaC_2H_3O_2$$

## Practice Problems

13.22 How many grams of $KNO_3$ can be dissolved in 800 mL of water at 20°C? at 100°C?

13.23 How many grams of $CaCO_3$ can be dissolved in 4 L of water at 20°C?

13.24 100 mL water at 100°C has 225 g of $Na_2S_2O_3$ dissolved in it. It is slowly cooled to 20°C. When a seed crystal is added, how many grams of solid fall out?

## 13.5 The Solution Process

When a solute is dissolved in a solvent, an opposing process, **crystallization** is also occurring.

Equation 13.1 is the dissolution process. Equation 13.2 is the crystallization process. Equation 13.3 is an equilibrium representing both processes occurring at once.

The degree that a substance dissolves is dependent on which attractive forces predominate. If the molecules in the solid are more attracted to the molecules of the solvent then EQ 13.1 will predominate over EQ 13.2. The opposite will occur if the attractive forces of the solid are stronger.

$$CaCO_3(s) \rightarrow Ca^{2+}(aq) + CO_3^{2-}(aq) \qquad \text{(EQ 13.1)}$$

$$Ca^{2+}(aq) + CO_3^{2-}(aq) \rightarrow CaCO_3(s) \qquad \text{(EQ 13.2)}$$

$$CaCO_3(s) \rightleftharpoons Ca^{2+}(aq) + CO_3^{2-}(aq) \qquad \text{(EQ 13.3)}$$

**Example 13.6:** Write the equilibrium equation for the mixture of $KNO_3$ in water.

**Solution:** This will look like EQ 13.3.

$$KNO_3(s) \rightleftharpoons K^+(aq) + NO_3^-(aq)$$

---

### Practice Problems

**13.25** Write the equilibrium equation for the mixture of sodium chloride in water.

**13.26** Write the equilibrium equation for the mixture of lead(II) chloride in water.

**13.27** Write the equilibrium equation for the mixture of glucose ($C_6H_{12}O_6$) in water.

---

## 13.6 Factors That Affect Solubility

Conditions that affect solubility include

- Nature of the solute and solvent
  Like dissolves like (see Table 13.3).
  Ionic compounds dissolve in polar solvents, if ions prefer interacting with solvent molecules over attraction to each other within the crystal.
- Temperature
  For most ionic substances or molecular solids in liquid, high temperatures increase solubility.
  For gases in liquids, high temperatures decrease solubility.
- Pressure
  For gases, high pressure increases solubility.
  For solids in liquids, pressure has very little affect.

Table 13.3 Summary of Factors Affecting Solubility in Types of Solvents

|  | Polar Solvent | Weakly Polar Solvent | Nonpolar Solvent |
|---|---|---|---|
| Examples | Water<br>Alcohol (low weight)<br>Liquid $NH_3$ | Acetone<br>$CH_2Cl_2$<br>Ether | $CCl_4$<br>$C_6H_{14}$ |
| Types of Solutes | Ionic<br>Polar molecules<br>Molecules with OH | Weak polar<br>Nonpolar | Nonpolar |

**Example 13.7:** $K_2SO_4$ is a blue ionic crystalline compound. $I_2$ is a silver-purple molecular crystalline element. NiS is an ionic crystalline compound. Predict the solubility of each in water and carbon tetrachloride.

**Solution:** $K_2SO_4$ is very *soluble* in water because it is ionic and water is polar. The $K^+$ ions are attracted to the electronegative oxygen of water, and the $SO_4^{2-}$ ions are attracted to the electropositive hydrogens of water.

$K_2SO_4$ is very *insoluble* in the nonpolar $CCl_4$.

$I_2$ is a nonpolar solid, so it is quite *insoluble* in water and quite *soluble* in $CCl_4$.

NiS is ionic but is *insoluble* in water because the $Ni^{2+}$ ions are more attracted to the $S^{2-}$ ions than to the polar water molecules.

NiS is *insoluble* in $CCl_4$ because it is nonpolar and won't dissolve ionic compounds.

---

## Practice Problems

**13.28** Ammonium chloride is very *soluble* in water but very *insoluble* in $CCl_4$. Why?

**13.29** Barium sulfate is very *insoluble* in water. Why?

**13.30** Bromine is very *insoluble* in water. Why?

**13.31** Oxygen is more *soluble* in cold water than in warm water. Why?

**13.32** More carbon dioxide will dissolve in soda under the pressure of canning (6 atm) than at 760 torr. Why?

---

## 13.7 Mass Percent of Solute

This section deals with two types of problems:

- Given quantities of solute and solvent, determine the mass percent solute.
- Given a quantity of solution for a desired mass percent solute, determine how much solute and solvent to use.

## 214   Chapter 13

The solutions for both types of problems involve the use of EQ 13.4 in some form.

$$\text{mass percent solute} = \frac{\text{mass solute}}{\text{mass solution}} \times 100\% \qquad \text{(EQ 13.4)}$$

**Example 13.8:** 20 g of sugar in 80 g of water has what concentration (mass %)?

**Solution:** Note that the mass of the solution (soln) must be determined first.

$$\text{mass soln} = \text{mass solute} + \text{mass solvent} = 20 \text{ g} + 80 \text{ g} = 100 \text{ g}$$

$$\text{mass percent solute} = \frac{20 \text{ g solute}}{100 \text{ g solution}} \times 100\% = 20\%$$

**Example 13.9:** You need to prepare 750 g of a 2.5% solution of glucose in water. How would you do this?

**Solution:** Rearrange EQ 13.4 to isolate *mass solute* as shown below:

$$\text{mass solute} = \frac{(\text{mass \% solute})(\text{mass solution})}{100\%} = \frac{(2.5\%)(750 \text{ g})}{100\%} = 18.8 \text{ g glucose}$$

To determine the mass of solvent, subtract: 750 g soln - 18.8g solute = 731g solvent

This section also covers two other measurements of concentration. These are volume percent solute (for liquid in liquid) and parts per million (ppm). These concentrations are described in EQ 13.5 and EQ 13.6, respectively. Two example problems using these solutions will show how their application is similar to mass percent problems.

$$\text{volume percent solute} = \frac{\text{volume solute}}{\text{volume solution}} \times 100\% \qquad \text{(EQ 13.5)}$$

$$\text{ppm} = \frac{\text{mass solute}}{\text{mass solution}} \times 1{,}000{,}000 \qquad \text{(EQ 13.6)}$$

**Example 13.10:** If 250. mL of acetone is dissolved in 600. mL of ethyl acetate, what is the volume percent of solute?

**Solution:** The first step is to determine the volume of solution.

$$\text{vol soln} = \text{vol solute} + \text{vol solvent} = 250 \text{ mL} + 600 \text{ mL} = 850 \text{ mL}$$

Next, use EQ 13.5 to determine the volume percent solute.

$$\text{volume percent solute} = \frac{250 \text{ mL acetone}}{850 \text{ mL solution}} \times 100\% = 29.4\%$$

**Example 13.11:** A solution contains 0.07 mg of lead(II) ion in 1000 g. What is its concentration?

**Solution:** Remember that you need to convert milligrams of solute to grams of solute (Use EQ 13.6).

$$\text{ppm} = \frac{0.07 \text{ mg}}{1000 \text{ g solution}} \times \frac{1 \text{ g}}{1000 \text{ mg}} \times 1{,}000{,}000 = 0.07 \text{ ppm}$$

---

**Practice Problems**

**13.33** What is the mass percent solute if 4.0 g of NaOH is dissolved in 64.0 g of solution?

**13.34** What is the mass percent solute if 4.0 g of KBr is dissolved in 64.0 g of solution?

**13.35** What is the mass percent solute if 4.0 g of KBr is dissolved in 64.0 g of water?

**13.36** How would you prepare 250 g of 0.850% sodium chloride solution?

**13.37** What is the volume percent solute if 20 mL of methanol is dissolved in 60 mL of water?

**13.38** If there are $6.5 \times 10^{-6}$ g $Pb^{2+}$ in 100 g of water solution, what is the concentration in ppm?

---

## 13.8 Molarity and Normality

The unit of **molarity** is the most common way of measuring concentration in chemistry. It is defined by EQ 13.7:

$$\text{molarity} = \frac{\text{mol solute}}{\text{L solution}} = M \qquad \text{(EQ 13.7)}$$

There are two types of concentration problems using this definition.

- Given the amount of solute and amount of solution, calculate $M$.
- Given the amount of a solution at a desired molarity, determine how to prepare it.

Some examples illustrate how to do both types of problems. (Note that solution is abbreviated soln in the problems below.)

**Example 13.12:** If there are 68.5 g of NaOH in 250. mL of solution, what is the molarity?

## Chapter 13

**Solution:** Keep the definition of **molarity** in mind. The mass of solute must be converted to moles solute and the volume of solution must be converted to liters. This is done formally in the equation below.

$$M = \frac{68.5 \text{ g NaOH}}{250 \text{ mL soln}} \times \frac{1000 \text{ mL soln}}{1 \text{ L soln}} \times \frac{1 \text{ mol NaOH}}{40.0 \text{ g NaOH}} = 6.85 \text{ mol/L} = 6.85 \, M$$

**Example 13.13:** If you need to prepare 300. mL of a 0.100 $M$ aqueous solution of silver nitrate, how would you do this?

**Solution:** When preparing a solution in the lab, you would find a volumetric flask which is calibrated for the desired volume. In this case it would be 300. mL. You would then fill this flask with approximately half the volume of solvent (water). Next, you need to determine how many grams of solute to dissolve in the water. The calculation for this is shown below.

$$300 \text{ mL AgNO}_3 \text{ soln} \times \frac{0.100 \text{ mol AgNO}_3}{1000 \text{ mL soln}} \times \frac{169.9 \text{ g AgNO}_3}{1 \text{ mol AgNO}_3} = 5.10 \text{ g AgNO}_3$$

> Notice that for the definitions of 0.100 $M$, the equality 0.100 mol/1000 mL was used instead of 0.100 mol/L. These three definitions are all equivalent.

The rest of the solution to this problem deals with technique. After the 5.10 g of $AgNO_3$ has been dissolved in the water, then fill the vessel to the 300 mL mark on the flask, and the job is finished.

Another time you will apply the use of concentration is in stoichiometry problems. An example of such a problem is Example 13.14.

**Example 13.14:** For the reaction of silver nitrate solution with sodium chloride solution, how many milliliters of 0.100 $M$ silver nitrate must be added to an excess of sodium chloride to produce $7.25 \times 10^{-2}$ g of silver chloride precipitate?

**Solution:** This is a stoichiometry problem, so the first order of business requires writing a balanced equation. After this is done, you can use the information in Figure 13.1 to follow a course for setting up the calculation.

The balanced equation is

$$AgNO_3(aq) + NaCl(aq) \rightarrow AgCl(s) + NaNO_3(aq)$$

**Figure 13.1 Scheme for Conversions Within a Balanced Equation**

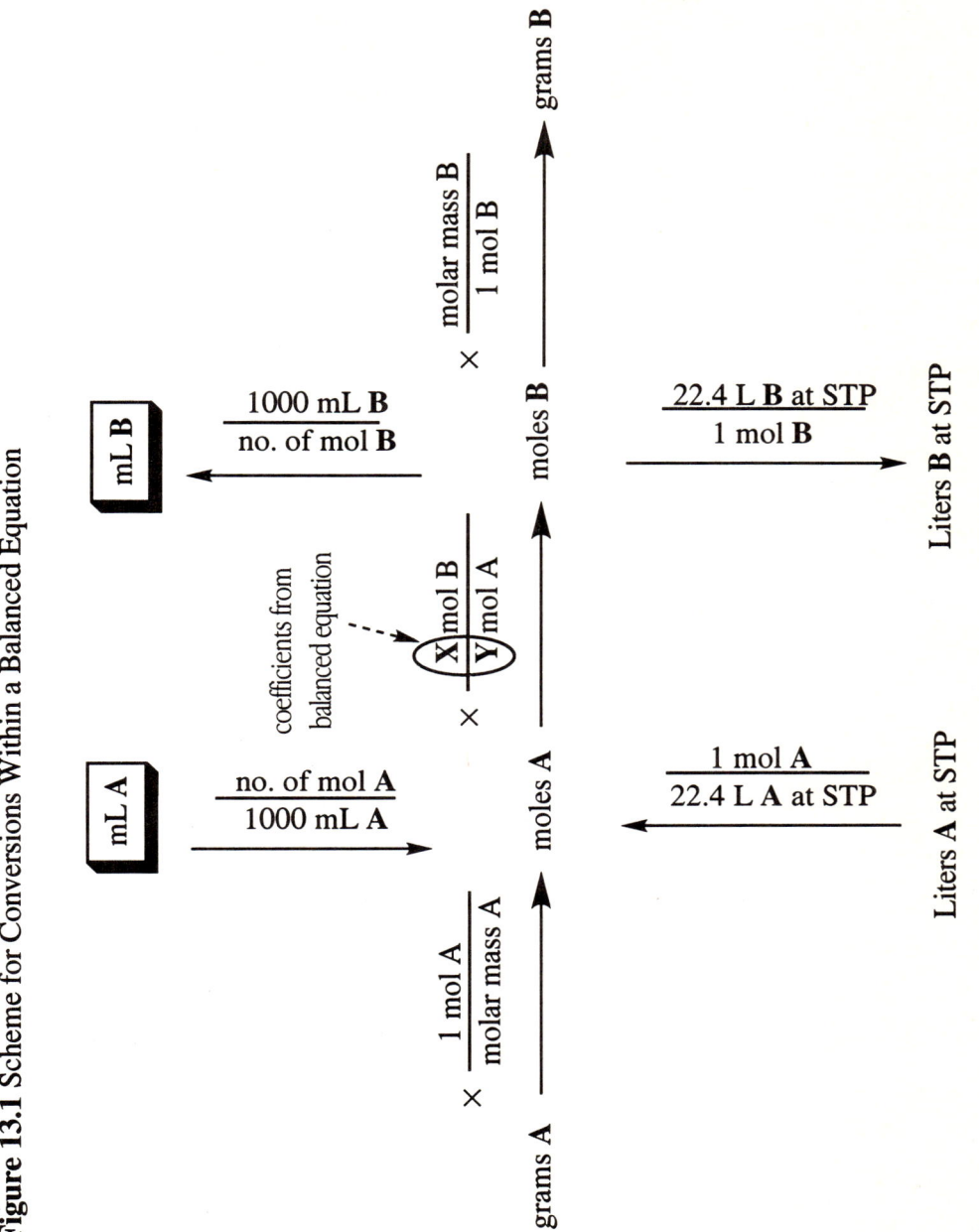

grams A $\xrightarrow{\times \dfrac{1 \text{ mol A}}{\text{molar mass A}}}$ moles A $\xrightarrow{\times \dfrac{\mathbf{X} \text{ mol B}}{\mathbf{Y} \text{ mol A}}}$ moles B $\xrightarrow{\times \dfrac{\text{molar mass B}}{1 \text{ mol B}}}$ grams B

(coefficients from balanced equation)

mL A $\xrightarrow{\dfrac{\text{no. of mol A}}{1000 \text{ mL A}}}$ moles A

moles B $\xrightarrow{\dfrac{1000 \text{ mL B}}{\text{no. of mol B}}}$ mL B

Liters A at STP $\xrightarrow{\dfrac{1 \text{ mol A}}{22.4 \text{ L A at STP}}}$ moles A

moles B $\xrightarrow{\dfrac{22.4 \text{ L B at STP}}{1 \text{ mol B}}}$ Liters B at STP

Solutions 217

Note from Figure 13.1 that you can incorporate milliliters of a substance in the balanced equation into stoichiometry.

In this current problem, you know grams of **A** (AgCl) and you want to find milliliters of **B** (0.100 $M$ AgNO$_3$). Try to follow how this is done using Figure 13.1.

Notice that

$$M = \text{mol/L} = \text{mol}/1000 \text{ mL} \text{ and } 1/M = M^{-1} = \text{L/mol} = 1000 \text{ mL/mol}$$

The solution for this problem follows:

$$7.25 \times 10^{-2} \text{ g AgCl} \times \frac{1 \text{ mol AgCl}}{143.4 \text{ g AgCl}} \times \frac{1 \text{ mol AgNO}_3}{1 \text{ mol AgCl}} \times \frac{1000 \text{ mL AgNO}_3}{0.100 \text{ mol AgNO}_3} =$$

$$0.506 \text{ mL AgNO}_3$$

**Example 13.15:** With excess NaCl solution, how many grams of AgCl will be formed from using 8.2 mL of 0.100 $M$ silver nitrate?

**Solution:** In this case you want to start with milliliters of **A** (AgNO$_3$ soln) and determine grams of **B** (AgCl). Use Figure 13.1.

$$8.2 \text{ mL AgNO}_3 \text{ soln} \times \frac{0.100 \text{ mol AgNO}_3}{1000 \text{ mL AgNO}_3 \text{ soln}} \times \frac{1 \text{ mol AgCl}}{1 \text{ mol AgNO}_3} \times \frac{143.4 \text{ g AgCl}}{1 \text{ mol AgCl}} =$$

$$0.12 \text{ g AgCl}$$

Another type of concentration calculation you will encounter is that found in **normality** problems. These problems use EQ 13.8. They use the term **equivalents**. This term refers to the amount of acid that will yield 1 mol of H$^+$ or the amount of base that will yield 1 mol of OH$^-$.

$$\text{normality} = \frac{\text{equivalents solute}}{\text{L solution}} = N \qquad \text{(EQ 13.8)}$$

**Example 13.16:** How many equivalents of H$^+$ are in 1 mol of HBr? in 1 mol H$_2$SO$_4$? in 1 mol H$_3$PO$_4$? in 2 mol of H$_3$PO$_4$?

**Solution:** Solve these by examining the formulas.

Each mole of HBr yields 1 mol of H$^+$.

Each mole of H$_2$SO$_4$ yields 2 mol of H$^+$.

Each mole of H$_3$PO$_4$ yields 3 mol of H$^+$. Therefore, 2 mol yields 6 mols H$^+$.

**Example 13.17:** What is the normality of 0.25 $M$ $H_2SO_4$?

**Solution:** Use EQ 13.8 to solve this.

$$N = \frac{0.50 \text{ equivalents solute}}{\text{L solution}} = 0.50\,N$$

**Example 13.18:** How many equivalents are in 75.5 mL of 0.035 $M$ $H_2SO_4$?

**Solution:** Rearrange EQ 13.8 to EQ 13.9. Then substitute and solve.

Equivalents = $N \times$ L solution  **(EQ 13.9)**

$$\text{equiv} = (0.070\,N)(0.755 \text{ L soln}) = \frac{0.070 \text{ equiv solute}}{\text{L solution}} \times 0.755 \text{ L soln} =$$

$$5.3 \times 10^{-3} \text{ equiv}$$

One common usage of the concept of equivalents is in calculations involving acid-base neutralization problems. These occur when a known concentration of one reactant is used to **titrate** a second reactant. The completion of the reaction is monitored by an indicator. Example 13.19 shows how this type of problem is solved.

**Example 13.19:** For the reaction of 0.126 $M$ sulfuric acid with an unknown concentration of potassium hydroxide, if it takes 17.9 mL of acid to neutralized the solution of potassium hydroxide, how many grams of base were present?

**Solution:** For a neutralization reaction, you know that when the reaction is complete:

equivalents of base = equivalents of acid

You can calculate the equivalents of acid using EQ 13.9.

$$\text{equiv} = \frac{0.252 \text{ equiv } H_2SO_4}{\text{L solution}} \times 0.0179 \text{ L soln} = 4.51 \times 10^{-3} \text{ equiv } H_2SO_4$$

At the reaction's completion, the equivalents of KOH are also $4.51 \times 10^{-3}$, and this can be inserted into a final dimensional analysis equation.

$$4.51 \times 10^{-3} \text{ equiv KOH} \times \frac{1 \text{ mol KOH}}{1 \text{ equiv KOH}} \times \frac{56.1 \text{ g KOH}}{1 \text{ mol KOH}} = 0.253 \text{ g KOH}$$

The final type of problem using molarity concentrations is one of **dilution**. The working equation for these types of problems is illustrated in EQ 13.10. This states that molarity of the concentrated solution ($M_i$) times the volume of the concentrated solution needed ($V_i$) is equal to the molarity of the diluted solution ($M_f$) times the volume of the diluted solution ($V_f$).

$$M_iV_i = M_fV_f \qquad \text{(EQ 13.10)}$$

**Example 13.20:** If you want to prepare 250 mL of 0.150 M H$_2$SO$_4$ from stock 6.0 M solution, how would you do this?

**Solution:** Rearrange EQ 13.10 to isolate the volume of stock ($V_i$) and then plug other values into the equation.

$$V_i = \frac{M_f V_f}{M_i} = \frac{(250 \text{ mL})(0.150 \, M)}{6.0 \, M} = 6.25 \text{ mL stock soln}$$

After determining the number of milliliters of stock, measure this amount into a 250 mL flask and dilute with solvent to the 250 mL mark.

---

**Practice Problems**

**13.39** What is the molarity of a solution where 10.6 g of NaCl is dissolved in 250 mL solution?

**13.40** How would you prepare 100 mL of 0.250 M MgCl$_2$?

**13.41** For the reaction of sulfuric acid and potassium hydroxide, how many grams of KOH will react with 6.5 mL of 0.50 M H$_2$SO$_4$?

**13.42** How much 3.0 M H$_2$SO$_4$ will it take to completely react with 1.25 g of KOH?

**13.43** How many equivalents are in 2.0 mol of H$_2$CO$_3$?

**13.44** What is the normality of 0.145 M Al(OH)$_3$?

**13.45** How many grams of AgCl will form from the reaction of 8.0 mL of 0.020 M AgNO$_3$ with an excess of NaCl?

**13.46** How many milliliters of 3.0 M HCl does it take to make 500. mL of 1.2 M HCl?

---

## 13.9 Molality

There are times when a concentration unit that is **not** temperature dependent is needed. This unit is called **molality**. Its formula is shown in EQ 13.11. Notice that it is similar to **molarity** except that instead of liters of solution, you measure kilograms of solvent. The abbreviation for it is lowercase *m*.

$$\text{molality} = \frac{\text{mol solute}}{\text{kg solution}} = m \qquad \text{(EQ 13.11)}$$

**Example 13.21:** If 17.6 g of sugar (M.W. = 342 g/mol) are dissolved in 74.0 g of water, what is the molality of the solution?

**Solution:** Use the given information to arrive at molality units. Verify that all units but those needed for molality cancel.

$$\frac{17.6 \text{ g sugar}}{74.0 \text{ g H}_2\text{O}} \times \frac{1 \text{ mol sugar}}{342 \text{ g sugar}} \times \frac{1000 \text{ g H}_2\text{O}}{1 \text{ kg H}_2\text{O}} = \frac{0.695 \text{ mol sugar}}{1 \text{ kg H}_2\text{O}} = 0.695 \, m$$

> **Practice Problems**
>
> **13.47** If 78.6 g of $C_2H_6O_2$ (ethylene glycol) is dissolved in 300. g of water, what is the molality?
>
> **13.48** If 68.5 g of $I_2$ is dissolved in 355 g of $CH_2Cl_2$, what is the molality?
>
> **13.49** If 575 mL ethylene glycol (density = 1.109 g/mL) is dissolved in 575 mL of water (density = 1.00 g/mL) what is the molality of the solution?

### 13.10 Freezing-Point Depression and Boiling-Point Elevation

This section looks at two of the three colligative properties listed below:

- Freezing-point depression (the f.p. of a solution is lower than that of the pure solvent).
- Boiling-point elevation (the b.p. of a solution is higher than that of the pure solvent).
- Osmotic pressure (pressure is exerted from solvent molecules crossing a membrane).

Every solvent has characteristic constants for depression of f.p. and elevation of b.p., which are noted in Table 13.4. These are symbolized as $K$ in the equations and Table 13.4 below.

$$\Delta T_f = K_f m \quad \text{(EQ 13.12)}$$

$$\Delta T_b = K_b m \quad \text{(EQ 13.13)}$$

**Table 13.4** Some Constants for Freezing Point Depression and Boiling-Point Elevation

| Name of Substance | Formula | Normal f.p. (°C) | $K_f$ (°C/m) | Normal b.p. (°C) | $K_b$ (°C/m) |
|---|---|---|---|---|---|
| Water | $H_2O$ | 0 | 1.86 | 100 | 0.52 |
| Cyclohexane | $C_6H_{12}$ | 6.6 | 20.0 | 80.7 | 2.79 |
| Acetic acid | $C_2H_4O_2$ | 16.6 | 3.90 | 117.9 | 2.93 |
| Ethanol | $C_2H_6O$ | -117 | 1.99 | 78.4 | 1.22 |
| Benzene | $C_6H_6$ | 5.455 | 5.12 | 80.1 | 2.53 |

The predictable behavior of f.p. and/or b.p. of solutions allows for methods that can determine the molality and the molecular mass of unknown compounds. Try to work with these properties in Examples 13.22-13.24.

**Example 13.22:** What is the boiling point of a 5.87 $m$ solution of glucose in water?

**Solution:** Use EQ 13.13 to determine the change in temperature. Add the change to the *normal* boiling point.

$$\Delta T_b = K_b m = \frac{0.52°C}{m} \times 5.87 \, m = 3.05°C$$

Solution b.p. = 100°C + 3.05°C = **103°C**

**Example 13.23:** What is the freezing point of a solution of benzene made by dissolving 10.6 g of $C_4H_8O_2$ in 315 g benzene?

**Solution:** First, the molality of the solution must be determined.

$$m = \frac{10.6 \text{ g } C_4H_8O_2}{0.315 \text{ kg benzene}} \times \frac{1 \text{ mol } C_4H_8O_2}{88.0 \text{ g } C_4H_8O_2} = 0.382 \, m$$

Use the calculated molality to determine the change in temperature. (Use EQ 13.12.)

$$\Delta T_f = K_f m = \frac{5.12°C}{m} \times 0.382 \, m = 1.96°C$$

Subtract the change in temperature from the normal freezing point (remember that freezing point is *depressed*).

5.455°C - 1.96°C = **3.50°C**

**Example 13.24:** An unknown substance is dissolved in acetic acid. If it freezes at 9.3°C, what is its molality?

**Solution:** This problem is slightly different because you need to determine *molality* and not the freezing point. The first step is to determine $\Delta T_f$ and then rearrange EQ 13.12 to solve for $m$.

$$\Delta T_f = 16.6°C - 9.3°C = 7.3°C$$

$$m = \frac{\Delta T_f}{K_f} = \frac{7.3°C}{3.90°C/m} = 1.87 \, m$$

When solving colligative property problems, always try to read the information carefully so that you know what parameter you are trying to determine.

---

**Practice Problems**

13.50 What is the freezing point of a 14.9 $m$ solution of ethylene glycol in water?

13.51 What is the boiling point of an 18.5 $m$ solution of sugar water?

**13.52** If 67.5 g $C_2H_6O_2$ is dissolved in 400 g of water, what is the freezing point of the solution?

**13.53** What is the boiling point of a 2.57 $m$ solution of cyclohexane?

**13.54** What is the molality of a solution of benzene that has a freezing point of 4.23°C?

## 13.11 Osmotic Pressure

When a semipermeable membrane is placed between two solutions having different concentrations, then the membrane will allow small solvent (usually water) molecules to pass from the dilute solution to the more concentrated solution. This is an attempt to equalize the concentrations.

The addition of extra solvent molecules to a closed system can exert a measurable pressure on that system. This is called **osmotic pressure** because it is derived from the *osmosis* of solvent molecules across a selective membrane.

These types of membranes exist in nature. The cell walls of bacteria, as well as cell membranes, are examples of structures which allow osmosis.

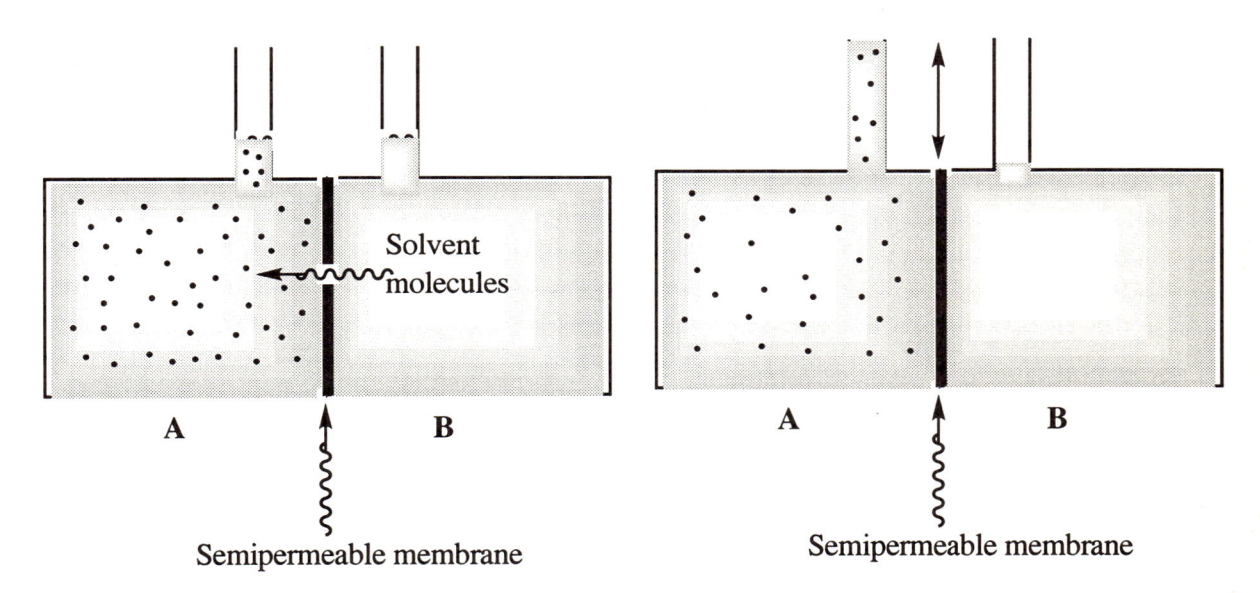

**Figure 13.2** Diagram of Osmotic Pressure

The double headed arrow in the second diagram indicates the amount of pressure created from osmosis.

**Example 13.25:** If an 0.85% solution of NaCl is isotonic with blood, what would happen to red blood cells placed in pure water?

**Solution:** Pure water is hypotonic to 0.85% saline. Therefore, water molecules would pass through the cell membrane to try to dilute the solution inside the cell. This would swell the cell to the point where it would eventually burst.

224    Chapter 13

## Practice Problems

**13.55** What would happen to red blood cells in 2.00% saline solution?

**13.56** Is 0.85% KBr solution isotonic, hypertonic, or hypotonic to 0.85% NaCl?

## Integration of Multiple Skills

This section gives you a chance to challenge yourself with problems that involve thinking about concepts you learned in this chapter and in previous chapters. You will see ways that they can be applied to problems you haven't encountered before.

**Table 13.5** Summary of Concentration Units

| Name of Concentration | Abbreviation | Formula |
|---|---|---|
| mass percent solute | g% | $\dfrac{\text{mass solute}}{\text{mass solution}} \times 100\%$ |
| volume percent solute | vol% | $\dfrac{\text{volume solute}}{\text{volume solution}} \times 100\%$ |
| parts per million | ppm | $\dfrac{\text{mass solute}}{\text{mass solution}} \times 1{,}000{,}000$ |
| molarity | M | $\dfrac{\text{mol solute}}{\text{L solution}}$ |
| normality | N | $\dfrac{\text{equivalents solute}}{\text{L solution}}$ |
| molality | m | $\dfrac{\text{mol solute}}{\text{kg solution}}$ |

**Example 13.26:** Concentrated sulfuric acid is 98.0 mass percent solute of $H_2SO_4$ in water. Its density is 1.83 g/mL. What is its molarity? What is its molality?

**Solution:** Use the definitions of each in Table 13.5. Try to solve each of these on your own before checking the following complete solutions. Examine each step to determine why the particular equality was used.

$$\frac{98 \text{ g H}_2\text{SO}_4}{100 \text{ g soln}} \times \frac{1.83 \text{ g soln}}{1 \text{ mL soln}} = \frac{1.7934 \text{ g H}_2\text{SO}_4}{1 \text{ mL soln}} \times \frac{1000 \text{ mL soln}}{1 \text{ L soln}} \times \frac{1 \text{ mol H}_2\text{SO}_4}{98.1 \text{ g H}_2\text{SO}_4} =$$

$$18.3 \text{ } M \text{ H}_2\text{SO}_4$$

$$\frac{98 \text{ g H}_2\text{SO}_4}{2 \text{ g H}_2\text{O}} \times \frac{1 \text{ mol H}_2\text{SO}_4}{98.1 \text{ g H}_2\text{SO}_4} \times \frac{1000 \text{ g H}_2\text{O}}{1 \text{ kg H}_2\text{O}} = 499 \text{ } m \text{ H}_2\text{SO}_4$$

**Example 13.27:** What is the molar mass of an unknown compound if 3.526 g dissolved in 115.0 g benzene has a freezing point of 4.387°C? If the compound is 49.0%C, 2.72%H, and 48.3% Cl, what is the molecular formula of the compound? What is the empirical formula?

**Solution:** To solve this problem, first determine the molality of the solution (see example 13.24). Next, use the definition of molality to determine how many grams are in 1 mol of the unknown.

The molality is 0.209m.

By definition this means that there 0.209 mols of solute in 1kg of benzene (solvent).

Use ratio and proportion to determine how many mols are in 115.0g of benzene.

$$\frac{0.209 \text{ mol}}{1 \text{ kg benzene}} = \frac{X}{0.115 \text{ kg benzene}}$$

Solving the above for X, you can determine that there are

**0.0240 mol in 0.115kg benzene.**

Use the definition of molar mass: **g/mol** to find that there are 3.526g/0.0240 mol =

**147g/mol = molar mass**

Finally you can take the relative % of each element to find the molecular formula.

$$0.490 \times 147\text{g} = 72\text{g C} \times 1\text{mol C}/12.0\text{g} = \textbf{6 mol C}$$
$$0.0272 \times 147\text{g} = 4\text{g H} \times 1\text{mol H}/1.00\text{g} = \textbf{4 mol H}$$
$$0.483 \times 147\text{g} = 71\text{g C} \times 1\text{mol C}/35.5\text{g} = \textbf{2 mol Cl}$$

Molecular formula: **C_6H_4Cl_2**

Empirical formula: **C_3H_2Cl**  (Remember this is just lowest common ratio of elements.)

# 226    Chapter 13

> **Practice Problems**
>
> **13.57** A solution of an unknown compound in acetic acid has a freezing point of 14.87°C. The solution is made up of 7.862 g of the compound in 145.0 g of solvent. What is the molality of the solution? What is the molar mass of the compound? If the compound is 68.9% C, 4.90% H, and 26.2% O, what is its molecular formula? What is its empirical formula?
>
> **13.58** 30.0 g of $NH_3$ is dissolved in 70.0 g of water and the solution has a density of 0.982 g/mL. What is the mass percent solute of the solution? What is the $M$? What is the $m$?

## Self-Exam

**I. (4 ea) Write T (true) or F (false) in each corresponding blank.**

_____1. A solution has NaCl solid dissolved in water. NaCl is the solvent.

_____2. Air is a solution of oxygen in nitrogen.

_____3. More oxygen can be dissolved in cold water than in hot water.

_____4. When you open a can of pop, the fizz is $CO_2$ escaping from solution due to a decrease in pressure.

_____5. $I_2$ is more soluble in $CCl_4$ than in water.

_____6. Gasoline won't dissolve in water, so these two liquids are miscible.

_____7. $BaSO_4$ is *insoluble* in water *because* it is nonpolar.

_____8. A solution of sugar water will have a b.p. lower than 100°C (at 1 atm).

_____9. A semipermeable membrane between pure water and a glucose solution will allow molecules of glucose to pass into the pure water.

_____10. Three moles of $H_3PO_4$ have nine equivalents of $H^+$.


**II. (4 ea) Multiple Choice. Place the letter of the BEST answer in the corresponding blank.**

_____1. 2.4 g of $C_6H_6O_2$ is dissolved in 1000 g of solution. Its concentration is
   A. 2.4%    B. 0.24%    C. 24%    D. 41.7%    E. 4.17%

_____2. 50 mL of a 2.5 $M$ solution of NaOH contains how many grams of NaOH?
   A. 50 g    B. 12.5 g    C. 2.0 g    D. 5.0 g    E. 20 g

_____3. 25 g of glucose, $C_6H_{12}O_6$, is dissolved in 200. mL of solution. The molarity is
   A. 0.125 $M$    B. 0.125%    C. 0.69 $M$    D. 5.0%    E. 0.140 $M$

_____4. The number of moles of solute dissolved in kilograms of solvent is what concentration unit?
   A. molarity         B. molality        C. mass % solute
   D. volume % solute  E. normality

Solutions    227

_____5. Which of the following is a polar solvent?
   A. $CH_3OH$    B. $CH_4$    C. $CCl_4$    D. $C_6H_{14}$    E. $C_6H_{12}$

_____6. 148.8 g $C_2H_6O_2$ is dissolved in 1000 g of water. What is the molality?
   A. 2.4 m    B. 0.23 m    C. 0.24 m    D. 0.00224 m    E. 0.00223 m

_____7. How many grams of glucose ($C_6H_{12}O_6$) are needed to make 150 mL of 1.5 M solution?
   A. 270 g    B. 180 g    C. 100 g    D. 81.0 g    E. 40.5 g

_____8. $2.50 \times 10^2$ mL of 0.450 M KOH needs to be prepared from 8.40 M stock. How many milliliters of stock are required?
   A. 945 mL    B. 466 mL    C. 134 mL    D. 13.4 mL    E. 0.015 mL

_____9. A solution is prepared by adding 128 mL of 2.00 M HCl to 178 mL of 5.75 M HCl. What is the final molarity? (Assume the volume is the sum of the two solutions.)
   A. 7.75 M    B. 5.75 M    C. 4.18 M    D. 3.75 M    E. 2.00 M

_____10. If 0.85% sodium chloride is the concentration of blood plasma, which of the following solutions is isotonic with this?
   A. 8.5 g NaCl in 100 g soln              B. 1.45 mol NaCl in 100 g soln
   C. 14.5 mol NaCl in 100 g of soln        D. 85 g NaCl in 1000 g of soln
   E. 0.145 mol NaCl in 1000 g soln

### III. Provide Complete Answers

1. (6 pts) How many milliliters of 0.200 M hydrochloric acid will completely react with 1.27 g of magnesium hydroxide?

2. (6 pts) For the precipitation reaction of 0.250 M silver nitrate with excess sodium chloride solution, how many grams of silver chloride will form from 4.8 mL of $AgNO_3$ solution?

3. (8 pts) 8.82 g of an unknown compound is dissolved in 215 g of benzene ($C_6H_6$), which has a normal freezing point of 5.455°C. The solution freezes as 3.911°C. What is the molar mass of the unknown compound?

## Answers to Practice Problems

| | | | | | |
|---|---|---|---|---|---|
| 13.1 | 23% | 13.2 | 12 g | | |
| 13.3 | 0.315 mol NaOH | 13.4 | 156 g HCl | 13.5 | 0.172 mol $H^+$ |
| 13.6 | 0.399 mol $OH^-$ | 13.7 | Solute | 13.8 | Solvent |
| 13.9 | Solution | 13.10 | Immiscible | 13.11 | Gas in gas |
| 13.12 | Solid in liquid | 13.13 | Gas in liquid | | |
| 13.14 | Liquid in liquid | 13.15 | Solid in solid | | |
| 13.16 | Solution | 13.17 | Solution | 13.18 | Pure substance |
| 13.19 | Pure substance | 13.20 | Solution | | |

**13.21** Mixture (heterogeneous)

**13.22** 106g at 20°C, 1980g at 100°C

**13.23** 0.056g

**13.24** 175g

**13.25** $NaCl(s) \rightleftharpoons Na^+(aq) + Cl^-(aq)$

**13.26** $PbCl_2(s) \rightleftharpoons Pb^{2+}(aq) + 2Cl^-(aq)$

**13.27** $C_6H_{12}O_6(s) \rightleftharpoons C_6H_{12}O_6(aq)$

**13.28** The ionic compound, $NH_4Cl$, will dissolve in polar solvents ($H_2O$) but not nonpolar solvents ($CCl_4$).

**13.29** The multi-charged ions $Ba^{2+}$ and $SO_4^{2-}$ are more attracted to each other than to $H_2O$.

**13.30** $Br_2$ is nonpolar and is not attracted to the polar water molecules.

**13.31** With an increase in temperature, the gas $O_2$ molecules have enough energy to leave the solution.

**13.32** Pressure can force more gas molecules into a solution.

| | | | | | |
|---|---|---|---|---|---|
| **13.33** | 6.3% | **13.34** | 6.3% | **13.35** | 5.9% |

**13.36** 2.13 g NaCl in 248 g $H_2O$

| | | | | | |
|---|---|---|---|---|---|
| **13.37** | 25% | **13.38** | 0.065 ppm | **13.39** | 0.72 $M$ |

**13.40** Dissolve 2.4 g $MgCl_2$ in 100 mL of solution

| | | | | | |
|---|---|---|---|---|---|
| **13.41** | 0.36g KOH | **13.42** | 3.7 mL | **13.43** | 4 equiv. |
| **13.44** | 0.435 $N$ | **13.45** | 0.023g AgCl | **13.46** | 200 mL |
| **13.47** | 4.23 $m$ | **13.48** | 0.760 $m$ | **13.49** | 17.9 $m$ |
| **13.50** | -27.7°C | **13.51** | 110°C | **13.52** | -5.1°C |
| **13.53** | 87.9°C | **13.54** | 0.239$m$ | | |

**13.55** They would shrink (water leaves cell).

**13.56** Isotonic

**13.57** 0.444 $m$; M.W. = 122 g/mol; $C_7H_6O_2$ is both molecular and empirical formula.

**13.58** 25.2 $m$ $NH_3$; 30%; 17.3 $M$

# Solutions 229

## Answers to Self-Exam

**I. T/F**
1. F  2. T  3. T  4. T  5. T  6. F  7. F  8. F  9. F  10. T

**II. MC**
1. B  2. D  3. C  4. B  5. A  6. A  7. E  8. D  9. C  10. E

**III. Complete Answers**

1. $2HCl + Mg(OH)_2 \rightarrow 2H_2O + MgCl_2$
   218 mL of 0.200 $M$ HCl

2. $AgNO_3 + NaCl \rightarrow AgCl + NaNO_3$
   0.17 g of AgCl

3. Solution molality = 0.302 $m$; M.W. = 136 g/mol

# 14
# Reaction Rates and Chemical Equilibrium

This chapter looks at factors that affect the speed of reactions. This is done from a qualitative point of view and does not involve exact calculations of rates. You will want to understand what conditions of a reaction determine how fast the reaction will occur.

The second part of this chapter looks at chemical equilibrium. This topic will be examined from both a qualitative and a quantitative perspective.

Finally, you will look a Le Châtelier's principle. This is a very powerful rule that allows the chemist who understands it to set reaction conditions that lead to desired products.

Keys for success in this chapter:

- Read the descriptions of reaction rates and equilibria.
- Understand how factors affect rates and equilibria.
- Learn new terms and definitions.
- Practice problems.

## Summary of Verbal Knowledge

**collision theory:** This says that reacting molecules must come so close that they collide. It also says that the energy of the collision must be greater than a certain minimum value in order for a reaction to occur.
**activation energy:** The minimum energy that leads to a reaction. A high activation energy means a slow reaction, and a low one means a fast reaction.
**catalyst:** A substance that increases the rate of a chemical reaction without being consumed in the reaction.
**chemical equilibrium:** The point where a reaction seems to stop because no more increase in products is observed.
**law of mass action:** This states that each reaction has an equilibrium constant with its own characteristic value at a given temperature.
**homogeneous equilibrium:** A reaction where all substances in the reaction are in the same state.
**heterogeneous equilibrium:** A reaction where substances are in more than one state.
**Le Châtelier's principle:** This states that when a reaction at equilibrium is disturbed by a change in volume, temperature, or the concentration of one of the components, the reaction will shift in a way that tends to counteract the change. This will reestablish equilibrium.
**exothermic:** This describes a reaction where heat is produced during a reaction.
**endothermic:** This describes a reaction where heat is absorbed during a reaction.

# Reaction Rates and Chemical Equilibrium

## Review of Mathematical and Calculator Skills

You will need to work with the $x^2$ functions and the square root ($\sqrt{\ }$) functions on your calculator. If you are not familiar with these, go through the following stepwise examples.

**Example 14.1:** What are the units for $M$, $M^{-1}$, $M^2$?

**Solution:** Remember from Chapter 13 that molarity ($M$) is mol/L.

The exponent -1 means 1/unit.
Thus: $M^{-1}$ means $1/M$. Mathematically, this means

$$\frac{1}{M} = \frac{1}{\text{mol/L}} = 1 \times \frac{L}{\text{mol}} = \frac{L}{\text{mol}}$$

Remember when dividing a fraction by a fraction, invert and multiply, as above.

The final unit is solved as follows:

$$M^2 = \left(\frac{\text{mol}}{L}\right)^2 = \frac{\text{mol}^2}{L^2}$$

**Example 14.2:** Solve $\dfrac{(1.25)^2}{(1.62)(0.17)^2}$

**Solution:** The solution to this problem involves using the $x^2$ function on your calculator. Two types of executions for this problem on calculators are shown. Check to see if your calculator matches one of these. It should follow one of the two.

| Operation | Display |
|---|---|
| 1. Press [1.25] | 1.25 |
| 2. Press [$x^2$] | 1.5625 |
| 3. Press [÷] [1.62] [÷] | 0.964506172 |
| 4. Press [ ( ] | ( 0 |
| 5. Press [0.17] | ( 0.17 |
| 6. Press [$x^2$] | ( 0.0289 |
| 7. Press [ ) ] | 0.0289 |
| 8. Press [ = ] | 33.37391602 |

If you have a *scientific* calculator, you will probably find the previous solution works for this problem. The number should be rounded to three significant figures, and the number displayed should not be reported. The correct number is **33.4**.

If you have a *graphic* calculator, you will probably find the solution below works. Again report the answer with the correct significant figures.

| | Operation | Display |
|---|---|---|
| 1. Press | [1.25] | 1.25_ |
| 2. Press | [shift] [√] | $1.25^2$ |
| 3. Press | [÷] [1.62] [÷] | $1.25^2 \div 1.62 \div$_ |
| 4. Press | [ ( ] [0.17] | $1.25^2 \div 1.62 \div (0.17$ |
| 5. Press | [shift] [√] [ ) ] | $1.25^2 \div 1.62 \div (0.17^2)$ |
| 6. Press | [EXE] | 33.37391602 |

**Example 14.3:** If $x^2 = 7.86 \times 10^{-2}$, what is the value of x?

**Solution:** To solve for x, take the square root of both sides.

$$\sqrt{x^2} = \sqrt{7.86 \times 10^{-2}}$$

$$x = 0.280$$

If you are not familiar with these calculations on your calculator, do the following if you have a *scientific* calculator.

| | Operation | Display |
|---|---|---|
| 1. Press | [√] | √ 0 |
| 2. Press | [7.86] [Exp] | √ $7.86^{00}$ |
| 3. Press | [2] [+/-] | √ $7.86^{-02}$ |
| 4. Press | [=] | 0.280356915 |

There are only three significant figures that can be reported, **0.280**.

# Reaction Rates and Chemical Equilibrium 233

If you have a *graphic* calculator, then this sequence of steps should work:

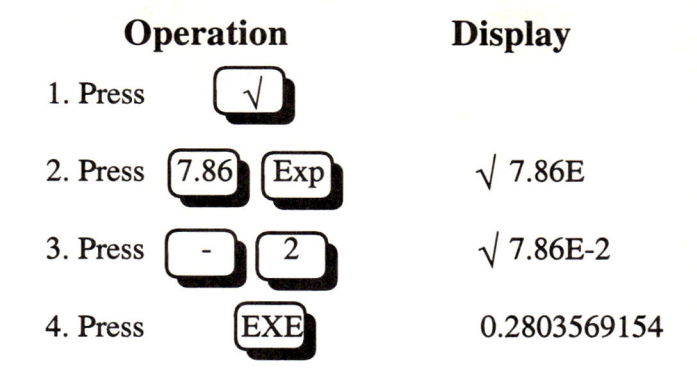

| Operation | Display |
|---|---|
| 1. Press $\sqrt{}$ | |
| 2. Press 7.86 Exp | $\sqrt{}$ 7.86E |
| 3. Press - 2 | $\sqrt{}$ 7.86E-2 |
| 4. Press EXE | 0.2803569154 |

Report your answer as **0.280**.

---

**Practice Problems**

**14.1** What is $M^3$?

**14.2** Solve for x: $x^2 = 8.23 \times 10^3$

**14.3** Solve $\dfrac{(0.25)^2}{(0.062)(3.17)^2}$

**14.4** Solve for x: $x^2 = 1.43 \times 10^{-5}$

**14.5** Solve $\dfrac{(0.68)(1.28)^2}{(1.42)^2(0.18)}$

---

## Application of Skills and Concepts

### 14.1 Collision Theory and Activation Energy

Review the definition of this theory at the beginning of the chapter. Note the following:

- Collisions are necessary but not sufficient for a reaction.
- Collisions must have a given minimum energy to react.

Figure 14.1 shows two pictorial representations of reactions. The height of the line from the energy level of the reactant to the top of the curve represents the minimum energy needed for a collision to result in a reaction. You should note that

- The higher the activation barrier, the slower the reaction.
- The lower the activation barrier, the faster the reaction.

Use the information in the Figure 14.1 to answer the practice problems.

**Figure 14.1** Two Reactions and Their Energy Barriers

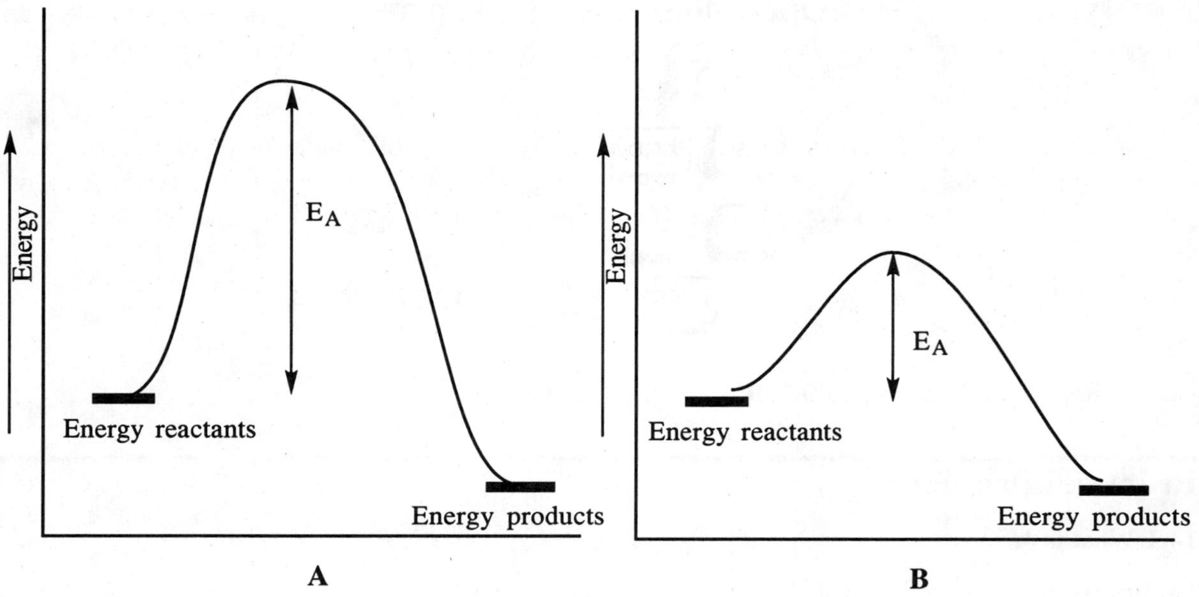

A

B

**Practice Problems**

**14.6** Which reaction (A or B) is faster?

**14.7** Which reaction (A or B) has a higher activation barrier?

**14.8** Which reaction (A or B) needs more energy in the collisions of molecules?

## 14.2 Factors Affecting Reaction Rates

These factors are summarized below:

- Identity of reactants
- Concentration of reactants
- Temperature of reaction
- Presence of a catalyst

The ways these factors affect rates are summarized in Table 14.1.

**Table 14.1** How Various Factors Affect Rates of Reactions

| Factor Affecting Reaction | Effect on Rate | Effect on Rate |
|---|---|---|
| Identity | This is based on experiment | This is based on experiment |
| Concentration of Reactants | If ↑, then rate ↑ | If ↓, then rate ↓ |
| Temperature of Reactions | If ↑, then rate ↑ | If ↓, then rate ↓ |
| Presence of Catalyst | If present, rate ↑ | It absent, rate ↓ |

**Example 14.3:** Why does food cook faster in a pressure cooker?

**Solution:** This question requires some application of the gas laws you learned in Chapter 11. From the ideal gas law (PV = nRT), you know that if you increase the pressure in a fixed volume, you will increase the temperature.

You also know that at a higher pressure, the boiling point of water (used to cook vegetables) will be higher. This is because the boiling point is equal to the temperature where the vapor pressure of water equals the ambient pressure.

From Table 14.1, you know that an increase in temperature increases the rate of the reaction; therefore, the food will cook faster.

---

## Practice Problems

Use the information in the following equation and Table 14.1 to answer the following questions.

$$3H_2(g) + N_2(g) \rightleftharpoons 2NH_3(g)$$

**14.9** Will the reaction be faster at 25°C or at 100°C?

**14.10** If the concentration of $H_2$ and $N_2$ are both 0.10 $M$, what will happen to the rate of the reaction when it is run when the concentrations of both are 0.05 $M$?

**14.11** What will happen to the rate of the reaction if the pressure is increased?

---

### 14.3 The Dynamic Nature of Chemical Equilibrium

Equilibrium is not a static state. Reaction mixtures that appear to be unchanging are constantly changing at the molecular level. However, at equilibrium the rate of the forward reaction is the same as the rate of the reverse reaction, so there appears to be no change.

To illustrate this, look at the dimerization of two molecules of nitrogen dioxide gas to one molecule of dinitrogen tetroxide gas.

**Example 14.4:** For the dimerization of nitrogen dioxide, write an equation for the forward reaction; for the reverse reaction; for the equilibrium reaction.

**Solution:** The forward reaction is written just as the words describe it. See EQ 14.1.

$$2NO_2 \rightarrow N_2O_4 \tag{EQ 14.1}$$

The reverse reaction involves writing the decomposition of $N_2O_4$ to two molecules of $NO_2$, as shown in EQ 14.2.

$$N_2O_4 \rightarrow 2NO_2 \tag{EQ 14.2}$$

## 236   Chapter 14

The equilibrium equation is the combination of the two equations. One arrow represents going from right to left (the traditional way of writing equations) and the other arrow from right to left. See EQ 14.3.

$$2NO_2 \rightleftharpoons N_2O_4 \qquad \text{(EQ 14.3)}$$

### Practice Problems

**14.12** Write the forward reaction for the equilibrium in problem 14.9.

**14.13** Write the reverse reaction for the equilibrium in problem 14.9.

**14.14** Write the reverse reaction for the combination of hemoglobin and oxygen:

$$Hb + 4O_2 \rightarrow Hb(O_2)_4$$

**14.15** Write the reverse reaction and equilibrium reaction for $S_8 + 8O_2 \rightarrow 8SO_2$.

## 14.4 The Equilibrium Expression

Any equation expressed as a balanced equilibrium can be used to find an equilibrium constant.

**The equilibrium constant is symbolized as K.**

Look at the generic equation in EQ 14.4.

$$aA + bB \rightleftharpoons cC + dD \qquad \text{(EQ 14.4)}$$

In this equation
- Upper case letters represent *substances* in the reaction.
- Lower case letters represent *coefficients* in the reaction.

Experiments show that for a given temperature, the equilibrium constant for the generic equation in EQ 14.4 is

$$K = \frac{[C]^c[D]^d}{[A]^a[B]^b}$$

The bracket symbols stand for concentration in molarity (*M*) or mol/L. However, the units are typically **not** used in the equilibrium expression. K is usually expressed as a unitless quantity.

**Example 14.5:** Write the equilibrium expression for the reaction of NOCl gas to form nitrogen monoxide gas and chlorine gas.

**Solution:** First, write a balanced equation.

$$2NOCl(g) \rightleftharpoons 2NO(g) + Cl_2(g)$$

Next, write the expression for K.

$$K = \frac{[NO]^2[Cl_2]}{[NOCl]^2}$$

---

**Practice Problems**

**14.16** Write the equilibrium expression for phosphorus trichloride gas plus chlorine gas to form phosphorus pentachloride gas.

**14.17** Write the equilibrium expression for sulfur dioxide gas plus oxygen gas to form sulfur trioxide gas.

**14.18** Write the equilibrium expression for nitrogen gas plus chlorine gas to form nitrogen trichloride gas.

---

## 14.5 Calculating an Equilibrium Constant

There are two concepts that are covered in this section.

- Learn to calculate K from given concentrations.
- Understand that K is the same at a given temperature regardless of the starting conditions.

The first task is learning to calculate K, given a reaction and the concentrations of all substances at equilibrium.

- Write the equilibrium expression from the balanced equation.
- Put the given concentrations into the equilibrium expressions.

**Example 14.6:** A 1.00-L vessel at 727°C has 1.00 mol each of sulfur dioxide and oxygen at the beginning of a reaction. After a time it forms sulfur trioxide (see problem 14.7). At equilibrium the concentration of sulfur dioxide is 0.075 $M$; the oxygen is 0.537 $M$; the sulfur trioxide is 0.925 $M$. Calculate K for the reaction.

**Solution:** Write the balanced equation.

$$2SO_2(g) + O_2(g) \rightleftharpoons 2SO_3(g)$$

Write the expression for $K = \dfrac{[SO_3]^2}{[SO_2]^2[O_2]}$

Substitute the given values into $K = \dfrac{(0.925)^2}{(0.075)^2(0.537)} = 280$

**Example 14.7:** The same 1.00-L vessel in Example 14.6 is filled with pure $SO_3$ gas at 727°C. After equilibrium is reached, the concentrations of each substance were $[SO_2] = 0.011$ $M$,

[O$_2$] = 1.05 $M$, and [SO$_3$] = 1.89 $M$. Calculate K. How does it compare with K in Example 14.6?

**Solution:** Use the same expression for K, and substitute the equilibrium concentrations.

$$K = \frac{(1.89)^2}{(0.11)^2(1.05)} = 280$$

This is the same value, even though the starting materials were different. At the same temperature, the value of K is the same, no matter what the starting conditions.

## Practice Problems

**14.19** Determine K for the reaction of carbon monoxide gas plus hydrogen gas to produce water vapor and CH$_4$ gas. The concentrations at equilibrium are: [CO] = 0.0613 $M$, [H$_2$] = 0.184 $M$, [H$_2$O] = 0.037 $M$, [CH$_4$] = 0.037 $M$.

**14.20** Determine K for the reaction where NOCl gas decomposes to nitrogen monoxide gas and chlorine gas. The concentrations at equilibrium are: [NOCl] = 0.44 $M$, [NO] = 0.06 $M$, [Cl$_2$] = 0.03 $M$.

**14.21** Determine K for the reaction of iodine gas plus hydrogen gas to produce HI vapor. The concentrations at equilibrium are: [I$_2$] = 0.42 $M$, [H$_2$] = 0.42 $M$, [HI] = 3.16 $M$.

**14.22** Determine K for the reaction of sulfur dioxide gas plus oxygen gas to produce sulfur trioxide gas at 877°C. The concentrations at equilibrium are: [SO$_2$] = 0.145 $M$, [O$_2$] = 0.0725 $M$, [SO$_3$] = 0.230 $M$.

**14.23** Compare the K in problem 14.22 to the K in Example 14.6.

**14.24** Write a balanced equation from the equilibrium expression: $K = \dfrac{[O_3]^2}{[O_2]^3}$.

## 14.6 Using an Equilibrium Constant

There are two skills you will master in this section.

- Use K to make predictions about reaction amounts.
- Use K and the equilibrium expression to determine unknown concentrations.

The value of K tells you if the reaction favors the left or the right side of the equation (or neither). Table 14.2 summarizes what K tells about a reaction.

Table 14.2 Values of K and the Reaction Side That Is Favored

|  | K < 1 | K = 1 | K > 1 |
|---|---|---|---|
| Side of reaction that predominates | Left | Neither | Right |

# Reaction Rates and Chemical Equilibrium

Remember that K is unique for a given reaction no matter what the initial conditions as long as the temperature is not changed. However, if T changes, K changes, as Table 14.3 shows.

**Table 14.3** Values of K at Different Temperatures

| Reaction | Temperature (°C) | K |
|---|---|---|
| $3H_2(g) + N_2(g) \rightleftharpoons 2NH_3(g)$ | 25 | $4.1 \times 10^8$ |
|  | 400 | 0.50 |
|  | 600 | $1.2 \times 10^{-2}$ |
| $2SO_2(g) + O_2(g) \rightleftharpoons 2SO_3(g)$ | 579 | 12,800 |
|  | 727 | 280 |
|  | 877 | 35 |
| $2NO_2 \rightleftharpoons N_2O_4$ | 25 | 1300 |
|  | 100 | 2.5 |

**Example 14.8:** Predict the direction of reactions **A** and **B**.
Reaction **A** has $K = 1.8 \times 10^3$
Reaction **B** has $K = 5.6 \times 10^{-4}$

**Solution:** Use Table 14.2 to solve this. Reaction **A** has K > 1; therefore, it favors products (reaction lies to the right).

Reaction **B** has K < 1; therefore, it favors reactants (reaction lies to the left).

**Example 14.9:** Given the reaction below, which has K = 0.22, [HI] = 0.85 $M$, [I$_2$] = 0.60 $M$, predict the [H$_2$].

$$2HI(g) \rightleftharpoons H_2(g) + I_2(g)$$

**Solution:** Use the definition for $K = \dfrac{[H_2][I_2]}{[HI]^2} = \dfrac{[H_2](0.60)}{(0.85)^2} = 0.22$

Now rearrange the equation to isolate [H$_2$].

$$[H_2] = \dfrac{(0.22)(0.85)^2}{(0.60)} = 0.26\ M$$

**Example 14.10:** Given the reaction below, which has K = 0.10, [N$_2$] = 0.0319 $M$, [O$_2$] = 0.0864 $M$, predict the [NO].

$$N_2(g) + O_2(g) \rightleftharpoons 2NO(g)$$

Solution: $K = \dfrac{[NO]^2}{[N_2][O_2]} = \dfrac{[NO]^2}{(0.0319)(0.0864)} = 0.10$

Solve the equation for $[NO]^2$.

$$[NO]^2 = (0.10)(0.0319)(0.0864)$$

Take the square root of both sides.

$$\sqrt{[NO]^2} = \sqrt{2.756 \times 10^{-4}}$$

$$[NO] = \mathbf{0.0166}\ M$$

---

## Practice Problems

**14.25** What might be a disadvantage of running a reaction at a lower temperature?

**14.26** What side of the reaction for the production of $NH_3$ from $H_2 + N_2$ is favored at 500°C?

**14.27** At a given temperature, K for the following is 78: $SO_2(g) + NO_2(g) \rightleftharpoons SO_3(g) + NO(g)$. Which side of the reaction predominates at this temperature?

**14.28** For $2NO_2 \rightleftharpoons N_2O_4$, if K = 7.5 and $[N_2O_4] = 0.0750\ M$, what is $[NO_2]$?

**14.29** For $CO_2(g) + H_2(g) \rightleftharpoons H_2O(g) + CO(g)$, if K = 1.60, $[H_2] = 0.087\ M$, $[CO] = 0.087\ M$, $[H_2O] = 0.11\ M$, what is $[CO]$?

---

## 14.7 Heterogeneous Equilibria

Not all equilibria have homogeneous substances (i.e., all in the same state). Some of the states that reactants and products can assume during equations are summarized in Table 14.4. One of the consequences of various states is the significance (or insignificance) of these states on concentration.

For most situations, pure solids or pure liquids (unless very dilute) do not change.

**Table 14.4** States of Substances and Their Relationship to Concentration

| Constant Concentration | Variable Concentration |
|---|---|
| Pure liquid | Gas |
| Pure solid | Aqueous solids |
|  | Aqueous ions |
|  | Dissolved liquids |

One of the consequences of heterogeneous equilibria is that the substances with constant concentration are not included in the expression for K.

Consider the reaction in EQ 14.5. Notice that ammonia is a gas; water is a pure liquid (the solvent); and $NH_4^+$ and $OH^-$ are dissolved (aqueous) ions.

# Reaction Rates and Chemical Equilibrium 241

$$NH_3(aq) + H_2O(l) \rightleftharpoons NH_4^+(aq) + OH^-(aq) \qquad \text{(EQ 14.5)}$$

If this were written with all the substances in the equilibrium expression, it would look like EQ 14.6.

$$K' = \frac{[NH_4^+][OH^-]}{[NH_3][H_2O]} \qquad \text{(EQ 14.6)}$$

In EQ 14.6 there are two constants, $K'$ and liquid water. These can be combined as in EQ 14.7.

$$K'[H_2O] = \frac{[NH_4^+][OH^-]}{[NH_3]} \qquad \text{(EQ 14.7)}$$

Both constants can be combined to equal K, which leads to the correct equilibrium expression, as shown in EQ 14.8.

$$K = \frac{[NH_4^+][OH^-]}{[NH_3]} \qquad \text{(EQ 14.8)}$$

**Example 14.11:** What is the equilibrium expression for the decomposition of solid potassium chlorate to form solid potassium chloride and oxygen gas?

**Solution:** First, write a balanced equation for the reaction

$$2KClO_3(s) \rightleftharpoons 2KCl(s) + 3O_2(g)$$

Now write the expression for K using the usual rules but omitting any solids or pure liquids. You can see that both $[KClO_3]$ and $[KCl]$ are not in the equation becasue they are constant.

$$K = [O_2]^3$$

**Example 14.12:** What is the equilibrium expression for the mixture of lead(II) iodide in water?

**Solution:** $PbI_2$ is one of those compounds classified as *insoluble*. This means that very few ions will be formed, but *some* will. The rules for writing K let you know that the solid is not included from the balanced equation

$$PbI_2 \rightleftharpoons Pb^{2+}(aq) + 2I^-(aq) \qquad \text{(EQ 14.9)}$$

Also note that the equilibrium constant has the subscript *sp* because this special case of K refers to **solubility product**.

$$K_{sp} = [Pb^{2+}][I^-]^2$$

As always, a small equilibrium constant means that the equation lies to the left. Small $K_{sp}$ values mean very low solubility. Some common $K_{sp}$ values are summarized in the following table. You can compare solubilities of compounds by examining $K_{sp}$ (if the number of ions is the same).

Table 14.5 Some Substances and Their Solubility Products

| Compound | $K_{sp}$ | Compound | $K_{sp}$ |
|---|---|---|---|
| AgI | $1.5 \times 10^{-16}$ | BaSO$_4$ | $1.1 \times 10^{-10}$ |
| AgBr | $3.3 \times 10^{-13}$ | SnI$_2$ | $1.0 \times 10^{-4}$ |
| AgCl | $1.8 \times 10^{-10}$ | PbSO$_4$ | $1.8 \times 10^{-8}$ |
| AgCN | $1.2 \times 10^{-16}$ | CoCO$_3$ | $2.5 \times 10^{-14}$ |
| NiCO$_3$ | $6.6 \times 10^{-9}$ | Mg(OH)$_2$ | $1.5 \times 10^{-11}$ |

**Example 14.13:** Put the following compounds, AgI, AgCN, and AgBr, in order of increasing solubilities.

**Solution:** Use Table 14.5 to answer this. Remember that a *large negative* exponent is a *small* number.

$$1.2 \times 10^{-16} < 1.5 \times 10^{-16} < 3.3 \times 10^{-13}$$

Therefore,
AgCN < AgI < AgBr

---

### Practice Problems

**14.30** What is the equilibrium expression for the decomposition of solid sodium hydrogen carbonate to form solid sodium carbonate, carbon dioxide gas, and water vapor?

**14.31** What is the equilibrium expression for the reaction of iron(III) oxide with hydrogen gas to form solid iron and water vapor?

**14.32** What is the equilibrium expression for the reaction of solid calcium hydroxide to form aqueous calcium and aqueous hydroxide ions?

**14.33** What is more soluble, CdS ($K_{sp} = 3.6 \times 10^{-29}$) or CdCO$_3$ ($K_{sp} = 2.5 \times 10^{-14}$)?

**14.34** What is more soluble, Ca(OH)$_2$ ($K_{sp} = 2.5 \times 10^{-14}$) or Mg(OH)$_2$ (See Table 14.5)?

---

## 14.8 Le Châtelier's Principle

This states that when a reaction at equilibrium is disturbed by a change in volume, temperature, or the concentration of one of the components, the reaction will shift in a way that tends to counteract the change. This will reestablish equilibrium.

To illustrate this principle, consider the generic reaction in EQ 14.10.

$$2A(g) \rightleftharpoons 3B(g) \quad\quad\quad (EQ\ 14.10)$$

If the reaction is at equilibrium and more A is added to the reaction mixture, then the reaction shifts to the right (forms more B) because this will reestablish K.

If A is removed from the reaction mixture, then the reaction will shift to the left to reestablish K.

Consider the reaction in EQ 14.10 if the volume is increased. This would decrease the pressure that the reaction is under. With reduced pressure, the number of gas molecules would increase. This would shift the reaction to the right (3 molecules of gas versus 2 on the left), and more product would be present in the vessel.

The opposite shift would occur if the volume were decreased (pressure increased).

Consider a change in temperature for the reaction. You remember from earlier data in this chapter that K will vary with temperature. Remember that heat can be either a product or a reactant in a reaction. This is summarized in Table 14.6.

**Table 14.6** Summary of Heat in Reactions

| Reaction Type | Heat | Heat in Reaction |
|---|---|---|
| Endothermic | Absorbed | Reactant |
| Exothermic | Liberated | Product |

These two types of reactions are shown pictorially below.

**Figure 14.2** Reaction Coordinates for Exothermic (**A**) and Endothermic (**B**) Reactions

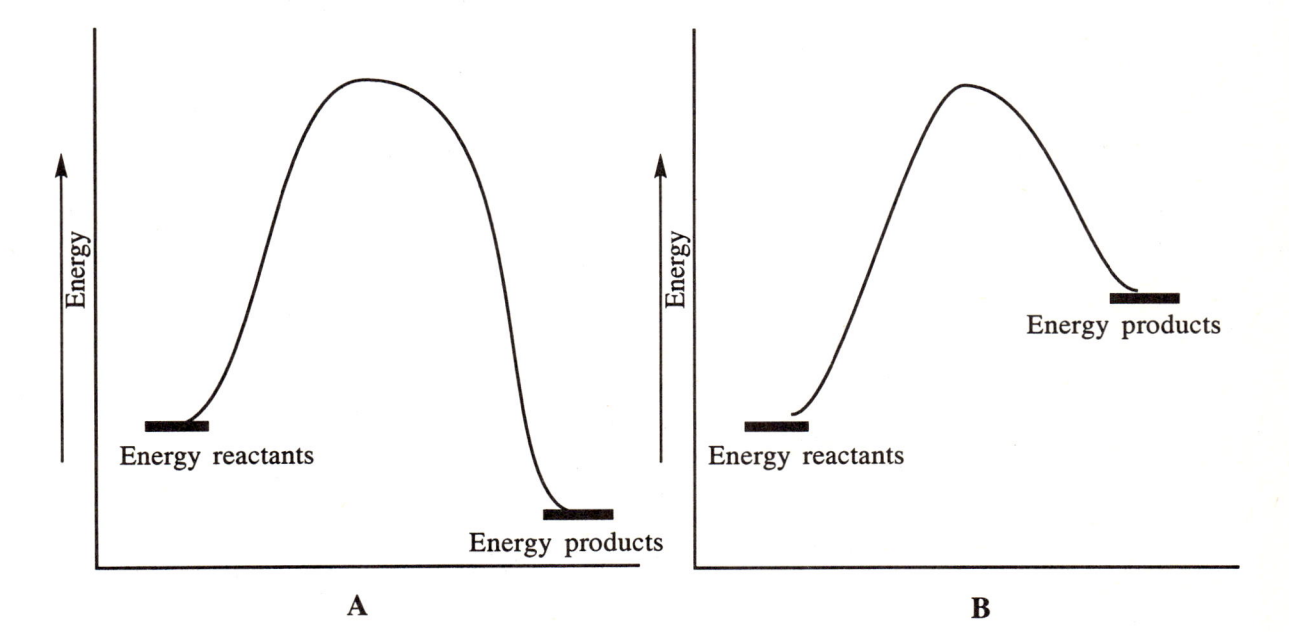

A

B

An increase in temperature for a reaction means that heat is added as a **reactant** for **endothermic** reactions and as a **product** for **exothermic** reactions. For an endothermic reaction, this will shift the reaction to the **right**. For an exothermic reaction, raising the temperature (adding heat) will shift the reaction to the **left**. The opposite occurs if heat is removed.

Lastly, consider the addition of a catalyst to the reaction. This substance will increase the rate of the reaction in both the forward and reverse directions but will not affect K.

**Table 14.7** Summary of Changes in Reaction Conditions

| Factor Changed | Direction of Change | Direction of Reaction | Direction of Change | Direction of Reaction |
|---|---|---|---|---|
| Pressure | Increase | Toward side with fewer gas molecules | Decrease | Toward side with more gas molecules |
| Volume | Increase | Toward side with more gas molecules | Decrease | Toward side with fewer gas molecules |
| Concentration | Increase | Away from side where increase occurred | Decrease | Toward side where decrease occurred |
| Catalyst | Increase | No change in K | Decrease | No change in K |
| Temperature | Increase | Away from side where heat was added. | Decrease | Toward side where heat was removed |

**Example 14.14:** For the reaction below (which is exothermic), what is the effect of
    A. Adding more oxygen?
    B. Removing ozone?
    C. Raising the temperature?
    D. Reducing the volume?

$$2O_3(g) \rightleftharpoons 3O_2(g)$$

**Solution:** Adding more oxygen would shift the reaction to the **left**.
Removing ozone ($O_3$) would shift the reaction to the **left**.
Raising the temperature would add more heat, and since this reaction is exothermic, it would add heat to the product side of the reaction, which would shift the reaction to the **left**.
Reducing the volume would favor fewer molecules (higher pressure), thus the reaction would shift to the **left**.

---

**Practice Problems**

For problems 14.35-14.37, use the following endothermic reaction to determine the effects of the action indicated.

$$2SO_3(g) \rightleftharpoons 2SO_2(g) + O_2(g)$$

**14.35** Adding sulfur trioxide.

**14.36** Raising the temperature.

> **14.37** Removing oxygen.
>
> For problems 14.38-14.40, use the following exothermic reaction to determine the effects of the action indicated.
>
> $3H_2(g) + N_2(g) \rightleftharpoons 2NH_3(g)$
>
> **14.38** Lowering the temperature.
>
> **14.39** Decreasing the volume.
>
> **14.40** Adding a catalyst.

## Integration of Multiple Skills

This section is not necessary to master Chapter 14. However, you may find the following type of problem an interesting challenge.

**Example 14.15:** How many grams of AgI can you dissolve in a liter of solution?

> **Solution:** Look at the $K_{sp}$ for AgI in Table 14.5. This means that you can calculate the concentration of each ion that is dissolved.
>
> $$K_{sp} = 1.5 \times 10^{-16} = [Ag^+][I^-]$$
>
> Since $[Ag^+] = [I^-]$, you can take the square root of $1.5 \times 10^{-16}$, which is $1.2 \times 10^{-8}$.
>
> This means that the concentration of dissolved ions is $1.2 \times 10^{-8} M$.
>
> Now look at the definition of molarity, which means there are $1.2 \times 10^{-8}$ mol/L of dissolved AgI. Use this with the molecular weight, which is 234.8 g/mol to determine the number of grams in 1 liter.
>
> $$\frac{1.2 \times 10^{-8} \text{ mol AgI}}{\text{L soln}} \times \frac{234.8 \text{ g AgI}}{\text{mol AgI}} = 2.8 \times 10^{-6} \text{ g/L soln}$$

> **Practice Problems**
>
> **14.41** How many grams of nickel(II) carbonate can you dissolve in 1 L of solution?
>
> **14.42** How many grams of silver chloride can you dissolve in 1 L of solution?

# Self-Exam

**I. (4 ea) Write T (true) or F (false) in each corresponding blank.**

_____ 1. Temperature increase and concentration increase can speed up reactions.

_____ 2. A catalyst increases the rate of a reaction by lowering the activation energy.

_____ 3. According to collision theory, molecules must collide with a minimum amount of energy to react.

_____ 4. An equilibrium constant changes when the concentration changes.

For questions 5-10, use

$$2NO_2(g) \rightleftharpoons N_2O_4(g)$$

_____ 5. The equilibrium constant expression for this reaction is $K = \dfrac{[N_2O_4]}{[NO_2]^2}$.

_____ 6. An increase in the temperature will not change K.

_____ 7. An increase in volume will shift the reaction to the right.

_____ 8. An increase in temperature will increase the rate of forward reaction and decrease the rate of the reverse reaction (the reaction is exothermic).

_____ 9. An increase in $N_2O_4$ concentration will shift the reaction to the right.

_____ 10. Removal of $N_2O_4$ from the mixture will shift the reaction to the right.

**II. (4 ea) Multiple Choice. Place the letter of the BEST answer in the corresponding blank.**
For problems 1, 2, and 9, use the following exothermic reaction:

$$3H_2(g) + N_2(g) \rightleftharpoons 2NH_3(g)$$

_____ 1. Which will cause more product to form?
  A. Increase P      B. Decrease T      C. Increase $[H_2]$
  D. Remove $NH_3$   E. All will

_____ 2. Which will cause product to form faster?
  A. Decrease P      B. Decrease T      C. Increase $[H_2]$
  D. Remove $N_2$    E. All will

_____ 3. Which of the following is an example of a homogeneous reaction?
  A. $2NO_2(g) \rightleftharpoons N_2O_4(g)$
  B. $CaCO_3(s) \rightleftharpoons CaO(s) + CO_2(g)$
  C. $2H_2O(l) \rightleftharpoons 2H_2(g) + O_2(g)$
  D. $PbCl_2(s) \rightleftharpoons Pb^{2+}(aq) + 2Cl^-(aq)$
  E. $NH_3(g) + 2H_2O(g) \rightleftharpoons NH_4^+(aq) + OH^-(aq)$

## Reaction Rates and Chemical Equilibrium 247

_____ 4. Which is an example of a heterogeneous reaction?

  A. $2SO_2(g) + O_2(g) \rightleftharpoons 2SO_3(g)$

  B. $CaCO_3(s) \rightleftharpoons CaO(s) + CO_2(g)$

  C. $2H_2O(g) + CO(g) \rightleftharpoons CH_3OH(g)$

  D. $2H_2(g) + O_2(g) \rightleftharpoons H_2O(g)$

  E. $PCl_3(g) + Cl_2(g) \rightleftharpoons PCl_5(g)$

_____ 5. The reaction to form ammonia has a value for K of 0.50 at 400°C and 0.012 at 600°C. From this information which of the following conclusions can you make?
  A. The reaction is exothermic.     B. The reaction is endothermic.
  C. The reaction must have a catalyst.     D. The rate of the reaction is very fast.
  E. The rate of the reaction is very slow.

_____ 6. For the reaction of carbon disulfide gas with hydrogen gas to produce dihydrogen monosulfide gas and methane ($CH_4$) gas, what is the correct expression of the equilibrium constant, K?

  A. $\dfrac{[CS_2][H_2]}{[CH_4][H_2S]}$   B. $\dfrac{[CH_4][H_2S]}{[CS_2][H_2]}$   C. $\dfrac{[CH_4][H_2S]^2}{[CS_2][H_2]^4}$

  D. $\dfrac{[CS_2][H_2]^4}{[CH_4][H_2S]^2}$   E. $\dfrac{[CH_4]}{[CS_2][H_2]^4}$

_____ 7. What is the value of K in the following if $[N_2O_4] = 0.82\ M$ and $[NO_2] = 0.57\ M$?

$$2NO_2(g) \rightleftharpoons N_2O_4(g)$$

  A. 2.5     B. 1.4     C. 0.70     D. 0.42     E. 0.40

_____ 8. Find the value for $[S^{2-}]$ given $K = 2.56 \times 10^{-10}$ and $[Mn^{2+}] = 1.6 \times 10^{-5}\ M$.

$$PCl_3(g) + Cl_2(g) \rightleftharpoons PCl_5(g)$$

  A. $2.6 \times 10^{-10}\ M$   B. $1.6 \times 10^{-5}\ M$   C. $1.3 \times 10^{-10}\ M$
  D. $1.3 \times 10^{-5}\ M$   E. $2.6 \times 10^{-5}\ M$

_____ 9. Which statement(s) is(are) true about exothermic reactions?
  I. Heat is a product.
  II. They *always* have a lower activation energy than endothermic reactions.
  III. They *always* are faster than endothermic reactions.
  IV. The reactions will yield more product at lower temperatures.

  A. All     B. I and II     C. I, II, and III     D. I and IV     E. None

_____ 10. For the reaction of iron(III) oxide solid with hydrogen gas to produce iron solid and water vapor, which of the following is the correct equilibrium expression?

  A. $\dfrac{[Fe][H_2O]}{[Fe_2O_3][H_2]}$   B. $\dfrac{[Fe_2O_3][H_2]}{[Fe][H_2O]}$   C. $\dfrac{[H_2O]}{[[H_2]}$

  D. $\dfrac{[Fe]^2[H_2O]^3}{[Fe_2O_3][H_2]^3}$   E. $\dfrac{[[H_2O]^3}{[H_2]^3}$

## 248  Chapter 14

**III. Provide Complete Answers**

1. (5 pts) A balloon is filled with 2 mol H₂ gas and 1 mol O₂ gas. It sits for 6 hours, and no apparent reaction occurs, even though the reaction of hydrogen gas with oxygen gas is known to be very exothermic. When a match is touched to the balloon, it immediately explodes. Explain these observations.

2. (5 pts) The reaction shown below is exothermic. Use a reaction diagram to show the differences in the course of reaction with a catalyst versus one without a catalyst.

$$3H_2(g) + N_2(g) \rightleftharpoons 2NH_3(g)$$

3. (5 pts) What is the concentration of HI given the following reaction and information about equilibrium constant and concentrations of reactants at equilibrium?
K = 56, [H₂] = [I₂] = 0.0037 M.

$$3H_2(g) + I_2(g) \rightleftharpoons 2HI(g)$$

4. (5 pts) Which is more soluble, AgBr ($K_{sp}$ = 3.3 × 10⁻¹³) or AgCl ($K_{sp}$ = 1.8 × 10⁻¹⁰)?

## Answers to Practice Problems

| | | | | | |
|---|---|---|---|---|---|
| **14.1** | mol³/L³ | **14.2** | 90.7 | **14.3** | 0.10 |
| **14.4** | 3.78 × 10⁻³ | **14.5** | 3.07 | **14.6** | B |
| **14.7** | A | **14.8** | A | **14.9** | 100°C |
| **14.10** | Slower rate | **14.11** | Faster rate | | |

**14.12**  $3H_2(g) + N_2(g) \rightarrow 2NH_3(g)$

**14.13**  $2NH_3(g) \rightarrow 3H_2(g) + N_2(g)$

**14.14**  $Hg(O_2)_4 \rightarrow Hb + 4O_2$

**14.15**  $8SO_2 \rightarrow S_8 + 8O_2$;  $S_8 + 8O_2 \rightleftharpoons 8SO_2$

**14.16**  $PCl_3(g) + Cl_2(g) \rightleftharpoons PCl_5(g)$;  $K = \dfrac{[PCl_5]}{[PCl_3][Cl_2]}$

**14.17**  $2SO_2(g) + O_2(g) \rightleftharpoons 2SO_3(g)$;  $K = \dfrac{[SO_3]^2}{[SO_2]^2[O_2]}$

**14.18**  $N_2(g) + 3Cl_2(g) \rightleftharpoons 2NCl_3(g)$;  $K = \dfrac{[NCl_3]^2}{[N_2][Cl_2]^3}$

**Reaction Rates and Chemical Equilibrium** 249

14.19 $K = \dfrac{[CH_4][H_2O]}{[CO][H_2]^3} = \dfrac{(0.037)(0.037)}{(0.0613)(0.184)^3} = 3.59$

14.20 $K = \dfrac{[NO]^2[Cl_2]}{[NOCl]^2} = \dfrac{(0.06)^2(0.03)}{(0.44)^2} = 5.6 \times 10^{-4}$

14.21 $K = \dfrac{[HI]^2}{[H_2][I_2]} = \dfrac{(3.16)^2}{(0.42)(0.42)} = 57$

14.22 $K = \dfrac{[SO_3]^2}{[SO_2]^2[O_2]} = \dfrac{(.230)^2}{(.145)^2(0.0725)} = 34.7$

14.23 K in Example 14.6 for this reaction was 280. K in problem 14.22 was 34.7. The difference is due to temperature. In the former it was 727°C, and in the latter was 877°C. Remember K is temperature dependent.

14.24 $3O_2(g) \rightleftharpoons 2O_3(g)$

14.25 Product forms too slowly.

| | | | | | |
|---|---|---|---|---|---|
| 14.26 | Favors reactants | 14.27 | Right | 14.28 | 0.10 M |
| 14.29 | 0.11 M | 14.30 | K = [CO$_2$][H$_2$O] | | |
| 14.31 | K = [H$_2$O]$^3$/[H$_2$]$^3$ | 14.32 | [Ca$^{2+}$][OH$^-$]$^2$ | | |
| 14.33 | CdCO$_3$ | 14.34 | Mg(OH)$_2$ | | |
| 14.35 | To right | 14.36 | To right | | |
| 14.37 | To right | 14.38 | To right | 14.39 | To right |
| 14.40 | No change in K | 14.41 | 0.0096 g NiCO$_3$ | 14.42 | 0.0019 g AgCl |

## Answers to Self-Exam

**I. T/F**
1. T  2. T  3. T  4. F  5. T  6. F  7. F  8. F  9. F  10. T

**II. MC**
1. E  2. C  3. A  4. B  5. A  6. C  7. A  8. B  9. D  10. E

**III. Complete Answers**
1. The match provides enough energy to get reacting molecules over the activation barrier. After the initial spark, there is enough energy released by the exothermic nature of the reaction to keep the reaction going.

## 250　Chapter 14

2.

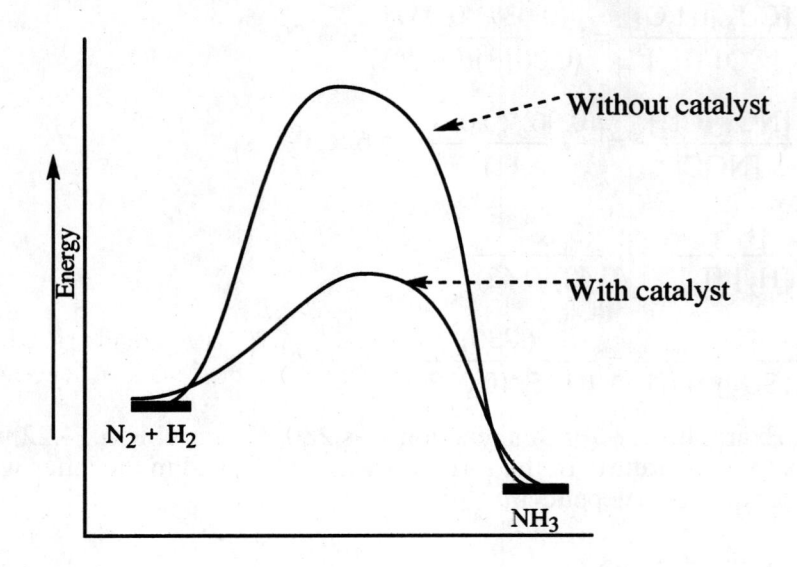

```
          Energy
            ↑
            |         ___
            |        /   \          ← Without catalyst
            |       /     \
            |      /  ___  \
            |     /  /   \  \       ← With catalyst
            |    /  /     \  \
            |   /__/       \__\
            |  N₂ + H₂        \___
            |                  NH₃
            |_____
```

3. [HI] = 0.028 $M$

4. AgCl is more soluble because $1.8 \times 10^{-10} > 3.3 \times 10^{-13}$

# 15
# Acids and Bases

This chapter illustrates the many everyday applications of acid-base chemistry. The chapter has important conceptual ideas that should be mastered. There are also many calculations that are slightly different from those you have encountered in the past.

Many examples in your textbook and in this *Study Guide* are provided to help you master the material.

Keys for success in this chapter:

- Learn the definitions.
- Learn to manipulate logarithmic functions.
- Apply your understanding of concepts to qualitative solutions.
- Practice many problems.

## Summary of Verbal Knowledge

**acid:** A substance that produces $H^+$ ions (protons) when it is dissolved in water.
**base:** A substance that produces $OH^-$ ions (hydroxide) when it is dissolved in water.
**salt:** An ionic compound containing the cation from a base and anion from an acid.
**molecular equation:** This is one in which each substance is written as if it were a molecular substance, even though some may actually exist as ions.
**total ionic equation:** This is one that shows all the ions in solution.
**net ionic equation:** This is one that shows only the ions that take part in the reaction.
**hydronium ion:** This is the ion that $H^+$ exists as in water, $H_3O^+$.
**conjugate acid-base pair:** Two substances (one an acid and the other a base) in an acid-base reaction that differ by gain or loss of a proton.
**acid strength:** This is a measure of the degree to which an acid will ionize. In other words, it is the degree to which the reaction lies to the right.
**monoprotic acid:** This is a molecule that produces one proton per molecule, which can be transferred to a base.
**diprotic acid:** This is a molecule that produces two protons per molecule, which can be transferred to a base.
**strong acid:** This is an acid that completely dissociates in water to $H_3O^+$ ions and anions.
**weak acid:** This is an acid that dissociates only to a small degree in water.
**strong base:** This is a hydroxide compound that totally dissociates in water.
**weak base:** This is a compound that does not totally ionize in water, and it leads to a small concentration of $OH^-$ ions.
**amphoteric substance:** One that can behave as either an acid or a base.

**self-ionization:** This is a process where two identical molecules react to give one anion and one cation.
**ion-product constant for water:** The equilibrium value of the ion product, $[H_3O^+][OH^-]$.
**pH:** The negative log of the $[H_3O^+]$.
**logarithm of a number:** The power to which 10 must be raised to equal that number.
**acid-base indicator:** A weak acid whose solution will change color within a small pH range and indicate the pH of the solution by the color.
**buffer solution:** A mixture that contains a weak acid and the salt of its conjugate base or a weak base and the salt of its conjugate acid. It is resistant to changes in pH when small amounts of acid or base are added to it.

## Review of Mathematical and Calculator Skills

The following examples illustrate some new applications that will be helpful in making calculations with acids and bases.

**Example 15.1:** What is the log of $3.56 \times 10^{-6}$? What is the -log of $3.56 \times 10^{-6}$?

**Solution:** Both these problems start the same way. If you have a *scientific* calculator, the solution will look like this:

| Operation | Display |
|---|---|
| 1. Press [log] | log  0 |
| 2. Press [3.56] [EXP] | log  $3.56^{00}$ |
| 3. Press [6] [+/-] | log  $3.56^{-06}$ |
| 4. Press [=] | -5.448550002 |

Notice that the calculator gives more than the correct significant figures. The number above should be rounded off to **-5.45**.

The negative log of $3.56 \times 10^{-6}$ simply equals $-(-5.45) = $ **5.45**.

If you are using a *graphic* calculator, the solution will look as follows.

| Operation | Display |
|---|---|
| 1. Press [log] | log_ |
| 2. Press [3.56] [EXP] | log 3.56E |
| 3. Press [-] [6] | log 3.56E-6 |
| 4. Press [EXE] | -5.448550002 |

Again, the calculator gives more than the correct significant figures. The number above should be rounded off to **-5.45**.

**Example 15.2:** What is the inverse log of −5.44?

**Solution:** Inverse log is equal to $\log^{-1}$. It is simply the "undoing" of the function in the first question. The solution on the *scientific* calculator looks as follows:

| Operation | Display |
|---|---|
| 1. Press [shift] [log] | 10ˣ   0 |
| 2. Press [5.45] [+/−] | 10ˣ   -5.45 |
| 3. Press [=] | 3.548133892⁻⁶ |

There are only three significant figures in the original data, so there can only be three in the solution. Written as scientific notation, the answer is $3.55 \times 10^{-6}$.

The solution on the *graphic* calculator looks as follows:

| Operation | Display |
|---|---|
| 1. Press [shift] [log] | 10_ |
| 2. Press [-] [5.45] | 10-5.45 |
| 3. Press [EXE] | 0.00000354813 |

Written as scientific notation, the answer is $3.55 \times 10^{-6}$.

**Example 15.3:** Solve: $1.00 \times 10^{-14} \div 6.25 \times 10^{-5}$.

**Solution:** When you enter exponents into the calculator, you don't need to enter the × 10. Refer to Example 2.4 if you forgot how to do this.

$$\frac{1.00 \times 10^{-14}}{6.25 \times 10^{-5}} = 1.60 \times 10^{-10}$$

---

**Practice Problems**

15.1 What is the log of $10^{-2}$?

15.2 What is the log of $10^{-13}$?

15.3 What is the -log of $10^{-2}$?

15.4 What is the -log of $10^{-13}$?

15.5 What is -log $4.52 \times 10^{-5}$?

15.6 What is the -log of $8.91 \times 10^{-12}$?

15.7 What is the inverse log of -7.25?

15.8 What is the inverse log of -1.29?

15.9 What is quotient of $1.00 \times 10^{-14} \div 4.29 \times 10^{-12}$?

---

## Application of Skills and Concepts

### 15.1 The Arrhenius Theory of Acids and Bases

This section provides some historical background of acids and bases and early theories of these types of compounds.

**Table 15.1** Characteristics of Acids and Bases

| Substance | Taste | Found in Nature | Arrhenius |
|---|---|---|---|
| Acid | Sour | Foods | Produces protons ($H^+$) |
| Base | Bitter | Soaps | Produces hydroxide ions ($OH^-$) |

**Example 15.4:** The juice of lemons tastes sour. Is it an acid or a base?

**Solution:** Use Table 15.1 to determine that lemons are acidic.

## Acids and Bases

**Practice Problems**

**15.10** Dish soap is bitter tasting. Is it an acid or a base?

**15.11** Vinegar is sour tasting. Is it an acid or a base?

**15.12** A solution of sodium hydroxide feels slippery. Is it an acid or base?

### 15.2 Neutralization, Salts, and Net Ionic Equations

Before beginning this section, review the definitions at the beginning of the chapter.

**Example 15.5:** Write a balanced equation for the reaction of hydrobromic acid solution with magnesium hydroxide solution. Label the substances in the reaction as acid, base, or salt.

**Solution:** Remember how to transform a word description of a reaction to formulas (see Chapter 5). After writing the correct formulas and their states, balance the reaction.

$$2HBr(aq) + Mg(OH)_2(aq) \rightarrow 2H_2O(l) + MgBr_2(aq)$$
$$\text{acid} \qquad \text{base} \qquad\qquad\qquad \text{salt}$$

**Example 15.6:** Write a total ionic and a net ionic equation for the reaction of hydrobromic acid with magnesium hydroxide.

**Solution:** The sample equations in Table 15.2 show how to transform a molecular equation into a total ionic equation and a net ionic equation.

**Total ionic equation**

$$2H^+(aq) + 2Br^-(aq) + Mg^{2+}(aq) + 2OH^-(aq) \rightarrow 2H_2O(l) + Mg^{2+}(aq) + 2Br^-(aq)$$

Ions that are the same on both sides of the equation are canceled to yield:

**Net ionic equation**

$$2H^+(aq) + 2OH^-(aq) \rightarrow 2H_2O(l)$$

**Table 15.2** Types of Equations

| Equation Type | Characteristics | Example |
|---|---|---|
| Molecular | All molecules are included and ionic compounds are written as molecules. | $2NaOH(aq) + H_2SO_4(aq) \rightarrow 2H_2O(l) + Na_2SO_4(aq)$ |
| Total ionic | All ions are written out in the condition they are in. | $2Na^+(aq) + 2OH^-(aq) + 2H^+(aq) + SO_4^{2-}(aq) \rightarrow 2H_2O(l) + 2Na^+(aq) + SO_4^{2-}(aq)$ |
| Net ionic | Only the species involved in the reaction are written out. | $2OH^-(aq) + 2H^+(aq) \rightarrow 2H_2O(l)$ |

> **Practice Problems**
>
> **15.13** Write molecular, total ionic, and net ionic equations for the reaction of aqueous sulfuric acid and aqueous potassium hydroxide.
>
> **15.14** Write molecular, total ionic, and net ionic equations for the reaction of aqueous acetic acid and aqueous aluminum hydroxide.
>
> **15.15** Write molecular, total ionic, and net ionic equations for the reaction of aqueous nitric acid and aqueous lithium hydroxide.

## 15.3 The Brønsted-Lowry Theory

This section will look at an expanded definition of acids and bases in water solutions. It will also look at acids and bases as conjugate pairs in a chemical reaction.

**Table 15.3** Summary of Acid-Base Definitions

| Subsistence | Arrhenius | Brønsted-Lowry |
|---|---|---|
| Acid | Substance that produces protons | Proton donor |
| Base | Substance that produces hydroxide ions | Proton acceptor |

The concept of an acid-base conjugate pair is easy to understand if you remember that when acids dissolve in water, they produce protons ($H^+$ ions). In aqueous solutions these ions are not isolated. Rather, they are attached to a water molecule, forming the **hydronium** ion ($H_3O^+$) as shown in EQ 15.1.

$$HA + H_2O \rightleftharpoons H_3O^+ + A^- \qquad \text{(EQ 15.1)}$$

Each side of EQ 15.1 has one acid molecule and one base molecule. They are different from their conjugate partners by only one proton ($H^+$). With this definition in mind, it is easy to pick out the conjugate pairs. Follow these simple rules:
- Identify the acid species on each side of the equilibrium (use definition in Table 15.3).
- Identify the base species on each side of the equilibrium (use definition in Table 15.3).
- Identify the conjugate pairs by noting the species that differ by a proton.

In EQ 15.1, HA and $H_3O^+$ are acids.
  Going from left to right, HA *donates* a proton to water.
  Going from right to left, $H_3O^+$ *donates* a proton to $A^-$.

In EQ 15.1, $H_2O$ and $A^-$ are bases.
  Going from left to right, $H_2O$ accepts a proton.
  Going from right to left, $A^-$ accepts a proton.

In EQ 15.1, HA and $A^-$ are a conjugate pair (they differ by a proton).
In EQ 15.1, $H_3O^+$ and $H_2O$ are a conjugate pair (they differ by a proton).

**Example 15.7:** What is the conjugate acid of $Cl^-$? What is the conjugate base of $HNO_3$?

**Solution:** Remember that conjugate pairs only differ by a proton. In a pair, the acid has one more proton than its conjugate base.

$HCl$ is the conjugate acid of $Cl^-$.
$NO_3^-$ is the conjugate base of $HNO_3$.

**Example 15.8:** Identify the acids and bases in the reaction below by their conjugate pairs.

$$H_3PO_4 + H_2O \rightleftharpoons H_2PO_4^- + H_3O^+$$

**Solution:** Remember in acid-base equilibria, there is an acid on each side of the equation and a base on each side of the equation.

In this equation, $H_3PO_4$ and $H_3O^+$ are acids.
In this equation, $H_2O$ and $H_2PO_4^-$ are bases.

$H_3PO_4$ is the conjugate acid of $H_2PO_4^-$.
$H_3O^+$ is the conjugate acid of $H_2O$.

**Example 15.9:** Identify the acids and bases in the reaction below by their conjugate pairs.

$$CH_3NH_2 + H_2O \rightleftharpoons CH_3NH_3^+ + OH^-$$

**Solution:** There is an acid on each side of the equation and a base on each side of the equation.

In this equation, $H_2O$ and $CH_3NH_3^+$ are acids.
In this equation, $CH_3NH_2$ and $OH^-$ are bases.

$H_2O$ is the conjugate acid of $OH^-$.
$CH_3NH_3^+$ is the conjugate acid of $CH_3NH_2$.

## Practice Problems

**15.16** What is the conjugate acid of $HSO_3^-$?

**15.17** What is the conjugate base of $HCO_3^-$?

**15.18** What is the conjugate acid of $HCO_3^-$?

**15.19** What is the conjugate base of $NH_3$?

**15.20** What is the conjugate acid of $NH_3$?

**15.21** Identify the acid and base in each side of the equilibrium for the reaction of hydrofluoric acid and water.

> **15.22** Write the equilibrium equation for the reaction of $NH_3$ and water. Identify the acid and base on each side of the equilibrium.
>
> **15.23** For the equation below, identify the acid and base on each side of the equilibrium.
>
> $HC_2H_3O_2 + NO_2^- \rightleftharpoons HNO_2 + C_2H_3O_2^-$
>
> **15.24** For the equation below, identify the acid and base on each side of the equilibrium.
>
> $F^- + NH_4^+ \rightleftharpoons HF + NH_3$

## 15.4 The Relative Strengths of Acids and Bases

This section examines the **strength** of acids and bases. This should not be confused with **concentration** of acids and bases.

- The **strength** of an acid or base depends on how completely it ionizes in water.
- The **concentration** of an acid or base depends on its molarity.

Very *strong* acids completely ionize to hydronium ions and anions. An example is hydrochloric acid. This is shown in EQ 15.2.

$$HCl(aq) + H_2O(l) \rightarrow H_3O^+(aq) + Cl^-(aq) \qquad \text{(EQ 15.2)}$$

Notice that only one arrow (in the left to right direction) is used for a strong acid, because the reverse reaction is so small that it is negligible. This reaction has a very, very large K.

*Weak* acids do not completely ionize, and so the opposing equilibrium arrows are drawn. An example is phosphoric acid, shown in EQ 15.3.

$$H_3PO_4(aq) + H_2O(l) \rightleftharpoons H_3O^+(aq) + H_2PO_4^-(aq) \qquad \text{(EQ 15.3)}$$

Weak acids vary in strength. The larger the K for the acid, the stronger it is.

Some acids and their conjugate bases are listed in Figure 15.1. The strong acids all have very large equilibrium constants, while the weaker ones are listed in order of *decreasing* strength.

Notice that as an acid becomes *weaker*, its conjugate base becomes *stronger*.

The same treatment can be used with bases. A *strong* base is shown in EQ 15.4.

$$NaOH(s) + H_2O(l) \rightarrow Na^+(aq) + OH^-(aq) + H_2O(l) \qquad \text{(EQ 15.4)}$$

A weak base in water only partially produces $OH^-$ ions, so it is written with opposing arrows. The most common example of this is a solution of ammonia in water, shown in EQ 15.5.

$$NH_3(g) + H_2O(l) \rightleftharpoons NH_4^+(aq) + OH^-(aq) \qquad \text{(EQ 15.5)}$$

**Figure 15.1 Acids and Bases and Their Relative Strengths**

| Strong Acids | Their Conjugate Bases |
|---|---|
| HCl | Cl⁻ |
| HBr | Br⁻ |
| HI | I⁻ |
| HNO₃ | NO₃⁻ |
| HClO₄ | ClO₄⁻ |
| H₂SO₄ | HSO₄⁻ |

| Weak Acids | Their Conjugate Bases |
|---|---|
| H₃O⁺ | H₂O |
| HSO₄⁻ | SO₄²⁻ |
| H₂SO₃ | HSO₃⁻ |
| HNO₂ | NO₂⁻ |
| HF | F⁻ |
| HC₂H₃O₂ | C₂H₃O₂⁻ |
| H₂CO₃ | HCO₃⁻ |
| HSO₃⁻ | SO₃²⁻ |
| NH₄⁺ | NH₃ |
| HCO₃⁻ | CO₃²⁻ |
| H₂S | HS⁻ |
| H₂O | OH⁻ |

(Left arrow: Increasing Strength — upward for acids; Right arrow: Increasing Strength — downward for bases)

**Example 15.10:** Use Figure 15.1 to determine which acid, HF or HCl, is stronger?

**Solution:** HCl is in the group of strong acids and HF is in the group of weak acids. Therefore hydrochloric acid is stronger than hydrobromic acid.

---

## Practice Problems

**15.25** Which is a stronger acid, acetic acid or sulfurous acid?

**15.26** Which is a stronger base, Br⁻ or NO₂⁻?

**15.27** Which is a stronger acid, water or NH₄⁺?

**15.28** Which is a stronger base, water or F⁻?

## 15.5 Water: An Acid and a Base

This section describes an **amphoteric** substance. This is demonstrated by EQ 15.6. Two molecules of water can combine to form one hydronium ion and one hydroxide ion. Notice that this equation is written with the opposing arrows. The K value for this equation is very small; thus the reaction favors the left side of the equation.

$$H_2O(l) + H_2O(l) \rightleftharpoons H_3O^+(aq) + OH^-(aq) \qquad \text{(EQ 15.6)}$$

One molecule of water in EQ 15.6 acts as a base and the other acts as an acid.

**Example 15.11:** Write an equation where water acts as an acid; one where it acts as a base.

**Solution:** Use the definitions of acids and bases. There are many examples of these types of reactions.

Water as an acid: $H_2O + CH_3O^- \rightleftharpoons OH^- + CH_3OH$

Water as a base: $H_2O + H_2CO_3 \rightleftharpoons H_3O^+ + HCO_3^-$

---

**Practice Problems**

**15.29** Show an example of water acting as an acid.

**15.30** Show an example of water acting as a base.

---

## 15.6 Acidic, Neutral, and Basic Solutions

Water autoionizes as in EQ 15.6. Remember that the equilibrium constant for heterogeneous systems omits pure solvents. In this case, water is not included in K. This equilibrium is special, so it has a subscript $w$ for water. For pure water and water solutions, $K_w$ is a constant as shown in EQ 15.7.

$$K_w = [H_3O^+][OH^-] = 1.00 \times 10^{-14} \qquad \text{(EQ 15.7)}$$

In pure water, $[H_3O^+] = [OH^-] = 1.00 \times 10^{-7} \, M$

Check with your calculator to verify that $(1.00 \times 10^{-7})(1.00 \times 10^{-7}) = 1.00 \times 10^{-14}$

**Example 15.12:** What is $K_w$ if $[H_3O^+] = 1.00 \times 10^{-2} \, M$ and $[OH^-] = 1.00 \times 10^{-12} \, M$?

**Solution:** Use your calculator to verify the product is $1.00 \times 10^{-14}$.

**Example 15.13:** An aqueous solution has $[H_3O^+] = 2.37 \times 10^{-5} \, M$. Is it an acid, a base, or neutral? What is $[OH^-]$?

**Solution:** Use the rules in Table 15.4 to arrive at the answer. Since the concentration of hydronium ion is $>1.00 \times 10^{-7}$, it is an acidic solution.

The second part of the solution involves using the definition of $K_w$ and solving for the unknown, [OH$^-$]. The solution is shown below.

$$[OH^-] = \frac{K_w}{[H_3O^+]} = \frac{1.00 \times 10^{-14}}{2.37 \times 10^{-5}} = 4.22 \times 10^{-10} \, M$$

**Table 15.4** Concentrations in Acidic, Basic, and Neutral Solutions

| Solution | [H$_3$O]$^+$ | [OH$^-$] |
|---|---|---|
| Acid | $> 1.00 \times 10^{-7}$ | $< 1.00 \times 10^{-7}$ |
| Base | $< 1.00 \times 10^{-7}$ | $> 1.00 \times 10^{-7}$ |
| Neutral | $= 1.00 \times 10^{-7}$ | $= 1.00 \times 10^{-7}$ |

## Practice Problems

**15.31** What is $K_w$ if [OH$^-$] = $2.60 \times 10^{-5}$?

**15.32** What is [H$_3$O$^+$] if [OH$^-$] = $1.7 \times 10^{-10}$?

**15.33** What is [OH$^-$] if [H$_3$O$^+$] = $2.34 \times 10^{-7}$?

**15.34** What type of solution has [OH$^-$] = $4.56 \times 10^{-9}$?

**15.35** What type of solution has [H$_3$O$^+$] = $3.79 \times 10^{-7}$?

**15.36** What type of solution has [OH$^-$] = $7.54 \times 10^{-2}$?

**15.37** What type of solution has [H$_3$O$^+$] = $8.92 \times 10^{-3}$?

## 15.7 pH

This section looks at practical ways to measure acidity or basicity. There are two skills you need to master for this section.

- Determine pH, given the [H$_3$O$^+$].
- Determine [H$_3$O$^+$], given the pH.

For the first of these you need to know the definition of pH. It is given in EQ 15.7.

$$\text{pH} = -\log[H_3O^+] \tag{EQ 15.7}$$

**Example 15.14:** What is the pH of a solution that has [H$_3$O$^+$] = $5.62 \times 10^{-6}$ $M$?

**Solution:** Determine the log of $5.62 \times 10^{-6}$ (which will be a negative number).

$$\log 5.62 \times 10^{-6} = -5.25$$

Then find the negative of this number.

$$-(-5.25) = 5.25 = \text{pH}$$

For the second skill, you can use EQ 15.8 to convert from concentration to pH.

$$10^{-\text{pH}} = [H_3O^+] \quad \quad (\text{EQ 15.8})$$

This is really the inverse of the pH. See the samples at the beginning of this chapter. Go through the steps in the next example.

**Example 15.15:** What is $[H_3O^+]$ of a solution with a pH = 4.2?

**Solution:** Determine the inverse log of -4.2. This is the same as $10^{-4.2}$.

$$10^{-4.2} = 6.31 \times 10^{-5} \, M$$

Table 15.5 Solutions and Their pH's

| Solution | $[H_3O]^+$ (M) | pH |
|---|---|---|
| Acid | $> 1.00 \times 10^{-7}$ | < 7 |
| Base | $< 1.00 \times 10^{-7}$ | > 7 |
| Neutral | $= 1.00 \times 10^{-7}$ | = 7 |

Notice the relationship between pH and $[H_3O^+]$. One important thing to understand about the pH scale is that each unit represents a factor of 10. A solution with pH = 6 is 10 times as acidic as one with pH = 7.

**Example 15.16:** How much more acidic is a solution of pH = 2 than one of pH = 5?

**Solution:** These two solutions differ by 3 pH units (5 - 2 = 3). Each of these units is a factor of 10. Therefore, pH of 2 is $10^3$ times (or **1000**) as acidic as pH of 5.

## Practice Problems

**15.38** What is the pH of a solution with $[H_3O^+] = 2.92 \times 10^{-3} \, M$?

**15.39** What is the pH of a solution with $[H_3O^+] = 1.65 \times 10^{-9} \, M$?

**15.40** What is [H$_3$O$^+$] of a solution with pH = 6.42? What is [OH$^-$]?

**15.41** A solution with pH = 8.00 is how much more basic than one of pH = 4.00?

## 15.8 Measuring pH

There are several ways of measuring pH.

- Electronically, with a pH meter.
- Chemically, with acid-base indicators.

There are many substances found in nature (which are weak acids) that can be used to determine pH. These substances have various colors at difference pH's.

Table 15.6 Some Indicators Used to Determine pH

| Indicator | Color | at pH | Color | at pH | Color | at pH |
|---|---|---|---|---|---|---|
| Crystal violet | Yellow | < 0.5 | Green | 0.5-1.5 | Blue | > 1.5 |
| Bromphenol blue | Yellow | < 3.5 | Green | 3.5-4.5 | Blue | > 4.5 |
| Methyl orange | Orange | < 4.2 | | | Yellow | > 4.2 |
| Bromcresol green | Yellow | < 4.2 | Green | 4.2-5.0 | Blue | > 5.0 |
| Methyl red | Red | < 5.5 | | | Yellow | > 5.5 |
| Bromcresol purple | Yellow | < 5.8 | | | Purple | > 5.8 |
| Bromthymol blue | Yellow | < 6.2 | Green | 6.2-7.2 | Blue | > 7.2 |
| Phenolphthalein | Colorless | < 8.8 | | | Pink | > 8.8 |
| Thymolphthalein | Colorless | < 9.5 | | | Blue | >9.5 |

Many common foods are acidic, and many household cleaners are basic. Some of these are listed in Table 15.7.

Table 15.7 Some Common Substances and Their pH's

| Substance | pH |
|---|---|
| Pure water | 7.0 |
| Vinegar solution | 2.8 |
| Soda pop | 3.1 |
| Stomach acid | 1.0 |
| Baking soda solution | 8.5 |
| Lye solution (NaOH) | 13.0 |
| Milk of magnesia | 10.5 |
| Tomato juice | 4.7 |
| Lemon juice | 2.5 |
| Ammonia solution | 12.2 |
| Milk | 6.8 |

## Chapter 15

**Example 15.17:** What color would a solution with a pH = 4.5 be with phenolphthalein indicator? With methyl red indicator?

**Solution:** Find the range in Table 15.6.
Since the pH of 4.5 is less than 8.8, it would be colorless with phenolphthalein.
Since the pH of 4.5 is less than 5.5, it would be red with methyl red.

---

### Practice Problems

**15.42** What color would vinegar be in bromthymol blue?

**15.43** What color would vinegar be in crystal violet?

**15.44** What color is NaOH solution with bromcresol purple indicator?

**15.45** What color is pure water with methyl orange indicator?

**15.46** What color is stomach acid with methyl red indicator?

**15.47** What color is baking soda solution with bromphenol blue indicator?

---

### 15.9 Buffer Solutions

These solutions resist changes in pH when small amounts of acid or base are added to them. They can be designed around many optimal pH's. They are made of two types.

- Weak acid and a salt of its conjugate base.
- Weak base and a salt of its conjugate acid.

**Example 15.18:** How would you make a buffer solution with $NH_3$?

**Solution:** This compound is a weak base in water solutions. Therefore, add a salt of its conjugate acid ($NH_4^+$) such as $NH_4Cl$.

---

### Practice Problems

**15.48** What salt might be added to HOCl (hypochlorous acid) to make a buffer solution?

**15.49** What acid might be added to sodium phosphate to make a buffer solution?

**15.50** Give an example of two components of a weak acid buffer using $C_6H_5CO_2^-$ ion.

**15.51** Give an example of a weak base buffer solution using $CH_3NH_2$.

---

### Integration of Multiple Skills

This section will look at three topics that integrate current subjects with each other and previous topics you have encountered in this book.

- Acid strength versus concentration

- Calculation of pH from [OH⁻]
- Calculation of pH from concentration of a strong acid

Examples will illustrate these three ideas.

**Example 15.19:** Compare the strength and concentration of 1 $M$ $H_2SO_4$ to 1 $M$ $H_2SO_3$.

**Solution:** Both of these are the same concentration (1 $M$).
The sulfuric acid is a *strong* acid (see Figure 15.1) and sulfurous is a *weak* acid.

**Example 15.20:** If [OH⁻] is $2.56 \times 10^{-5}$ $M$, what is the pH?

**Solution:** First determine $[H_3O^+] = \dfrac{K_w}{[OH^-]} = \dfrac{1 \times 10^{-14}}{2.56 \times 10^{-5}} = 3.91 \times 10^{-10}$ $M$

Find the pH = -log $3.91 \times 10^{-10}$ = **9.41**

**Example 15.21:** If 2.86 g of HCl is dissolved in 250 mL of solution, what is the pH?

**Solution:** First, calculate the molarity of $[H_3O^+]$. Remember that HCl is a strong acid. Therefore, it completely dissociates, and the molarity of the acid = molarity of $[H_3O^+]$.

$$\dfrac{2.86 \text{ g HCl}}{0.250 \text{ L soln}} \times \dfrac{1 \text{ mol HCl}}{36.5 \text{ g HCl}} = 0.313 \text{ mol/L} = 0.313 \, M$$

Find the pH = -log 0.313 = **0.500**

---

**Practice Problems**

**15.52** Compare the strength and concentration of 1 $M$ $HClO_4$ to 2 $M$ $HClO_4$.

**15.53** Compare the strength and concentration of 2 $M$ HF to 2 $M$ HI.

**15.54** What is the pH of a solution where the [OH⁻] is $7.56 \times 10^{-7}$ $M$?

**15.55** What is the pH of a solution where the [OH⁻] is $3.59 \times 10^{-13}$ $M$?

**15.56** If 2.35 g of NaOH is dissolved in 300. mL of solution, what is the pH?

**15.57** If 0.872 g of $HNO_3$ is dissolved in 150. mL of solution, what is the pH?

---

## Self-Exam

**I. (4 ea) Write T (true) or F (false) in each corresponding blank.**

_____ 1. The conjugate base of water is OH⁻.

_____ 2. Bases have a sour taste.

# 266   Chapter 15

_____3. The salt from hydrochloric acid reacting with sodium hydroxide is NaCl.

_____4. When acid dissolves in water, hydronium ions form.

_____5. HNO$_2$ produces hydroxide ions when dissolved in water.

_____6. A base is a proton acceptor according to Brønsted-Lowry definition.

_____7. Acid-base indicators are weak acids.

_____8. Phosphoric acid is a monoprotic acid.

_____9. The concentration of [OH$^-$] in pure water is $1.00 \times 10^{-14} M$.

_____10. A solution with a pH = 2 is 100,000 times as acidic as one with a pH = 7.

## II. (4 ea) Multiple Choice. Place the letter of the BEST answer in the corresponding blank.

_____1. Which is a base?
   A. HCl   B. H$_3$PO$_4$   C. Ca(OH)$_2$   D. HC$_2$H$_3$O$_2$   E. H$_3$O$^+$

_____2. Which of the following is the conjugate acid of NH$_3$?
   A. NH$_4^+$   B. NH$_3$   C. H$_2$O   D. H$_3$O$^+$   E. NH$_2^-$

_____3. For the reaction shown below, which is(are) an acid?

$$CH_3OH + OH^- \rightleftharpoons CH_3O^- + H_2O$$

   A. OH$^-$   B. CH$_3$O$^-$   C. H$_2$O   D. None   E. All

_____4. For the reaction shown below, which is(are) a base?

$$H_2O + CN^- \rightleftharpoons OH^- + HCN$$

   A. CN$^-$   B. H$_2$O   C. HCN   D. OH$^-$   E. A & D

_____5. Which of the following is a weak acid?
   A. HCl   B. HF   C. HI   D. OH$^-$   E. NH$_2^-$

_____6. Which of the following solutions is most basic?
   A. pH = 2.0   B. pH = 5.2   C. pH = 7.0   D. pH = 10.5   E. pH = 13.5

_____7. Which of the following is the strongest acid?
   A. H$_2$O   B. NH$_2^-$   C. NH$_3$   D. H$_2$SO$_4$   E. H$_2$SO$_3$

_____8. What is [OH$^-$] in a solution where [H$_3$O$^+$] = $2.79 \times 10^{-9} M$?
   A. $2.72 \times 10^{-9} M$   B. $1.00 \times 10^{-13} M$   C. $3.58 \times 10^{-6} M$
   D. $1.00 \times 10^{-7} M$   E. $2.72 \times 10^{-5} M$

_____9. What is [OH$^-$], if pH = 4.78?
   A. 9.22 M   B. $6.03 \times 10^{-10} M$   C. $4.78 \times 10^{-5} M$
   D. $1.66 \times 10^{-5} M$   E. $4.78 \times 10^{-7} M$

_____10. Which of the following is a buffer solution?
   A. HC$_2$H$_3$O$_2$ + NaC$_2$H$_3$O$_2$   B. HCl + NaCl   C. HC$_2$H$_3$O$_2$ + HCl
   D. NaOH + KOH   E. NH$_3$ + KOH

### III. Provide Complete Answers

1. (5 pts) What is the pH of a solution where $[H_3O^-] = 2.35 \times 10^{-10}\,M$.

2. (5 pts) What is the $[H_3O^-]$ of a solution with pH = 5.72?

3. (5 pts) Write a balanced molecular equation for the reaction of $HC_2H_3O_2$ with $NH_3$. Identify the conjugate acid base pairs.

4. (5 pts) Write the total ionic equation and the net ionic equation for the reaction of calcium hydroxide reacting with hydrobromic acid.

## Answers to Practice Problems

| | | | | | |
|---|---|---|---|---|---|
| 15.1 | -2 | 15.2 | -13 | 15.3 | 2 |
| 15.4 | 13 | 15.5 | 4.34 | 15.6 | 11.1 |
| 15.7 | $5.62 \times 10^{-8}$ | 15.8 | $5.13 \times 10^{-2}$ | | |
| 15.9 | $2.33 \times 10^{-3}$ | | | | |
| 15.10 | base | 15.11 | acid | 15.12 | base |

15.13    $H_2SO_4(aq) + 2KOH(aq) \rightarrow 2H_2O(l) + K_2SO_4(aq)$
$2H^+(aq) + SO_4^{2-}(aq) + 2K^+(aq) + 2OH^-(aq) \rightarrow 2H_2O(l) + 2K^+(aq) + SO_4^{2-}(aq)$
$2H^+(aq) + 2OH^-(aq) \rightarrow 2H_2O(l)$

15.14    $3HC_2H_3O_2(aq) + Al(OH)_3(aq) \rightarrow 3H_2O(l) + Al(C_2H_3O_2)_3(aq)$
$3H^+(aq) + 3C_2H_3O_2^-(aq) + Al^{3+}(aq) + 3OH^-(aq) \rightarrow$
$\qquad\qquad\qquad\qquad\qquad\qquad\qquad 3H_2O(l) + Al^{3+}(aq) + 3C_2H_3O_2^-(aq)$
$3H^+(aq) + 3OH^-(aq) \rightarrow 3H_2O(l)$

15.15    $HNO_3(aq) + LiOH(aq) \rightarrow H_2O(l) + LiNO_3(aq)$
$H^+(aq) + NO_3^-(aq) + Li^+(aq) + OH^-(aq) \rightarrow H_2O(l) + Li^+(aq) + NO_3^-(aq)$
$H^+(aq) + OH^-(aq) \rightarrow H_2O(l)$

| | | | | | |
|---|---|---|---|---|---|
| 15.16 | $H_2SO_3$ | 15.17 | $CO_3^{2-}$ | 15.18 | $H_2CO_3$ |
| 15.19 | $NH_2^-$ | 15.20 | $NH_4^+$ | | |

15.21    $HF + H_2O \rightleftharpoons H_3O^+ + F^-$
          acid   base        acid   base

15.22    $NH_3 + H_2O \rightleftharpoons NH_4^+ + OH^-$
          base   acid        acid   base

15.23    $HC_2H_3O_2 + NO_2^- \rightleftharpoons HNO_2 + C_2H_3O_2^-$
          acid       base        acid     base

**15.24** $F^- + NH_4^+ \rightleftharpoons HF + NH_3$
        base  acid        acid  base

**15.25** $H_2SO_3$        **15.26** $NO_2^-$        **15.27** $NH_4^+$

**15.28** $F^-$        **15.29** $SO_4^{2-} + H_2O \rightleftharpoons OH^- + HSO_4^-$

**15.30** $H_2O + HF \rightleftharpoons H_3O^+ + F^-$

**15.31** $1.00 \times 10^{-14}\ M$      **15.32** $5.88 \times 10^{-5}\ M$      **15.33** $4.27 \times 10^{-8}\ M$

**15.34** Acidic        **15.35** Acidic        **15.36** Basic

**15.37** Acidic        **15.38** 2.53        **15.39** 8.78

**15.40** $3.80 \times 10^{-7}\ M = [H_3O^+];\ 2.63 \times 10^{-8}\ M = [OH^-]$      **15.41** 10,000

**15.42** Yellow        **15.43** Blue        **15.44** Purple

**15.45** Yellow        **15.46** Red        **15.47** Blue

**15.48** NaOCl        **15.49** $H_3PO_4$

**15.50** $C_6H_5CO_2H / C_6H_5CO_2^-Na^+$

**15.51** $CH_3NH_2 / CH_3NH_3^+Cl^-$

**15.52** Both are strong 2 $M$ is twice as concentrated as 1 $M$

**15.53** Same concentration, but HF is weak and HI is strong      **15.54** 7.88

**15.55** 1.56        **15.56** 13.4        **15.57** 0.81

## Answers to Self-Exam

**I. T/F**
1. T    2. F    3. T    4. T    5. F    6. T    7. T    8. F    9. F    10. T

**II. MC**
1. C    2. A    3. C    4. E    5. B    6. E    7. D    8. C    9. B    10. A

**III. Complete Answers**

1. pH = 9.63            2. $1.91 \times 10^{-6}$

3. $HC_2H_3O_2 + NH_3 \rightleftharpoons C_2H_3O_2^- + NH_4^+$
    acid      base        base     acid

4. Total ionic equation:
     $Ca^{2+}(aq) + 2OH^-(aq) + 2H^+(aq) + 2Br^-(aq) \rightarrow 2H_2O(l) + Ca^{2+}(aq) + 2Br^-(aq)$

     Net ionic equation: $2OH^-(aq) + 2H^+(aq) \rightarrow 2H_2O(l)$

# 16
# Oxidation-Reduction Reactions

This chapter will examine reactions where electrons are transferred. You actually were introduced to some of these reactions back in Chapter 6, when you learned to balance single-replacement reactions.

Keys for success in this chapter:

- Review single replacement reactions.
- Practice problems for assigning oxidation numbers.
- Apply the rules for balancing oxidation-reduction reactions by doing problems.
- Learn how oxidation-reduction principles are applied in electrochemistry.

## Summary of Verbal Knowledge

**oxidation-reduction reaction:** A reaction in which electrons are transferred from one reactant to another.
**oxidation:** The loss of electrons.
**reduction:** The gain of electrons.
**oxidizing agent:** The reactant that causes oxidation (the one that is reduced).
**reducing agent:** The reactant that causes reduction (the one that is oxidized).
**activity series:** This is an experimental summary of elements listed in order of their potential to be oxidized. The metals on top are more oxidizable than the ones below.
**oxidation number:** This is also known as the oxidation state. It is the charge assigned to an atom in a compound.
**spontaneous process:** This is a physical or chemical change that occurs by itself.
**electricity:** The flow of electrons.
**electrochemistry:** The study of the conversion of stored chemical energy into electrical energy and vice versa.
**electrochemical cell:** An apparatus that either generates or uses an electric current.
**voltaic (galvanic) cell:** An electrochemical cell in which a spontaneous chemical reaction generates an electric current.
**half-cell:** That portion of an electrochemical cell in which a half-reaction (either oxidation or reduction) takes place.
**anode:** The electrode where oxidation occurs.
**cathode:** The electrode where reduction occurs.
**volt:** The SI unit describing the energy per unit charge.
**electrolytic cell:** An electrochemical cell in which an external electric current drives a nonspontaneous oxidation-reduction reaction.
**battery:** A single voltaic cell or a series of voltaic cells.

# Review of Mathematical and Calculator Skills

You should review the following algebraic skill. Look at Example 16.1 to recall how to use algebra to find the value of an unknown.

**Example 16.1** Exmine EQ 16.1.

$$7y + 2x = z \qquad \text{(EQ 16.1)}$$

Solve for x if y = -2 and z = -2.

**Solution:** Substitute the given values into EQ 16.1, then solve by isolating x.

$$7(-2) + 2x = -2$$
$$-14 + 2x = -2$$
$$2x = 12$$
$$x = 6$$

## Practice Problems

16.1 Solve for x: $4y + x = z$ (where y = -2, z = -1).

16.2 Solve for x: $2y + x + 3z = 0$ (where y = 1, z = -2).

# Application of Skills and Concepts

## 16.1 Oxidation and Reduction

Oxidation and reduction reactions are found in

- Biological systems
- Nature and the environment
- Batteries
- Electrolysis reactions

In this first section you will learn how to identify the following from a balanced chemical equation.

- Element being oxidized
- Element being reduced
- Oxidizing agent
- Reducing agent

Review the definitions of these identities at the beginning of the section. The following rules will help you to identify simple oxidizing and reducing agents.

# Oxidation-Reduction Reactions

1. If the charge on a monatomic substance increases, the substance has been **oxidized**. The substance is the **reducing agent**.

2. If the charge on a monatomic substance decreases, the substance has been **reduced**. The substance is the **oxidizing agent**.

These rules can be demonstrated by an example.

**Example 16.2:** For the equation shown below, identify the substance that is **oxidized**; the substance that is **reduced**; the **oxidizing agent**; the **reducing agent**.

$$Cu + 2Ag^+ \rightarrow Cu^{2+} + 2Ag$$

**Solution:** Using the rules above, notice that metallic copper went from neutral (0) to +2. This is an increase in charge. Therefore:

Cu is **oxidized**.
Cu is the **reducing agent**.

Using the rules above, notice that the silver ions went from +1 charge to neutral (0). Therefore:

$Ag^+$ is **reduced**.
$Ag^+$ is the **oxidizing agent**.

Notice that the equation in Example 16.2 is a *net ionic* equation. Even if this equation were written as a *molecular* equation (EQ 16.2), you would still reach the same conclusions as above. The spectator ions (in this case $NO_3^-$) do not change oxidation states during the equation.

$$Cu + 2AgNO_3 \rightarrow Cu(NO_3) + 2Ag \qquad (EQ\ 16.2)$$

Table 16.1 is a summary of the nomenclature introduced in this section.

**Table 16.1 Identity of Substances in Oxidation-Reduction Reactions**

| Substance Changes by | Term for What Happens to Substance | Name for Substance | Examples |
|---|---|---|---|
| Loss of electrons | Oxidation | Reducing Agent | $Na \rightarrow Na^+$<br>$H_2 \rightarrow 2H^+$<br>$S^{2-} \rightarrow S$ |
| Gain of electrons | Reduction | Oxidizing Agent | $Mg^{2+} \rightarrow Mg$<br>$O_2 \rightarrow O^{2-}$<br>$Cl_2 \rightarrow 2Cl^-$ |

## Practice Problems

For each reaction, identify the substance **oxidized; reduced; oxidizing agent; reducing agent**. When necessary, identify the **spectator ions**.

16.3 $Ba + Mg^{2+} \rightarrow Ba^{2+} + Mg$

16.4 $Cd + 2H^+ \rightarrow Cd^{2+} + H_2$

16.5 $2Al + Fe_2O_3 \rightarrow 2Fe + Al_2O_3$

16.6 $Ca + 2HCl \rightarrow CaCl_2 + H_2$

## 16.2 Activity Series

This is derived from experimental evidence. Neutral metals in the series are listed in order of their decreasing potential for oxidation. In an oxidation-reduction reaction, a neutral metal listed higher in the series will react with the ion of a metal below in the series.

A neutral metal will *not* react with the *ion* of a metal above it.

**Table 16.2** Activity Series of Metals and Hydrogen

| Metal (including hydrogen) | Metal Ion | Metal (including hydrogen) | Metal Ion |
|---|---|---|---|
| Li | $Li^+$ | Cd | $Cd^{2+}$ |
| K | $K^+$ | Co | $Co^{2+}$ |
| Ba | $Ba^{2+}$ | Ni | $Ni^{2+}$ |
| Ca | $Ca^{2+}$ | Sn | $Sn^{2+}$ |
| Na | $Na^+$ | Pb | $Pb^{2+}$ |
| Mg | $Mg^{2+}$ | $H_2$ | $H^+$ |
| Al | $Al^{3+}$ | Cu | $Cu^{2+}$ |
| Zn | $Zn^{2+}$ | Hg | $Hg^{2+}$ |
| Cr | $Cr^{2+}$ | Ag | $Ag^+$ |
| Fe | $Fe^{2+}$ | Pt | $Pt^{2+}$ |
| Cd | $Cd^{2+}$ | Au | $Au^{3+}$ |

**Example 16.3:** Which reaction(s) will occur?
  Zinc metal with HCl
  Copper(II) nitrate with cadmium metal
  HCl with copper metal

**Solution:** Use the activity series. Zinc is *above* $H_2$ in the activity series. Therefore it is a better *reducing agent* (will be oxidized) than $H_2$. Therefore, the following reaction occurs, where zinc is oxidized to $Zn^{2+}$ and $H^+$ is reduced to $H_2$.

$$Zn + 2HCl \rightarrow ZnCl_2 + H_2$$

Cadmium metal is *above* copper in the activity series. Therefore metallic, cadmium will be oxidized (it is a better *reducing agent*) by copper(II).

$$Cd + Cu(NO_3)_2 \rightarrow Cd(NO_3)_2 + Cu$$

Copper is *below* hydrogen in the activity series. Therefore copper metal is too weak a *reducing agent* to become *oxidized* when reacted with HCl.

$$Cu + HCl \rightarrow NO\ RXN$$

## Practice Problems

For each problem below write products when a reaction occurs; NO RXN when none occurs.

**16.7** Silver metal plus copper(II) nitrate

**16.8** Copper metal plus silver nitrate

**16.9** Zinc metal plus lead(II) chloride

**16.10** Lead metal plus hydrochloric acid

**16.11** Magnesium metal plus calcium chloride

**16.12** Magnesium metal plus hydrobromic acid

## 16.3 Oxidation Numbers

You need to learn to assign oxidation numbers to two types of atoms.

- Atom or monatomic ion
- Atom in a substance (molecule or polyatomic ion) with covalent bonds

The previous two sections alluded to the oxidation numbers on monatomic ions and neutral metals. Oxidation numbers are a way of keeping track of electrons in oxidation-reduction reactions. The rules for this system of bookkeeping are listed below. Note that the first two rules were applied in problems you have already encountered.

**Rules for Assigning Oxidation Numbers**

1. The oxidation number of an atom in a pure element is zero. This includes single atoms or atoms in elemental molecules, such as $H_2$, $S_8$, etc.

2. The oxidation number of a monatomic ion is equal to the charge on the ion.

3. The oxidation number of fluorine is -1 in all compounds (ionic and molecular).

4. The oxidation number of chlorine, bromine, and iodine in binary compounds containing only one type of Group VIIA element is -1.

5. The oxidation number of hydrogen is +1 in all compounds with nonmetals. It is -1 when in a compound with a metal in Group IA or Group IIA (see Rule 7).

6. In all compounds, oxygen is -2, except peroxides (e.g., $H_2O_2$), where it is -1.

7. In all compounds, Group IA metals are +1. In all compounds, Group IIA metals are +2.

8. The sum of the oxidation numbers of the atoms in a compound is always zero. In a polyatomic ion, the sum of the oxidation numbers of the constituent atoms will add up to the charge on the ion.

Examples will help demonstrate the application of the rules.

**Example 16.4:** What are the oxidation states on pure hydrogen, on pure oxygen, on each of these elements in water?

**Solution:** Rule 1 can be used to apply the oxidation numbers for $H_2$ and $O_2$, which are in their pure state, and therefore **0**.

Use Rules 5 and 6 to assign the oxidation numbers in $H_2O$.
Hydrogen is **+1** (in all compounds with nonmetals).
Oxygen is **-2** (in all compounds except peroxides).

**Example 16.5:** What are the oxidation numbers for each atom in $K_2Cr_2O_7$?

**Solution:** The oxidation state for K is **+1** (Rule 2).
The charge of the polyatomic group, $Cr_2O_7$, must be **-2** (to make a neutral compound).
The charge on oxygen is **-2** (Rule 6).
You need to apply Rule 8, to determine the oxidation number on chromium.

Let the charge on Cr be x. The rule says the sum of the charges must equal the charge on the polyatomic ion. Therefore,

$$2x + 7(-2) = -2$$
$$2x = -2 + 14$$
$$2x = 12$$
$$x = 6$$

Each chromium atom has a **+6** charge.

**Example 16.6:** What atom in the following equation is being oxidized? What atom is being reduced?

$$16H^+(aq) + 2\, MnO_4^-(aq) + 5C_2O_4^{2-}(aq) \rightarrow 2Mn^{2+}(aq) + 10CO_2(g) + 8H_2O(l)$$

**Solution:** Take each atom separately. Hydrogen is in a **+1** state on both sides of the equation.

Manganese is in a **+7** state on the left and in a **+2** state on the right. It has been *reduced* and therefore is an *oxidizing* agent.

Carbon is in a **+3** state on the left and in a **+4** state on the right. Carbon is *oxidized*. It is the *reducing agent*.

Oxygen is in a **-2** state on both sides of the equation.

**Example 16.7:** Which one of the following equations is an oxidation-reduction reaction? Identify the substances being oxidized and reduced.

$$NaOH(aq) + HBr(aq) \rightarrow NaBr(aq) + H_2O(l)$$
$$2Na(s) + 2 H_2O(l) \rightarrow 2NaOH(aq) + H_2(g)$$
$$NaCl(aq) + AgNO_3(aq) \rightarrow AgBr(s) + NaNO_3(aq)$$

**Solution:** The first equation is a neutralization reaction. Notice that no atoms change oxidation states in this reaction.

The second reaction is an oxidation-reduction reaction. Sodium has been *oxidized* from 0 to +1. Hydrogen has been *reduced* from +1 to 0.

The last reaction is a precipitation reaction. None of the atoms have changed oxidation states in this reaction.

---

## Practice Problems

**16.13** What is the oxidation number on each atom in $Na_2O$?

**16.14** What is the oxidation number on each atom in $Na_2O_2$?

**16.15** What is the oxidation number on each atom in $NaHSO_3$?

**16.16** What is the oxidation number on each atom in $Na_2SO_4$?

**16.17** What is the oxidation number on copper metal?

**16.18** What is the oxidation number on copper in CuCl?

**16.19** What is the oxidation number on copper in $CuCl_2$?

**16.20** What is the oxidation number on each atom in calcium fluoride?

**16.21** What is the oxidation number on each atom in fluorine?

**16.22** What is the oxidation number on each atom in $NH_3$?

**16.23** What is the oxidation number on each in NaH?

**16.24** Which (there may be more than one) of the following are oxidation-reduction equations? For those which are, name the atom being reduced, the atom being oxidized.

$H_2O(l) \rightarrow 2H_2(g) + O_2(g)$

$NH_3(g) + H_2O(l) \rightarrow NH_4^+(aq) + OH^-(aq)$

$2H_2O(l) + Ca(s) \rightarrow Ca(OH)_2(s) + H_2(g)$

---

## 16.4 Balancing Oxidation-Reduction Equations by the Half-Reaction Method

This section will help you use a system for balancing oxidation-reduction reactions. These are often very difficult to balance by traditional trial and error methods. The following set of rules will help you balance these types of reactions. The rules are followed by two examples, one for acidic conditions and one for basic conditions.

## Rules for Using the Half-Reaction Method to Balance Equations

1. Assign oxidation numbers to each atom in the skeletal equation. Decide which atoms were oxidized and which were reduced.

2. Split the skeletal equation into two half reactions, labeled **oxidation** and **reduction**.

3. Balance all of the atoms (except oxygen and hydrogen) in each half-reaction.

4. Balance the oxygen in each half-reaction by adding water.

5. In acidic solutions, balance the hydrogens by adding $H^+$. In basic solutions, balance the hydrogens by adding $H^+$ and then adding as many $OH^-$ ions to each side of the equation as there are hydrogens.

6. Balance the charge by adding electrons to account for oxidation or reduction electron transfers.

7. Multiply the half-reactions by factors that will cancel electrons when they are added together.

8. Add the two half-reactions.

9. Check the equation to make sure it is balanced.

**Example 16.8:** For the reaction of aqueous $MnO_4^-$ and aqueous $C_2O_4^{2-}$ which produces aqueous $Mn^{2+}$ and $CO_2$ gas in acidic solution, balance the equation using the rules above.

   **Solution:** Solve this using the rules above. $MnO_4^- + C_2O_4^{2-} \rightarrow Mn^{2+} + CO_2$

   | | | |
   |---|---|---|
   | **Rule 1:** | Mn is +7 on left, +2 on right | (Each Mn gains 5 e-) |
   | | O is -2 on left, -2 on right | (No change in e-) |
   | | C is +3 on left, +4 on right | (Each C loses 1 e-) |
   | **Rule 2:** | $C_2O_4^{2-} \rightarrow CO_2$ | oxidation |
   | | $MnO_4^- \rightarrow Mn^{2+}$ | reduction |
   | **Rule 3:** | $C_2O_4^{2-} \rightarrow 2CO_2$ | |
   | | $MnO_4^- \rightarrow Mn^{2+}$ | ok as is |
   | **Rule 4:** | $C_2O_4^{2-} \rightarrow 2CO_2$ | ok as is |
   | | $MnO_4^- \rightarrow Mn^{2+} + 4H_2O$ | |
   | **Rule 5:** | $C_2O_4^{2-} \rightarrow CO_2$ | ok as is |
   | | $8H^+ + MnO_4^- \rightarrow Mn^{2+} + 4H_2O$ | |
   | **Rule 6:** | $C_2O_4^{2-} \rightarrow CO_2 + 2e^-$ | |
   | | $5e^- + 8H^+ + MnO_4^- \rightarrow Mn^{2+} + 4H_2O$ | |

## Oxidation-Reduction Reactions

**Rule 7:** $5C_2O_4^{2-} \rightarrow 5CO_2 + 10e^-$ (multiplied times 5)

$10e^- + 16H^+ + 2MnO_4^- \rightarrow 2Mn^{2+} + 8H_2O$ (multiplied times 2)

**Rule 8:** $5C_2O_4^{2-} + 16H^+ + 2MnO_4^- \rightarrow 2Mn^{2+} + 8H_2O + 5CO_2$

**Rule 9:**

| Left | Right |
|------|-------|
| 16 H | 16 H |
| 10 C | 10 C |
| 28 O | 28 O |

**Example 16.9:** For the reaction of aqueous $Cl_2$ and aqueous $IO_3^-$ which produces $Cl^-$ and $IO_4^-$ in basic solution, balance the equation using the rules above.

**Solution:** This solution looks like that in Example 16.7, except for Rule 5.

$$Cl_2 + IO_3^- \rightarrow Cl^- + IO_4^-$$

**Rule 1:** 
- Cl is 0 on left; -1 on right    (Each Cl gains 1 e-)
- O is -2 on left, -2 on right    (No change in e-)
- I is +5 on left, +7 on right    (Each I loses 2 e-)

**Rule 2:**
- $Cl_2 \rightarrow Cl^-$    reduction
- $IO_3^- \rightarrow IO_4^-$    oxidation

**Rule 3:**
- $Cl_2 \rightarrow 2Cl^-$
- $IO_3^- \rightarrow IO_4^-$    ok as is

**Rule 4:**
- $Cl_2 \rightarrow 2Cl^-$    ok as is
- $H_2O + IO_3^- \rightarrow IO_4^-$

**Rule 5:**
- $Cl_2 \rightarrow 2Cl^-$    ok as is
- $H_2O + IO_3^- \rightarrow IO_4^- + 2H^+$    added protons
- $2OH^- + H_2O + IO_3^- \rightarrow IO_4^- + 2H^+ + 2OH^-$    added hydroxides
- $2OH^- + IO_3^- \rightarrow IO_4^- + H_2O$    canceled water

**Rule 6:**
- $Cl_2 + 2e^- \rightarrow 2Cl^-$
- $2OH^- + IO_3^- \rightarrow 2e^- + IO_4^- + H_2O$

**Rule 7:** ok as is

**Rule 8:** $2OH^- + IO_3^- + Cl_2 \rightarrow IO_4^- + H_2O + 2Cl^-$

**Rule 9:**

| Left | Right |
|------|-------|
| 2 H | 2 H |
| 2 Cl | 2 Cl |
| 5 O | 5 O |
| 1 I | 1 I |

## 278  Chapter 16

> **Practice Problems**
>
> **16.25** Balance the equation for the reaction of copper metal and silver nitrate.
>
> **16.26** Balance the equation for the reaction of aluminum metal and iron(III) oxide to form iron metal and aluminum oxide.
>
> **16.27** Balance the equation for the reaction of $Br_2$ and aqueous $HSO_3^-$ to form bromide ion and sulfate ion in basic solution.
>
> **16.28** Balance the equation for the reaction of zinc metal and manganese(IV) oxide to form zinc hydroxide and manganese(III) oxide (under basic conditions).
>
> **16.29** Balance the equation for the reaction of $C_2H_6O$ with $Cr_2O_7^{2-}$ under acidic conditions to produce $Cr^{3+}$ and $C_2H_4O_2$.

## 16.5  Voltaic Cells

These are methods that use oxidation and reduction reactions in such a way that the electron movement can be captured as electricity.

Voltaic cells:

- Use spontaneous reactions connected by a salt bridge and metal wire.
- Use reactions **not** at equilibrium.
- Can be designed by using activity series.
- Convert chemical energy to electrical energy.

Refer to Figure 16.9 in your textbook, *Introductory Chemistry*. This shows how you might set up a voltaic cell using equipment and reagents found in a typical chemistry laboratory. This cell has the following components.

- Beaker 1: Solution of zinc sulfate with a piece of zinc metal
- Beaker 2: Solution of copper(II) sulfate with a piece of copper metal
- Wire connecting the two pieces of metal
- Salt bridge connecting the two beakers

One of these beakers will have an oxidation reaction, and the other will have a reduction reaction. In order to determine which is which, look at the activity series. Zinc metal is higher in the series than copper metal; therefore, it will be oxidized in this system. This means that reduction must occur in the copper system, where $Cu^{2+}$ becomes Cu. $Cu^{2+}$ is lower in the activity series than $Zn^{2+}$, so it is more readily reduced.

The reactions that will occur (until equilibrium is reached):

- $Zn \rightarrow Zn^{2+}$    oxidation
- $Cu^{2+} \rightarrow Cu$    reduction

The electrons will flow throughout the circuit from the anode to the cathode. A summary of the voltaic cell is contained in Table 16.3.

# Oxidation-Reduction Reactions

Table 16.3 Summary of Action in Voltaic Cells

|  | Process | Example | Example | Example |
|---|---|---|---|---|
| **Anode** | Oxidation | $Cu \to Cu^{2+}$ | $Cd \to Cd^{2+}$ | $Cr \to Cr^{2+}$ |
| **Cathode** | Reduction | $Ag^+ \to Ag$ | $Cu^{2+} \to Cu$ | $Cd^{2+} \to Cd$ |
| **Electron flow** |  | $Cu \to Ag$ | $Cd \to Cu$ | $Cr \to Cd$ |

**Example 16.10:** Describe how to make a voltaic cell using copper, lead, and solutions.

**Solution:** Lead is higher in the activity series, so it will be oxidized. This means that lead metal is the anode. Copper(II) will be reduced, and copper metal will be the cathode. Use solutions with spectator ions such as $Zn(NO_3)_2$ and $Cu(NO_3)_2$.

## Practice Problems

**16.30** Describe how to make a voltaic cell using Al/AlCl$_3$ and Fe/FeCl$_2$. Name the anode and cathode.

**16.31** Describe how to make a voltaic cell using Zn/Zn(NO$_3$)$_2$ and Co/Co(NO$_3$)$_2$. Name the anode and cathode.

**16.32** Describe how to make a voltaic cell using Mg/MgSO$_4$ and Sn/SnSO$_4$. Name the anode and cathode

## 16.6 Electrolytic Cells

The oxidation-reduction reactions in electrolytic cells are **not** spontaneous. Reactions in voltaic cells are **spontaneous**. See Table 16.4 for a summary.

Table 16.4 Reactions in Electrolytic Cells

| Reaction | Anode (Oxidation) | Cathode (Reduction) |
|---|---|---|
| Electrolysis of molten NaBr | $2Br^- \to Br_2(g)$ | $Na^+ \to Na(s)$ |
| Electrolysis of molten CaCl$_2$ | $2Cl^- \to Cl_2(g)$ | $Ca^{2+} \to Ca(s)$ |

**Example 16.11:** For the electrolysis of water, at which electrode will each gas form?

**Solution:** First write the reaction.

$$2H_2O \to 2H_2 + O_2 \qquad \text{(EQ 16.3)}$$

Next, determine what happens with each half-reaction. By definition, oxidation occurs at the anode and reduction occurs at the cathode.

H goes from +1 to 0     **Reduction**
O goes from -2 to 0     **Oxidation**

You can predict that $H_2$ will form at the **cathode** and $O_2$ will form at the **anode**.

---

## Practice Problems

**16.33** What is produced (at each electrode) from the electrolysis of molten KI?

**16.34** What is produced (at each electrode) from the electrolysis of molten $AlCl_3$?

---

### 16.7 Three Important Batteries

The oxidation-reduction reactions of three batteries are summarized in this section. In general, batteries are voltaic cells. Those which can be recharged become electrolytic cells during the recharging process. The three types of batteries discussed in this section are

- Lead storage battery
- Nickel-cadmium battery
- Mercury battery

The oxidation-reduction reactions in each battery are summarized below. Remember, oxidation is occurring at the anode and reduction at the cathode.

For the lead storage battery, the reactions are summarized in EQ 16.4 and EQ 16.5.

**Anode:**
$$Pb(s) + H_2SO_4(aq) + 2H_2O(l) \rightarrow PbSO_4(s) + 2H_3O^+(aq) + 2e^- \quad \text{(EQ 16.4)}$$
**Cathode:**
$$PbO_2(s) + H_2SO_4(aq) + 2H_3O^+(aq) + 2e^- \rightarrow PbSO_4(s) + 4H_2O(l) \quad \text{(EQ 16.5)}$$

For the nickel-cadmium battery, the reactions are summarized in EQ 16.6 and EQ 16.7:

**Anode:**    $Cd(s) + 2OH^-(aq) \rightarrow Cd(OH)_2(aq) + 2e^-$     **(EQ 16.6)**
**Cathode:**   $NiO(OH) + H_2O + e^- \rightarrow Ni(OH)_2 + OH^-$     **(EQ 16.7)**

For the mercury battery, the reactions are summarized in EQ 16.8 and EQ 16.9:

**Anode:**    $Zn + 2OH^- \rightarrow Zn(OH)_2 + 2e^-$     **(EQ 16.8)**
**Cathode:**   $HgO + H_2O + 2e^- \rightarrow Hg + 2OH^-$     **(EQ 16.9)**

When batteries are designed in such a way that the spontaneous oxidation-reduction reactions can be reversed (by supplying an electric current), the battery is said to be **rechargeable**. Of the three batteries listed here, only the lead storage and nickel-cadmium are rechargeable.

**Example 16.12:** For the lead storage battery, what is being oxidized? What is being reduced? What are the reactions for the recharging of the battery?

**Solution:** At the anode, lead is being oxidized from 0 to +2.
At the cathode, lead is being reduced from +4 (in $PbO_2$) to +2.

The reactions for the recharging reverse those in the voltaic cell.

**Anode:** $PbSO_4(s) + 4H_2O(l) \rightarrow PbO_2(s) + H_2SO_4(aq) + 2H_3O^+(aq) + 2e^-$

**Cathode:** $PbSO_4(s) + 2H_3O^+(aq) + 2e^- \rightarrow Pb(s) + H_2SO_4(aq) + 2H_2O(l)$

---

## Practice Problems

**16.35** What is being oxidized in the nickel-cadmium battery? What is being reduced?

**16.36** What is being oxidized in the mercury battery? What is being reduced?

**16.37** What are the reactions for recharging the nickel-cadmium battery?

---

## Integration of Multiple Skills

There are several ideas from earlier chapters that can merge with the topics in this chapter.

**Example 16.13:** Consider the reaction of aluminum metal with iron(III) oxide. This reaction produces heat. Label the oxidizing and reducing reagents. If 1 mol of $Fe_2O_3$ produces 852 kJ of heat, how much heat would 186 g of iron(III) oxide produce?

**Solution:** Reaction is $2Al + Fe_2O_3 \rightarrow 2Fe + Al_2O_3$
Al is oxidize; therefore, it is the reducing agent.
Fe is reduced; therefore, it is the oxidizing agent.

You can use dimensional analysis to determine the heat produced.

$$186 \text{ g } Fe_2O_3 \times \frac{1 \text{ mol } Fe_2O_3}{159.7 \text{ g } Fe_2O_3} \times \frac{852 \text{ kJ}}{\text{mol } Fe_2O_3} = 992 \text{ kJ heat produced}$$

---

## Practice Problem

**16.38** How much heat is produced from reacting 48.7 g of $Fe_2O_3$ with an excess of Al?

---

## Self-Exam

**I. (4 ea) Write T (true) or F (false) in each corresponding blank.**

_____1. A substance that is *oxidized* is an *oxidizing* agent.

_____2. All oxidation reactions involve electron transfers.

282  Chapter 16

_____3. The lower a neutral metal in the activity series, the easier it is oxidized.

_____4. Reduction is the loss of electrons.

_____5. Voltaic cells operate spontaneously, while electrolytic cells require electric current from an outside source.

_____6. Batteries are voltaic (galvanic) cells.

_____7. All batteries can be recharged.

_____8. For molecular oxygen, one atom has a -2 charge and the other a +2 charge.

_____9. The oxidation number for metallic sodium is +1.

_____10. The oxidation numbers for group VIIA elements in binary ionic compounds is -1.

**II. (4 ea) Multiple Choice. Place the letter of the BEST answer in the corresponding blank.**

_____1. Which reaction below is an oxidation-reduction reaction?
   A. $Pb(NO_3)_2(aq) + KI(aq) \rightarrow PbI_2(s) + 2KNO_3(aq)$
   B. $NaOH(aq) + HCl(aq) \rightarrow H_2O(l) + NaCl(aq)$
   C. $HC_2H_3O_2(aq) + NaHCO_3(aq) \rightarrow NaC_2H_3O_2(aq) + H_2O(l) + CO_2(g)$
   D. $Zn(s) + Cu(NO_3)_2(aq) \rightarrow Cu(s) + Zn(NO_3)_2(aq)$
   E. $2NaOH(aq) + MgCl_2(aq) \rightarrow Mg(OH)_2(s) + 2NaCl(aq)$

_____2. Which substance below has oxygen with a "zero" oxidation?
   A. $H_2O$     B. $H_2O_2$     C. $O_2$     D. $CaO$     E. $Na_2O$

For questions 3-5, use $Zn(s) + Cu(NO_3)_2(aq) \rightarrow Cu(s) + Zn(NO_3)_2(aq)$.

_____3. What is oxidized?
   A. Zn     B. Cu     C. $Zn^{2+}$     D. $Cu^{2+}$     E. $NO_3^-$

_____4. What is the oxidizing agent?
   A. Zn     B. Cu     C. $Zn^{2+}$     D. $Cu^{2+}$     E. $NO_3^-$

_____5. What is the reducing agent?
   A. Zn     B. Cu     C. $Zn^{2+}$     D. $Cu^{2+}$     E. $NO_3^-$

_____6. Using the activity series, which of the following substances will have no reaction?
   A. $Mg(s) + HCl(aq) \rightarrow$
   B. $Cu(s) + HCl(aq) \rightarrow$
   C. $Ca(s) + Zn(NO_3)_2(aq) \rightarrow$
   D. $Zn(s) + Cu(NO_3)_2(aq) \rightarrow$
   E. $Zn(s) + AgNO_3(aq) \rightarrow$

_____7. Using the activity series, which of the following is the strongest reducing agent?
   A. Cu     B. Ag     C. $Na^+$     D. $Cu^{2+}$     E. Na

_____ 8. What is the oxidation state of manganese in $KMnO_4$?
   A. +1    B. +3    C. +5    D. +7    E. -4

_____ 9. What is the oxidation state of chromium in $Cr_2O_7^{2-}$?
   A. +7    B. +6    C. +5    D. -2    E. -1

_____ 10. What is used to balance hydrogens (acidic conditions) in the standard oxidation-reduction balancing using half-reactions?
   A. $H_2$    B. $H_2O$    C. $H^+$    D. $H_2O_2$    E. $H^-$

**III. Provide Complete Answers**

1. (6 pts) Write the half reactions for zinc metal reacting with oxygen in water solution to produce zinc hydroxide.

2. (7 pts) Balance by half reaction method (assume acidic solution).
$$H_2C_2O_4(aq) + MnO_4^-(aq) \rightarrow Mn^{2+}(aq) + CO_2(g)$$

3. (7 pts) Balance by half reaction method (assume acidic solution).
$$C_2H_5OH(aq) + Cr_2O_7^{2-}(aq) \rightarrow CH_3CO_2H(aq) + Cr^{3+}(aq)$$


## Answers to Practice Problems

| | | | |
|---|---|---|---|
| 16.1 | x = 7 | 16.2 | x = 4 |
| 16.3 | Ba is oxidized; is reducing agent. $Mg^{2+}$ is reduced; is oxidizing agent. | 16.4 | Cd is oxidized; is reducing agent. $H^+$ is reduced; is oxidizing agent. |
| 16.5 | Al is oxidized; is reducing agent. $Fe^{3+}$ is reduced; is oxidizing agent. $O^{2-}$ is spectator ion. | 16.6 | Ca is oxidized; is reducing agent. $H^+$ is reduced; is oxidizing agent. $Cl^-$ is spectator ion. |
| 16.7 | No RXN | 16.8 | Ag metal and $Cu(NO_3)_2$ |
| 16.9 | Pb metal and $ZnCl_2$ | 16.10 | $H_2(g)$ and $PbCl_2$ |
| 16.11 | No RXN | 16.12 | $H_2(g)$ and $MgBr_2$ |
| 16.13 | +1 on Na, -2 on O | 16.14 | +1 on Na, -1 on O |
| 16.15 | +1 on Na, +1 on H, +4 on S, -2 on O | | |
| 16.16 | +1 on Na, +6 on S, -2 on O | 16.17 | 0 |
| 16.18 | +1 | 16.19 | +2 |
| 16.20 | +2 on Ca, -1 on F | 16.21 | 0 |
| 16.22 | -3 on N, +1 on H | 16.23 | +1 on Na, -1 on H |

**16.24** $H_2O(l) \rightarrow 2H_2(g) + O_2(g)$  Hydrogen reduced, oxygen oxidized
$NH_3(g) + H_2O(l) \rightarrow NH_4^+(aq) + OH^-(aq)$  Not oxidation-reduction
$2H_2O(l) + Ca(s) \rightarrow Ca(OH)_2(s) + H_2(g)$  Hydrogen reduced, calcium oxidized

**16.25** $Cu + 2AgNO_3 \rightarrow Cu(NO_3) + 2Ag$  **16.26** $2Al + Fe_2O_3 \rightarrow 2Fe + Al_2O_3$

**16.27** $3OH^- + Br_2 + HSO_3^- \rightarrow SO_4^{2-} + 2Br^- + 2H_2O$

**16.28** $2MnO_2 + Zn + H_2O \rightarrow Zn(OH)_2 + Mn_2O_3$

**16.29** $16H^+ + 3C_2H_6O + 2Cr_2O_7^{2-} \rightarrow 4Cr^{3+} + 3C_2H_4O_2 + 11H_2O$

**16.30** $Al \rightarrow Al^{3+}$ (anode); $Fe^{2+} \rightarrow Fe$ (cathode)

**16.31** $Zn \rightarrow Zn^{2+}$ (anode); $Co^{2+} \rightarrow Co$ (cathode)

**16.32** $Mg \rightarrow Mg^{2+}$ (anode); $Sn^{2+} \rightarrow Sn$ (cathode)

**16.33** $2KI(s) \rightarrow 2K(s) + I_2(s)$  **16.34** $2AlCl_3 \rightarrow 2Al(s) + 3Cl_2(g)$
  cathode  anode       cathode  anode

**16.35** Cd oxidized, Ni reduced  **16.36** Zn oxidized, Hg reduced

**16.37** **Anode:** $Ni(OH)_2 + OH^- \rightarrow NiO(OH)_2 + H_2O + e^-$
**Cathode:** $Cd(OH)_2(aq) + 2e^- \rightarrow Cd(s) + 2OH^-(aq)$

**16.38** 260 kJ

## Answers to Self-Exam

**I. T/F**
1. F  2. T  3. F  4. F  5. T  6. T  7. F  8. F  9. F  10. T

**II. MC**
1. D  2. C  3. A  4. D  5. A  6. B  7. E  8. D  9. B  10. C

**III. Complete Answers**
1. $Zn \rightarrow Zn^{2+}(aq) + 2e^-$  (oxidation)
   $4e + O_2 + H_2O \rightarrow 4OH^-(aq)$  (reduction)

2. $6H^+ + 5H_2C_2O_4 + 2MnO_4^- \rightarrow 2Mn^{2+} + 10CO_2 + 8H_2O$

3. $16H^+ + 3C_2H_5OH + 2Cr_2O_7^{2-} \rightarrow 3CH_3CO_2H + 4Cr^{3+} + 11H_2O$

# 17
# Nuclear Chemistry

The first sixteen chapters of this course have dealt with the chemistry of electron changes. This chapter will look at those chemical reactions where there are changes in the nuclei of atoms. The terminology and writing of equations for these types of reactions are significantly different from those you have encountered so far. Review new terms, and try to incorporate them into your memory.

By examining the examples given here and in your textbook, you will find the reactions are fairly easy to follow.

Keys for success in this chapter:

- Review the structure of the atom in Chapter 4.
- Learn new terms and phrases.
- Practice problems.
- Study figures and tables.

## Summary of Verbal Knowledge

**radioactivity:** The radiation in the form of particles or energy coming from the nucleus of an atom undergoing spontaneous disintegration.
**nucleons:** Particles (neutrons or protons) found in the tiny, dense nucleus.
**nuclide:** The nucleus of a specific isotope.
**radioactive decay:** The spontaneous disintegration of a nucleus.
**alpha emission:** The emission of a $^{4}_{2}He$ nucleus (known as an *alpha particle*) from the decaying nucleus.
**nuclear equation:** A symbolic representation of a nuclear reaction.
**beta emission:** The emission of a high-speed electron (known as a *beta particle*) from a decaying nucleus.
**positron emission:** The emission of a positron from a decaying nucleus.
**positron:** This is a particle of the same mass as an electron, but with a positive charge.
**gamma emission:** The emission of pure energy (a photon) with a single wavelength from an excited nucleus.
**radioactive decay series:** A sequence of decay steps that continues until a stable nucleus is reached.
**transmutation:** The change of one element into another.
**cosmic rays:** Radiation that originates from nuclear reactions in the sun and other stars.
**transuranium elements:** The elements with atomic numbers greater than that of uranium.
**half-life:** This is the time required for half of *any amount* of a radioactive nuclide to decay.
**radioactive dating:** A technique for determining the age of certain old objects that relies on the known decay rate of radioactive nuclides in the object.

**curie:** This is a unit for measuring radioactive decay. It is equal to 37 billion disintegrations per second.
**radioactive tracer:** A radioactive isotope that is added to a chemical or biological system to trace the path of a nonradioactive isotope that is normally used by the system.
**free radical:** A molecule with an unpaired valence electron.
**nuclear fission:** A process in which a heavy nucleus splits to form two lighter and more stable nuclei.
**nuclear fusion:** A process in which light nuclei combine to give a heavier, more stable nucleus.
**chain reaction:** A self-sustaining series of nuclear fissions caused by the absorption of neutrons from previous nuclear fissions.
**critical mass:** The minimum mass that will allow fission to become a chain reaction.

## Review of Mathematical and Calculator Skills

There are only two simple math skills you need to review for this section. The first two examples illustrate these.

**Example 17.1:** What is $(\frac{1}{2})^4$?

> **Solution:** To solve this, remember the rules for exponents. The 4 simply means to multiply the value inside the parentheses by itself 4 times.
>
> $$(\tfrac{1}{2})^4 = \tfrac{1}{2} \times \tfrac{1}{2} \times \tfrac{1}{2} \times \tfrac{1}{2} = \tfrac{1}{16}$$

**Example 17.2:** What is the mass number for an isotope of carbon that has 6 protons and 7 neutrons? Show two ways to represent this symbolically.

> **Solution:** The atomic mass of an isotope is equal to the sum of the protons and neutrons. This isotope has a mass number = **13**.
> From Chapter 4, this can be represented as carbon-13 or $^{13}_{6}C$.

---

**Practice Problems**

**17.1** What is $(\frac{1}{2})^2$?

**17.2** What is $(\frac{1}{2})^5$?

**17.3** How many protons and neutrons are present in nitrogen-15?

**17.4** An isotope of sulfur has 16 protons and 18 neutrons. What is its atomic mass?

---

## Application of Skills and Concepts

### 17.1 The Nuclear Model Revisited

Recall from Chapter 4 that the nucleus of an atom is made up of protons and neutrons.

- The **atomic number** is equal to the number of protons.
- The **mass number** is equal to the sum of neutrons plus protons.
- **Isotopes** are atoms with the same atomic number, but different mass numbers.

When dealing with the study of nuclear chemistry there are additional terms used.

- A **nucleon** is either a proton or neutron.
- The nucleus of a specific isotope is called a **nuclide**.

Look at Figure 17.1 to help recall how the symbols for isotopes are written.

**Figure 17.1** Nuclear Symbols

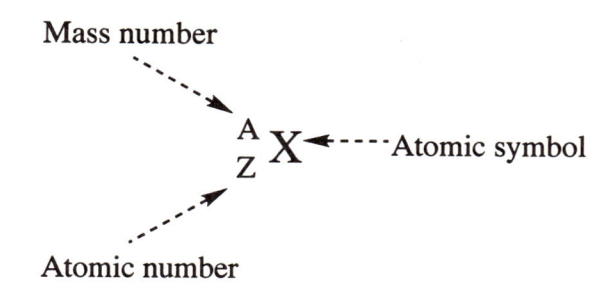

$$^A_Z X$$

Mass number → A
Atomic number → Z
Atomic symbol → X

**Example 17.3:** What is the symbol for the nuclide phosphorus-31? How many protons and neutrons does this nuclide have? How many total nucleons does it have?

**Solution:** The notation *phosphorus-31* tells you that the atom is P.
The mass number is **31**.
Since phosphorus is atomic number 15, there are **15** protons.
Since the number of nucleons is the sum of protons and neutrons, there are

$$31 - 15 = 16 \text{ neutrons}$$

Look at Figure 17.1, to arrive at the notation $^{31}_{15}P$.

---

**Practice Problems**

17.5 Draw the symbol for the vanadium-51 nuclide.

17.6 Draw the symbol for the hydrogen-3 nuclide.

17.7 Draw the nuclide for the carbon-14 nuclide.

17.8 Draw the nuclide for the uranium-235 nuclide.

---

## 17.2 Radioactive Decay

There are four types of radioactive decay that will be studied in this section.

- Alpha emission (emits $^4_2He$)
- Beta emission (emits $^0_{-1}e$)
- Positron emission (emits $^0_1e$)
- Gamma emission (emits $^0_0\gamma$)

Before examining the details of each type of radioactive decay, look at a summary of why some nuclides decay and others are stable. Table 17.1 has some observations about nuclides.

Table 17.1 Stability of Nuclides

| Atomic Number | Stable Ratio of Neutrons: Protons | Examples | Unstable Ratio of Neutrons: Protons | Examples |
|---|---|---|---|---|
| 1-20 | n = p or n = p + 1 | Hydrogen-1 Sodium-23 | n > p or n > p + 1 | Silicon-32 Hydrogen-3 |
| 20-83 | n = 1.5p | Bismuth-126 Copper-63 | n > 1.5p | Titanium-45 Strontium-90 Cobalt-60 |
| > 83 | None | None | All | Uranium-235 Uranium-238 Polonium-209 Plutonium-239 |

Each one of the above types of radioactive decay is unique and will be looked at separately. In this section, the methods for writing nuclear reactions will be studied. The next four examples illustrate how to write these nuclear equations.

**Example 17.4:** Americium-243 decays by alpha emission to form neptunium-239 and an alpha particle. Write a reaction showing how this decay occurs.

**Solution:** To start this problem, write the symbols for americium-243 and neptunium-239. Also write the symbol for an alpha particle (see definitions at the beginning of the chapter).

Americium-243 has the symbol: $^{243}_{95}Am$

Neptunium-239 has the symbol: $^{239}_{93}Np$

Alpha particle has the symbol: $^4_2He$

Now put the nuclides into equation form.

$$^{243}_{95}Am \rightarrow {}^{239}_{93}Np + {}^4_2He \tag{EQ 17.1}$$

Do some addition to verify conservation of mass and conservation of charge.

**Nuclear Chemistry** 289

|  | Left Side | Right Side | Both Sides |
|---|---|---|---|
| **Mass number** | 243 | 239 + 4 | 243 |
| **Atomic number** | 95 | 93 + 2 | 95 |

For all nuclear reactions, the mass number and atomic number must be equal on both sides of the equation.

**Example 17.5:** Cobalt-60 decays by beta emission to form nickel-60 and a beta particle (electron). Write a reaction showing how this decay occurs.

  **Solution:** To start this problem, write the symbols for cobalt-60 and nickel-60. Also write the symbol for a beta particle.

$$\text{Cobalt-60 has the symbol: } ^{60}_{27}\text{Co}$$

$$\text{Nickel-60 has the symbol: } ^{60}_{28}\text{Ni}$$

$$\text{Electron has the symbol: } ^{0}_{-1}\text{e}$$

Now put the nuclides into equation form.

$$^{60}_{27}\text{Co} \rightarrow {}^{60}_{28}\text{Ni} + {}^{0}_{-1}\text{e} \qquad \text{(EQ 17.2)}$$

Do some addition to verify conservation of mass and conservation of charge.

|  | Left Side | Right Side | Both Sides |
|---|---|---|---|
| **Mass number** | 60 | 60 + 0 | 60 |
| **Atomic number** | 27 | 28 - 1 | 27 |

**Example 17.6:** Nitrogen-13 decays by positron emission to form carbon-13 and a positron. Write a reaction showing how this decay occurs.

  **Solution:** To start this problem, look at the definition of a **positron** at the beginning of the chapter. See that the symbol below is consistent with this definition. Look also at the symbols for the nuclides in this reaction.

$$\text{Nitrogen-13 has the symbol: } ^{13}_{7}\text{N}$$

$$\text{Carbon-13 has the symbol: } ^{13}_{6}\text{C}$$

$$\text{Positron has the symbol: } ^{0}_{1}\text{e}$$

Now put the nuclides into an equation form.

$$^{13}_{7}\text{N} \rightarrow {}^{13}_{6}\text{C} + {}^{0}_{1}\text{e} \qquad \text{(EQ 17.3)}$$

Do some addition to verify conservation of mass and conservation of charge.

|  | Left Side | Right Side | Both Sides |
|---|---|---|---|
| **Mass number** | 13 | 13 + 0 | 13 |
| **Atomic number** | 7 | 6 + 1 | 7 |

**Example 17.7:** Metastable technetium-99 decays by gamma emission. Write an equation for this.

**Solution:** To solve this you need to know that gamma radiation is a photon, but has no mass. The symbol for a metastable compound is shown below.

Metastable technetium-99 has the symbol: $^{99m}_{43}Tc$

technetium-99 has the symbol: $^{99}_{43}Tc$

Gamma rays have the symbol: $^{0}_{0}\gamma$

Write the equation.

$$^{99m}_{43}Tc \rightarrow {}^{99}_{43}Tc + {}^{0}_{0}\gamma \qquad \text{(EQ 17.4)}$$

Do some addition to verify conservation of mass and conservation of charge.

|  | Left Side | Right Side | Both Sides |
|---|---|---|---|
| **Mass number** | 99 | 99 + 0 | 99 |
| **Atomic number** | 43 | 43 + 0 | 43 |

When writing nuclear equations, you will sometimes be given the unstable nucleus and the method of decay. You will then have to determine the products of the decay. An example illustrates this.

**Example 17.8:** Strontium-90 decays by beta emission. What is the nuclide in the product? Write a balanced equation.

**Solution:** To solve this, rationalize that the mass and charge are conserved. Strontium-90 has 38 protons (charge +38) and a mass of 90. A beta particle (electron) has a mass of 0 charge of -1. To balance both sides of the equation:

Mass: 90 - 0 = 90
Charge: 38 - (-1) = 39

This tells you that the mass number is **90** and the number of protons is **39**. This means that the identity of the nuclide is yttrium, Y. The equation for the reaction is:

$$^{90}_{38}Sr \rightarrow {}^{90}_{39}Y + {}^{0}_{-1}e$$

Notice that the total mass and total charge on the left are equal to those on the right.

## Practice Problems

**17.9** Radon-222 decays by alpha emission to form polonium-218. Write the nuclear equation.

**17.10** Polonium-218 decays by alpha emission to form lead-214. Write the nuclear equation.

**17.11** Thorium-234 decays by beta emission to form palladium-234. Write the nuclear equation.

**17.12** Palladium-234 decays by beta emission to form uranium-234. Write the nuclear equation.

**17.13** Oxygen-15 decays by positron emission to form nitrogen-15. Write the nuclear equation.

**17.14** Sodium-22 decays by positron emission to form neon-22. Write the nuclear equation.

**17.15** Lead-210 decays by alpha and beta emission. Write the nuclear equation for alpha decay.

**17.16** Write the nuclear equation for the beta decay of lead-210.

**17.17** Bismuth-214 decays by alpha and beta emission. Write the nuclear equation for alpha decay.

**17.18** Write the nuclear equation for the beta decay of bismuth-214.

**17.19** Neon-22 is formed from positron decay of what nuclide?

Table 17.2 Summary of Radioactive Decay

| Type | Particle Emitted | General Description | Examples |
|---|---|---|---|
| Alpha | $^{4}_{2}He$ | Nuclides with atomic number greater than 83 | $^{238}_{92}U \rightarrow {}^{234}_{90}Th + {}^{4}_{2}He$ |
| Beta | $^{0}_{-1}e$ | Wide range of nuclides High speed electron is emitted | $^{234}_{91}Pa \rightarrow {}^{234}_{92}U + {}^{0}_{-1}e$ |
| Positron | $^{0}_{1}e$ | Equivalent of converting a proton to a neutron | $^{207}_{84}Po \rightarrow {}^{207}_{83}Bi + {}^{0}_{1}e$ |
| Gamma | $^{0}_{0}\gamma$ | Emitted from high energy metastable nuclei | $^{99m}_{43}Tc \rightarrow {}^{99}_{43}Tc + {}^{0}_{0}\gamma$ |

## 17.3 Nuclear Transmutation

This type of nuclear reaction is similar to decay reactions. Once again, you will need to balance both the mass number and the charge (atomic number).

**Transmutation** is the conversion of one element into another. This type of reaction involves nuclear chemistry and is not similar to attempts by alchemists to convert cheap metals into gold, etc., by use of chemical means. An example will demonstrate this type of reaction.

**Example 17.9:** Aluminum-27 can be bombarded with hydrogen-3 to produce magnesium-27 and what other particle?

**Solution:** To solve this, set up a nuclear equation from the word description, and calculate the mass number and atomic number on each side.

$$^{27}_{13}Al + {}^{3}_{1}H \rightarrow {}^{27}_{12}Mg + \_\_$$

|  | Left Side | Right Side |
|---|---|---|
| Mass number | 27 + 3 = 30 | 27 + __ = 30 |
| Atomic number | 13 + 1 = 14 | 12 + __ = 14 |

Now you can fill the numbers in the empty spaces on the right side.

The unknown particle must have a mass of **3** (27 + 3 = 30).
The atomic number must be **2** (12 + 2 = 14).
Since the atomic number is **2** the particle must be helium (atomic number defines element of particle).

The missing particle on the right is: $^{3}_{2}He$

---

## Practice Problems

**17.20** Nitrogen-14 combines with a neutron to form a proton ($^{1}_{1}H$) and what other nuclide?

**17.21** An alpha particle and nitrogen-14 react to form a proton and what other nuclide?

**17.22** An alpha particle and beryllium-9 react to form carbon-12 and what other particle?

**17.23** A neutron ($^{1}_{0}n$) and uranium-238 combine to form what nuclide?

---

### 17.4 Rate of Radioactive Decay and Half-Life

The understanding of radioactive decay is very important. One new concept that you will encounter in this section is that of a **half-life**. This is simply the amount of time it takes for half of a given amount (mass) of a radioactive substance to decay.

Table 17.3 (page 294) lists some radioactive materials and their respective half-lives (in decreasing order). The half-lives vary from years to fractions of a second. To understand half-lives, look at Figures 17.2 and 17.3 (page 293). Each represents the decay of 20.0 g of a radioactive substance.

Notice from Figure 17.2 that at time = zero there are 20.0 g of phosphorus-32. From Table 17.3 you will notice that this isotope has a half-life of 14.3 days. From the graph you can see that after 14.3 days, **10.0 g** of phosphorus-32 (half the original mass) remains. After another 14.3 days passes (two half-lives) half of 10.0 g or **5.00 g** of phosphorus-32 remains. The other points on the graph can be verified in a similar manner.

The decay of carbon-14 (as shown in Figure 17.3) has a similar appearance. The only difference is the total amount of time it takes for the mass to disappear. If there are 20.0 g of carbon-14 at time equals zero, it will take one half-life (5730 years) for the mass to decay to 10.0 g, and 11,460 years to decay to 5.00 g.

**Figure 17.2 Decay of 20.0 g of Phosphorus-32**

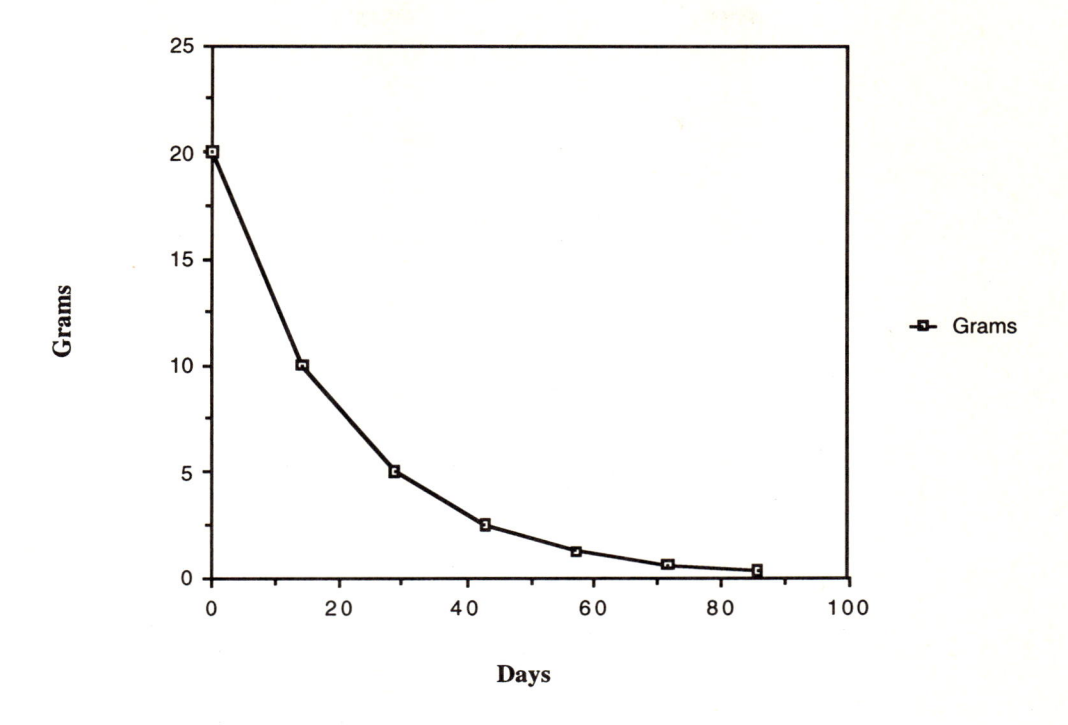

**Figure 17.3 Decay of 20.0 g Carbon-14**

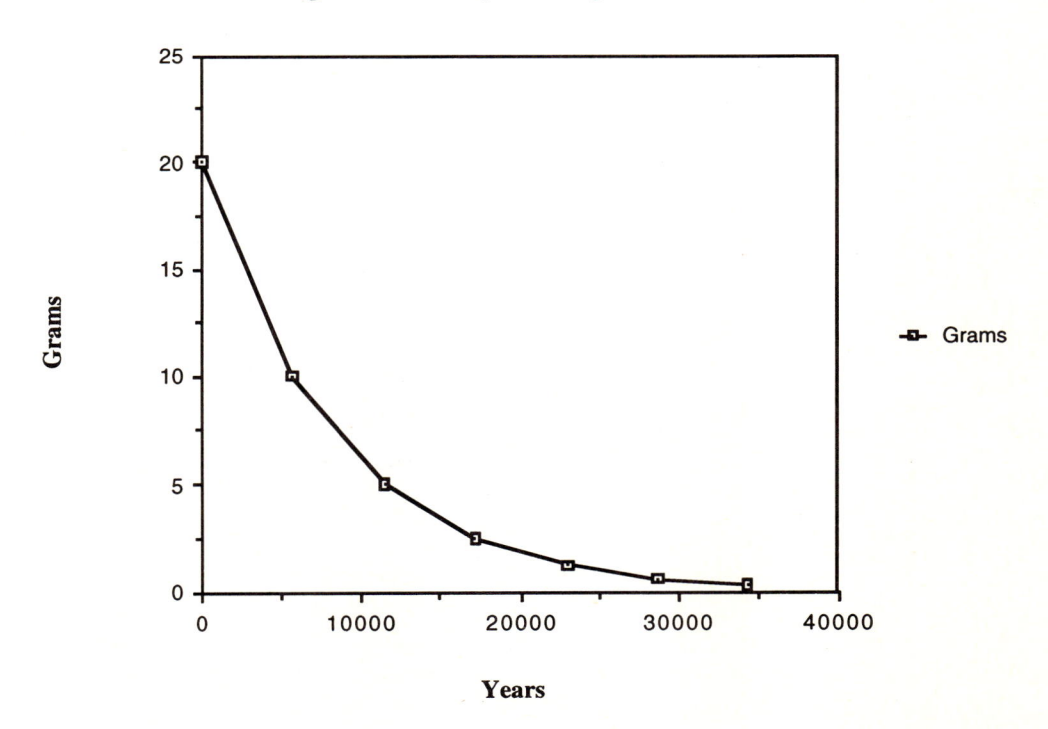

**Example 17.10:** If you start with a 2.00-g sample of phosphorus-32, how much will remain after 57.2 days?

**Solution:** To solve this problem, first determine how many half-lives have elapsed.

$$57.2 \text{ days} \times \frac{1 \text{ half-life}}{14.3 \text{ days}} = 4 \text{ half-lives}$$

Next, you can determine the mass left after 4 half-lives. This really means that you will multiply the original mass by 1/2 four times or $(1/2)^4$.

$$2.00 \text{ g} \times \left(\frac{1}{2}\right)^4 = 0.125 \text{ g}$$

**Table 17.3** Isotopic Half-Lives

| Isotope | Type of Decay | Half-Life |
|---------|---------------|-----------|
| U-238   | Alpha         | $4.51 \times 10^9$ years |
| U-235   | Alpha         | $7.1 \times 10^8$ years |
| Th-230  | Alpha         | 80,000 years |
| C-14    | Beta          | 5730 years |
| Ra-226  | Alpha         | 1600 years |
| Pu-238  | Alpha         | 86 years |
| Sr-90   | Beta          | 28 years |
| Pb-210  | Beta and Alpha | 21 years |
| H-3     | Beta          | 12.26 years |
| Co-60   | Not available | 5.26 years |
| Th-234  | Beta          | 24.1 days |
| P-32    | Beta          | 14.3 days |
| I-131   | Beta          | 8.05 days |
| Rn-222  | Alpha         | 3.82 days |
| Pb-214  | Beta          | 26.8 minutes |
| Bi-214  | Beta and Alpha | 19.7 minutes |
| Tl-206  | Beta          | 4.19 minutes |
| Po-218  | Alpha and Beta | 3.05 minutes |
| Tl-210  | Beta          | 1.3 minutes |
| O-15    | Positron      | 124 seconds |
| Po-214  | Alpha         | $1.64 \times 10^{-4}$ seconds |

The pattern of decay can be used for many practical purposes. One of these is radioactive dating. Carbon-14 is a good isotope for this technique. Carbon-14 is produced in the atmosphere on a regular basis by the reaction shown in EQ 17.5.

$$^{14}_{7}N + ^{1}_{0}n \rightarrow ^{1}_{1}H + ^{14}_{6}C \qquad \text{(EQ 17.5)}$$

Because the carbon-14 is produced on a regular rate, it is incorporated into all living material at a regular ratio (this is a very small amount of the total carbon, but it is measurable). When material ages the amount of carbon-14 decays over time by the process shown in EQ 17.6.

$$^{14}_{6}C \rightarrow {}^{14}_{7}N + {}^{0}_{-1}e \qquad \text{(EQ 17.6)}$$

The next example illustrates how carbon-14 can be used to measure the age of objects made from matter that incorporated organic (compounds with carbon) material at one time.

**Example 17.11:** A sample of aged material containing carbon-14 can be measured for radioactive decay. It is determined to have 2 beta emissions per minute per gram of total carbon. A new sample of the same material has 8 beta emissions per minute per gram of total carbon. How old is the sample?

**Solution:** The solution to this problem involves determining the amount of carbon-14 that has decayed in the old sample. To do this, note that there are 2 beta emissions to 8 beta emissions, which is the ratio of carbon-14 in the old sample versus one that is new.

$$\frac{2}{8} = \frac{1}{4} = \left(\frac{1}{2}\right)^2$$

The exponent in the above equation is the number of half-lives. The sample is 2 half-lives old.

$$2 \text{ half-lives} \times \frac{5730 \text{ years}}{1 \text{ half–life}} = 11,460 \text{ years}$$

---

**Practice Problems**

**17.24** What is the mass of a 0.15-g sample of oxygen-15 after 248 seconds?

**17.25** What is the mass of a 1.25-mg sample of hydrogen-3 after 36.78 years?

**17.26** What is the mass of a 27.0-mg sample of iodine-131 after 40.25 days?

**17.27** What is the mass of 2.9 g of plutonium-238 after 86 years?

**17.28** How old is a wood frame if it has 4 beta emissions per minute per gram of carbon, when a new sample of the same wood has 8 beta emissions per minute per gram of carbon?

**17.29** How old is a sample of cotton cloth if it has 2 beta emissions per minute per gram of carbon when a new sample of the cloth has 16 beta emissions per minute per gram of carbon?

**17.30** A sample is 5730 years old. How many beta emissions per minute per gram will it have if a new sample of the same material has 3.0 beta emissions per minute per gram of carbon?

## 17.5 Radioactivity Detection and Measurement

The unit for measuring radioactivity is the **curie**. This is defined as $37 \times 10^9$ disintegrations per second. This is the amount of radiation from a 1.0-g sample of radium.

Table 17.4 Instruments for Detecting Radioactive Decay

| Instrument | Particles of Radiation Detected | Type of Detection |
|---|---|---|
| Geiger-Müller counter | Alpha Beta Gamma | Radiation induced conductivity |
| Scintillation counter | Alpha Beta Gamma | Light emission induced by radiation |

### Practice Problems

**17.31** What type of emissions are detected by a Geiger-Müller counter?

**17.32** How does a scintillation counter work?

**17.33** How many disintegrations per second are in $1.0 \times 10^{-4}$ g of radium?

**17.34** How many disintegrations per second are in $1.0 \times 10^{-6}$ curies?

## 17.6 Medical Applications

Some of the common uses of radioactivity are summarized in Table 17.5. These are just a sample of the many applications currently used in medical therapy.

You should read this section of the chapter for an understanding of how radioactivity is used in medicine.

Table 17.5 Medical Applications of Radioactive Decay

| Name of Application | Method of Operation | Examples |
|---|---|---|
| Cancer cells irradiation | Emission of radioactive material destroys cells | Cobalt-60 |
| Biological tracers | Radioactive nuclides | Iodine-131 |
| PET | Positron emission | Carbon-11, oxygen-15 |

### Practice Problems

**17.35** Give an example of using radiation to destroy cancer cells.

**17.36** If a thyroid gland has an affinity for iodine, what substance might be used to trace activity in this organ?

## 17.7 Everyday Sources and Biological Effects of Radiation

The energy from radioactive decay is potentially harmful for two major reasons.

- Energy can ionize molecular bonds.
- Energy causes free radical formation.

Some of the qualities of radiation are summarized in Table 17.6.

**Table 17.6** Radiation Summary

| Type Radiation | Potential Damage | Depth of Penetration |
|---|---|---|
| Gamma | Can ionize bonds | Deepest penetration |
| Beta | Less damage than alpha, but over large area | Moderate penetration |
| Alpha | Damages biological tissues in small area | Shallow penetration |

The unit for measuring radiation exposure is called a **rem.** Some of the important measurements are found in Table 17.7.

**Table 17.7** Samples of Dosages of Radiation Important to Human Beings

|  | Rem Measurement |
|---|---|
| Dental x-ray | 0.005 |
| Causes decrease in white blood cells | 25-200 |
| Causes death | > 500 |
| Yearly average | 0.360 |

## Practice Problems

**17.37** What is the equation for the decay of radon-222?

**17.38** Why is radon gas dangerous for humans?

## 17.8 Nuclear Fission

This type of reaction releases huge amounts of energy. It is characterized by the breakup of a high-atomic-mass nuclide to form two smaller nuclides plus high-energy neutrons which carry the reaction to other atoms. There are two types that will be covered.

- Spontaneous fission
- Outside bombardment with high energy neutrons

An example of the first type is the decay of californium-252, as shown in EQ 17.7. Notice that the larger californium-252 is broken up to barium-142 and molybdenum-106. There are also 4 neutrons released which can bombard other atoms of this compound.

$$^{252}_{98}\text{Cf} \rightarrow {}^{142}_{56}\text{Ba} + {}^{106}_{42}\text{Mo} + 4\,{}^{1}_{0}\text{n} \qquad \text{(EQ 17.7)}$$

## 298    Chapter 17

An example of the bombardment type of fission is shown in EQ 17.8. When a uranium-235 sample (of the right mass) is hit with a high energy neutron, it will split apart in several known ways, one of which is shown in EQ 17.8. Again, notice that more neutrons are released by the reaction, which makes it easy to carry the reaction further.

$$_0^1n + {}_{92}^{235}U \rightarrow {}_{52}^{144}Cs + {}_{37}^{90}Rb + 2{}_0^1n \qquad \text{(EQ 17.8)}$$

**Example 17.12:** When uranium-235 is bombarded with a neutron, it forms xenon-142 plus four more high-energy neutrons plus what other nuclide?

**Solution:** To solve this problem, write out the information that is given. Then add up the mass numbers and atomic numbers on both sides, and solve the same way you solved in Example 17.9.

$$_0^1n + {}_{92}^{235}U \rightarrow {}_{54}^{142}Xe + \underline{\phantom{X}} + 4{}_0^1n$$

|  | Left Side | Right Side |
| --- | --- | --- |
| Mass number: | 235 + 1 = 236 | 142 + __ + 4 = 236 |
| Atomic number: | 0 + 92 = 92 | 54 + __ + 0 = 92 |

From the above you can see that the atomic number of the missing nuclide must be

$$92 - 54 = 38$$

The mass number for the missing nuclide must be

$$236 - 146 = 90$$

The nuclide can be identified as strontium (atomic number = 38). Therefore the missing nuclide from this fission reaction is **strontium-90**.

---

**Practice Problems**

**17.39** What other product is produced from the bombardment of uranium-235 with a neutron if three neutrons and barium-139 are the other products?

**17.40** What is the difference between uranium-235 and uranium-238 with respect to fission and radioactive decay?

**17.41** Which is the most likely to be fissionable, plutonium, lithium, or carbon? Why?

---

### 17.9 Nuclear Fusion

This is a reaction that also results in the production of energy. It differs from fission in that the energy is a result of the combination of light nuclides to form heavier ones.

**Example 17.13:** One type fusion is the reaction of two protons to form deuterium (hydrogen-2) and a positron. What is the nuclear equation for this reaction?

**Solution:** To solve this, simply transform the word description to symbols. Check to see if the atomic number and atomic mass are balanced.

$$2\,{}^{1}_{1}H \rightarrow {}^{0}_{1}e + {}^{2}_{1}H$$

---

**Practice Problems**

**17.42** Write the equation for the formation of an alpha particle and a positron from the fusion of helium-3 with a proton.

**17.43** Write the equation for the fusion of four protons to produce two positrons plus another particle.

**17.44** What particle fuses with a neutron to form tritium (hydrogen-3) plus an alpha particle?

---

## Integration of Multiple Skills

The problems below blend concepts of radioactive decay and concentration (see Chapter 13).

**Example 17.14:** A 450.-mL sample of a 0.00125 $M$ solution of iodine-131 would contain how many grams of the nuclide after 16.1 days?

**Solution:** To solve this problem, you need to review problems from Chapter 13 on determining mass from molarity.

$$450.\ \text{mL soln} \times \frac{0.00125\ \text{mol Iodine-131}}{1000\ \text{mL soln}} \times \frac{131\text{g Iodine-131}}{\text{mol Iodine-131}} = 0.0737\ \text{g Iodine-131}$$

---

**Practice Problems**

**17.45** What is the mass percent of the solution in Example 17.14?

**17.46** How long will it take the solution in Example 17.14 to have only 0.00230 g of the nuclide left?

---

## Self-Exam

**I. (4 ea) Write T (true) or F (false) in each corresponding blank.**

_____ 1. A *nuclide* is another term for an *isotope*.

_____ 2. A *nucleon* is either a proton or a neutron.

_____3. Alpha particles are equivalent to protons.

_____4. A positron has the same mass as an electron.

_____5. Transmutation can be used to prepare many transuranium elements.

_____6. The half-life of 1.00 g of radon-222 is shorter than that of a 2.00 g of radon-222.

_____7. Cobalt-60 is a radioactive substance that can be used to suppress cancer cell growth.

_____8. A Geiger-Müller counter can detect alpha, beta, and gamma radiation.

_____9. Fission is an endothermic reaction.

_____10. Fusion is the bombardment of large nuclides to form smaller ones.

II. (4 ea) Multiple Choice. Place the letter of the BEST answer in the corresponding blank.

_____1. Exposure to radioactivity is measured by what unit?
    A. Curie           B. Microcurie      C. Millicurie
    D. Rem           E. Beta emission

_____2. Gamma rays, which are produced from decay of metastable nuclides, have a mass of
    A. -1      B. +1      C. +2      D. +4      E. 0

_____3. A proton is identical to a(an)
    A. Alpha particle    B. Gamma ray
    C. Hydrogen-1 without its electron
    D. Beta particle     E. Positron

_____4. If nitrogen-13 decays by positron decay, the product nuclide is
    A. Carbon-13    B. Carbon-12    C. Nitrogen-15
    D. Nitrogen-14   E. Oxygen-17

_____5. When an atom of Be-9 is bombarded with an alpha particle, the products are one neutron and
    A. Carbon-13    B. Carbon-12    C. Lithium-18
    D. Boron-13     E. Boron-12

_____6. When a nuclide decays by alpha decay, the product nuclide
    A. Is one atomic number lower than the reactant.
    B. Is two atomic numbers lower than the reactant.
    C. Is one atomic number higher than the reactant.
    D. Is two atomic numbers higher than the reactant.
    E. Is four atomic numbers lower than the reactant.

_____7. A 10.0-g sample of an unknown isotope has a mass of 2.50 g after 39.4 min. How long is the half-life of this isotope?
    A. 78.8 min   B. 59.1 min   C. 39.4 min   D. 19.7 min   E. 9.9 min

_____ 8. The emission from an aged sample of wood has 2 beta emissions per min per gram of total carbon, while a new sample of the same compound has 8 beta emissions. How old is the sample (half-life for carbon-14 is 5730 years)?
   A. 22,920 years    B. 17,190 years    C. 11,460 years
   D. 5730 years      E. 1433 years

_____ 9. Radioactive decay of tritium (hydrogen-3) has a half-life of 12.5 years. How long will it take for 20.0-g sample of this compound to be reduced to 1.25 g of hydrogen-3?
   A. 1.25 years      B. 12.5 years      C. 25.0 years
   D. 37.5 years      E. 50 years

_____ 10. Which type of particle is penetrates the deepest?
   A. Alpha           B. Beta            C. Gamma
   D. Uranium-235     E. Uranium-238

### III. Provide Complete Answers

1. (5 pts) Write a nuclear equation for the alpha decay of uranium-238.

2. (5 pts) Write a nuclear equation for the positron decay of sodium-22.

3. (5 pts) What nuclide is formed in the spontaneous fission of californium-238, if the other products are molybdenum-106 and four neutrons?

4. (5 pts) What does aluminum-27 have to be bombarded with to produce magnesium-27 and helium-3?

## Answers to Practice Problems

| | | | | | |
|---|---|---|---|---|---|
| 17.1 | 1/4 | 17.2 | 1/32 | 17.3 | 7p; 8n |
| 17.4 | 34 | 17.5 | $^{51}_{23}V$ | 17.6 | $^{3}_{1}H$ |
| 17.7 | $^{14}_{6}C$ | 17.8 | $^{235}_{92}U$ | | |
| 17.9 | $^{222}_{86}Rn \rightarrow {}^{218}_{84}Po + {}^{4}_{2}He$ | 17.10 | $^{218}_{84}Po \rightarrow {}^{4}_{2}He + {}^{214}_{82}Pb$ | | |
| 17.11 | $^{234}_{90}Th \rightarrow {}^{0}_{-1}e + {}^{234}_{91}Pa$ | 17.12 | $^{234}_{91}Pa \rightarrow {}^{234}_{92}U + {}^{0}_{-1}e$ | | |
| 17.13 | $^{15}_{8}O \rightarrow {}^{15}_{7}N + {}^{0}_{1}e$ | 17.14 | $^{22}_{11}Na \rightarrow {}^{0}_{1}e + {}^{22}_{10}Ne$ | | |
| 17.15 | $^{210}_{82}Pb \rightarrow {}^{4}_{2}He + {}^{206}_{80}Hg$ | 17.16 | $^{210}_{82}Pb \rightarrow {}^{0}_{-1}e + {}^{210}_{83}Bi$ | | |
| 17.17 | $^{214}_{83}Bi \rightarrow {}^{4}_{2}He + {}^{210}_{81}Tl$ | 17.18 | $^{214}_{83}Bi \rightarrow {}^{0}_{-1}e + {}^{214}_{84}Po$ | | |
| 17.19 | Sodium-22 | 17.20 | Carbon-14 | | |
| 17.21 | Oxygen-17 | 17.22 | A neutron ($^{1}_{0}n$) | | |

**17.23**  Uranium-239

**17.24**  0.0375 g

**17.25**  0.156 mg

**17.26**  0.844 mg

**17.27**  1.45 g

**17.28**  5,730 years

**17.29**  17,190 years

**17.30**  1.5 emissions

**17.31**  Alpha, beta, and gamma

**17.32**  Radiation causes light emission.

**17.33**  $3.7 \times 10^6$ disintegrations/s

**17.34**  $3.7 \times 10^4$ disintegrations/s

**17.35**  Irradiation with cobalt-60

**17.36**  Iodine-131

**17.37**  $^{286}_{86}Rn \rightarrow {}^{4}_{2}He + {}^{218}_{84}Po$

**17.38**  Alpha particles damage tissues

**17.39**  Krypton-94

**17.40**  Both nuclides undergo alpha decay, but only uranium-235 undergoes fission.

**17.41**  Plutonium is fissionable. The others are too small.

**17.42**  ${}^{3}_{2}He + {}^{1}_{1}H \rightarrow {}^{0}_{1}e + {}^{4}_{2}He$

**17.43**  $4 {}^{1}_{1}H \rightarrow 2 {}^{0}_{1}e + {}^{4}_{2}He$

**17.44**  Lithium-6

**17.45**  0.00409 g%

**17.46**  $\dfrac{0.00230 \text{ g}}{0.07369 \text{ g}} = \dfrac{1}{32} = \left(\dfrac{1}{2}\right)^5$  This means 5 half-lives, or 40.25 days.

## Answers to Self-Exam

**I. T/F**
   1. T   2. F   3. F   4. T   5. T   6. F   7. T   8. T   9. F   10. F

**II. MC**
   1 D.   2. E   3. C   4. A   5. B   6. B   7. D   8. C   9. E   10. C

**III. Complete Answers**

   1. ${}^{238}_{92}U \rightarrow {}^{234}_{90}Th + {}^{4}_{2}He$

   2. ${}^{22}_{11}Na \rightarrow {}^{0}_{1}e + {}^{22}_{10}Ne$

   3. Barium-132

   4. Hydrogen-3

# 18
# Organic Chemistry

Like Chapter 17, this chapter takes a change in direction from the first sixteen chapters of the book. Organic chemistry is a large subdiscipline of chemistry that has a unique language. However, you will notice many applications from previous topics you have studied in this course. Topics about bonding, acidity, calculation of molar mass, balancing equations, etc., are all used in the study of organic chemistry.

Keys to success in this chapter:

- Learn new terminology.
- Relate new topics to ones previously encountered in this course.
- Learn the rules for nomenclature of organic compounds.
- Memorize the important *functional groups* introduced in this chapter.
- To help reduce the need for memorization, look for patterns in the nomenclature and reactivity of organic compounds.
- Learn the significant reactions of organic molecules.
- Be able to compare and contrast reactions of various functional groups.

## Summary of Verbal Knowledge

**organic chemistry:** The chemistry of carbon compounds.
**isomers:** These are compounds that have the same molecular formula but different structural formulas (that is, molecules with the same number of each kind of atom but with the atoms bonded in different ways).
**condensed structural formula:** A structural formula that uses established abbreviations for various groups of atoms.
**hydrocarbon:** A compound containing only carbon and hydrogen.
**alkane:** A hydrocarbon containing only single bonds and having the general formula $C_nH_{2n+2}$.
**alkyl group:** This is a group of atoms obtained by removing one hydrogen atom from an alkane.
**substitution reaction:** A reaction in which one atom (or atom group) substitutes for another atom (or group) on a molecule.
**alkene:** A hydrocarbon containing a carbon-carbon double bond and having the general molecular formula $C_nH_{2n}$.
**alkyne:** A hydrocarbon containing a carbon-carbon triple bond and having the general molecular formula $C_nH_{2n-2}$.
**addition reaction:** A reaction in which parts of a reactant molecule are added to each carbon atom of a carbon-carbon multiple bond; the carbon-carbon double bond becomes a single bond and a carbon-carbon triple bond becomes a double bond.

**polymer:** This is a very large molecule consisting of many repeating units of low molecular weight.

**monomer:** This is a compound used to prepare a polymer; the monomer gives rise to the polymer's repeating unit.

**addition (chain-growth) polymer:** A polymer formed by linking together many monomer molecules through addition reactions.

**aromatic hydrocarbon:** A hydrocarbon that has a structure based on the benzene ring.

**functional group:** A reactive portion of a molecule that undergoes predictable reactions.

**alcohol:** A compound with an —OH group bonded to a tetrahedral carbon atom.

**carbonyl compound:** A compound containing a C=O (carbonyl group).

**aldehyde:** A compound containing a carbonyl group bonded to a hydrocarbon group $R$ and a hydrogen atom.

**ketone:** A compound of formula RCOR' that contains a carbonyl group bonded to two hydrocarbon groups.

**carboxylic acid:** A compound containing the carboxyl group —COOH.

**ester:** A compound formed from a carboxylic acid, RCOOH, and an alcohol, R'OH, to form a compound with the general formula RCOOR'.

**amine:** A compound that is structurally derived by replacing one or more hydrogen atoms of ammonia ($NH_3$) with hydrocarbon groups.

**amide:** A compound derived from the reaction of ammonia or an amine with a carboxylic acid.

**condensation (step-growth) polymer:** A polymer formed from monomer molecules by condensation reactions.

## Review of Mathematical and Calculator Skills

Take time to review bonding concepts from Chapter 10. Bonding concepts in organic chemistry begin with an understanding of Lewis structures.

**Example 18.1:** How many bonds can carbon and nitrogen have in neutral molecules? How many lone pairs (l.p.) does each have in neutral molecules?

**Solution:** Carbon is in Group IVA. This means that it has four valence electrons. If it forms covalent bonds in neutral molecules in compliance with the octet rule, it can form 4 bonds. It will have no lone pairs.

Nitrogen is in Groups VA. This means that it has five valence electrons. If it forms covalent bonds in neutral molecules in compliance with the octet rule, it can form 3 bonds. It will have one lone pair.

| Atom | Bonds | Lone Pairs |
|------|-------|------------|
| C    | 4     | 0          |
| N    | 3     | 1          |

**Example 18.2:** Balance the equation for the reaction of glucose ($C_6H_{12}O_6$) to form ethanol ($CH_3CH_2OH$) and carbon dioxide gas.

**Solution:** You may need to review techniques for balancing equations (Chapter 6). For this particular equation, it will help to notice that all the hydrogen in the product is in the ethanol, so balance this product first. Also notice that the atoms in ethanol appear more than once. It might be easier to balance if you write ethanol as $C_2H_6O$.

$$C_6H_{12}O_6 \rightarrow 2CH_3CH_2OH + 2CO_2$$

## Practice Problems

**18.1** How many bonds and lone pairs will oxygen have in neutral molecules?

**18.2** How many bonds and lone pairs will hydrogen have in neutral molecules?

**18.3** How many bonds and lone pairs will Group VII atoms have in neutral molecules?

**18.4** Balance the equation for the reaction of $C_2H_2$ with $Br_2$ to form $C_2H_2Br_4$.

**18.5** Balance the equation for the combustion of $CH_3CH_2OH$ to form carbon dioxide and water.

## Application of Skills and Concepts

### 18.1 Carbon Atom Bonding

This chapter will cover the structure, nomenclature, functional groups and reactions of organic compounds. Like the chemistry of inorganic compounds, organic molecules can form ions. However, only neutral compounds will be addressed in this chapter.

Before beginning this study you will want to recall the following.

- Each covalent bond is represented by a line that denotes 2 electrons.
    Bonding electrons are shared by the two atoms they are between.
- Lone pairs (l.p.) are represented as 2 dots.
    For bookkeeping purposes, lone pairs belong to the atom they are on.
- Atoms in period 2, tend to form bonds that comply with the **octet rule**.
- Hydrogen, in period 1, will form bonds with 2 electrons around it.

Look at Table 18.1 to see a summary of the types of carbon atoms you will find in organic molecules. This table also reviews the shape of these molecules (about the central carbon atom). Try some problems which review the topics encountered in Chapter 10.

**Example 18.3:** Draw the Lewis structure for a molecule containing one carbon and four bromine atoms.

**Solution:** The first step for solving this problem involves counting the valence electrons. The carbon comes with 4 and each bromine with 7. Carbon will be the central atom, forming four single bonds, and each bromine (which can form one bond) will be attached to the central carbon atom. The structure is tetrahedral, and the Lewis structure looks like the structure following. Note the total of 32 total electrons.

```
        :Br:
         |
   :Br—C—Br:
         |
        :Br:
```

**Table 18.1** Specific Ways that Carbon Can Form Four Bonds

| Type | Bonds | Shape About C |
|---|---|---|
| —C— (with vertical bonds) | Four Single Zero Double | Tetrahedron |
| \C= / | Two Single One Double | Planar |
| =C= | Zero Single Two Double | Linear |
| —C≡ | One Single One Triple | Linear |

## Practice Problems

**18.6** Write the Lewis structure for a molecule with 1 carbon and 4 hydrogens.

**18.7** Write the Lewis structure for a molecule with 1 carbon, 2 hydrogens, and 1 oxygen.

**18.8** Write the Lewis structure for a molecule with 1 carbon, 2 hydrogens, and 2 chlorines.

**18.9** Write the Lewis structure for a molecule with 1 carbon, 4 hydrogens, and 1 oxygen.

### 18.2 Structural Formulas and Isomers

This section introduces some new terms dealing specifically with organic molecules. It will help to relate these new terms to ones you are already familiar with.

Look at Figure 18.1. This demonstrates the difference between a Lewis structure and a **structural formula**. The major difference is the omission of the lone pairs in the structural formula. This is done for brevity's sake, and the lone pairs still exist. It is assumed that the person writing and/or reading these formulas knows that the lone pairs are there.

**Organic Chemistry** 307

**Figure 18.1** Two Ways of Depicting Methanol

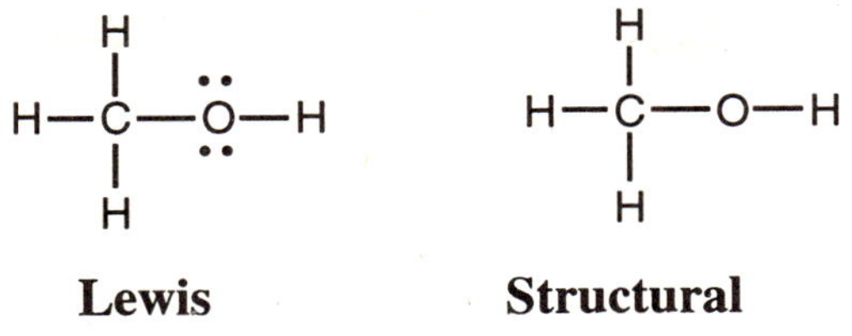

$$\text{Lewis} \qquad \text{Structural}$$

The molecular formula for methanol is CH$_4$O. Organic molecules are centered around carbon and usually contain hydrogen. Molecular formulas for organic molecules are therefore written with carbon first (followed by a subscript, which is omitted if it is 1), followed by hydrogen and its subscript, followed by other atoms in the molecules listed in alphabetical order with their subscripts.

Notice that there is only one way that the atoms in CH$_4$O can be connected together and still obey the **octet** rule. Try to prove this to yourself.

For most molecular formulas of organic molecules, this is not the case. There is a specific name for different structural formulas which have the same component atoms (i.e., same molecular formula). These molecules are **isomers** of each other.

Whereas structural formulas show every bond in the molecule, there is another way of depicting these molecules which incorporates abbreviations. These are called **condensed formulas**. They employ many abbreviations. Some of the more common ones are shown in Figure 18.2.

**Figure 18.2** Some Common Abbreviations Used in Condensed Organic Formulas

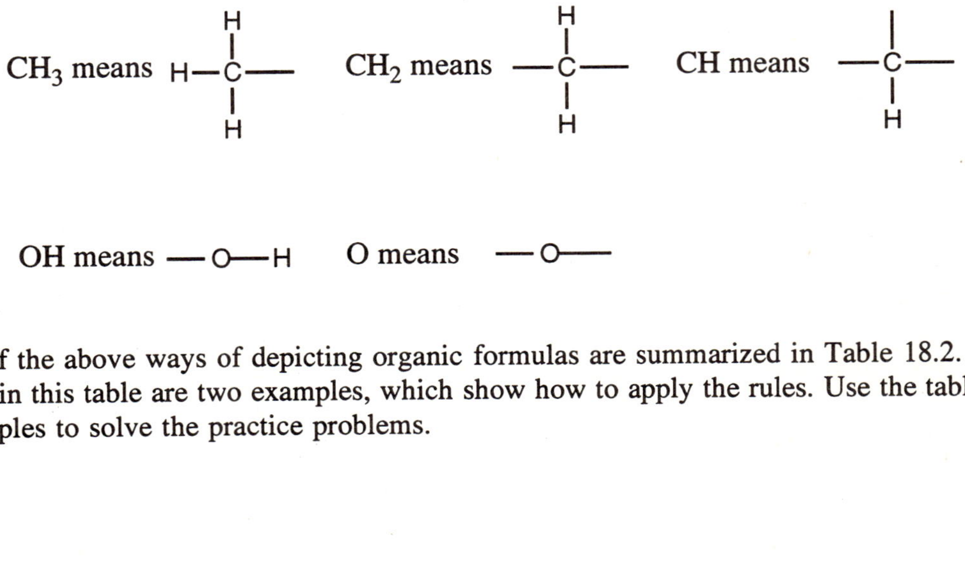

CH$_3$ means  H—C—  ;  CH$_2$ means  —C—  ;  CH means  —C—

OH means  —O—H    O means  —O—

All of the above ways of depicting organic formulas are summarized in Table 18.2. Also included in this table are two examples, which show how to apply the rules. Use the tables and the examples to solve the practice problems.

## Table 18.2 Organic Formulas

| Formula Type: | Molecular | Structural | Condensed |
|---|---|---|---|
| Rules | Write in the following order with appropriate subscripts: C, H, other atoms alphabetically. | Show all atoms and bonds between them. Lone pairs can be omitted. | Show bonding order of atoms using abbreviations. Actual lines depicting electrons in bonds may or may not be shown. |
| Example 18.4 (two isomers are shown) | $C_2H_6O$ | H-C(H)(H)-C(H)(H)-O-H  <br><br> H-C(H)(H)-O-C(H)(H)-H | $CH_3CH_2OH$ <br><br> $CH_3OCH_3$ |
| Example 18.5 (two isomers are shown) | $C_2H_4BrCl$ | H-C(H)(H)-C(H)(Br)-Cl <br><br> Br-C(H)(H)-C(H)(H)-Cl | $CH_3CHBrCl$ <br><br> $CH_2BrCH_2Cl$ |

## Practice Problems

**18.10** Write the condensed formulas for two isomers of $C_3H_8Cl$.

**18.11** Write the condensed formulas for three isomers of $C_3H_6O$.

**18.12** What is the molecular formula for $CH_3CH_2CH_2BrCHO$?

**18.13** What is the molecular formula for $CH_3CH_2OCH_2CH_3$?

### 18.3 Alkanes

There are three major skills you will want to master in this section.

- Understand general properties and uses of alkanes.
- Learn the system of nomenclature for alkanes.
- Understand the products of reactions of alkanes, and how to balance these reactions.

Some important facts about alkanes will make it easy to understand this class of organic molecules.

- They contain only carbon and hydrogen.
- All the bonds are single bonds.
- They are found in petroleum products and are purified by distillation.
- They undergo only two types of reactions.
- They have the general formula of $C_nH_{2n+2}$

A summary of the nomenclature for straight chain hydrocarbons is compiled in Table 18.3. Notice the pattern of the language of naming these organic compounds.

Table 18.3 Nomenclature for Straight Chain Alkanes, Alkenes, and Alkynes

| No. of C | Stem Name | Alkane | Molecular Formula $C_nH_{2n+2}$ | Alkene | Molecular Formula $C_nH_{2n}$ | Alkyne | Molecular Formula, $C_nH_{2n-2}$ |
|---|---|---|---|---|---|---|---|
| 1 | Meth- | Methane | $CH_4$ | None | None | None | None |
| 2 | Eth- | Ethane | $C_2H_6$ | Ethene | $C_2H_4$ | Ethyne | $C_2H_2$ |
| 3 | Prop- | Propane | $C_3H_8$ | Propene | $C_3H_6$ | Propyne | $C_3H_4$ |
| 4 | But- | Butane | $C_4H_{10}$ | Butene | $C_4H_8$ | Butyne | $C_4H_6$ |
| 5 | Pent- | Pentane | $C_5H_{12}$ | Pentene | $C_5H_{10}$ | Pentyne | $C_5H_8$ |
| 6 | Hex- | Hexane | $C_6H_{14}$ | Hexene | $C_6H_{12}$ | Hexyne | $C_6H_{10}$ |
| 7 | Hep- | Heptane | $C_7H_{16}$ | Heptene | $C_7H_{14}$ | Heptyne | $C_7H_{12}$ |
| 8 | Oct- | Octane | $C_8H_{18}$ | Octene | $C_8H_{16}$ | Octyne | $C_8H_{14}$ |
| 9 | Non- | Nonane | $C_9H_{20}$ | Nonene | $C_9H_{18}$ | Nonyne | $C_9H_{16}$ |
| 10 | Dec- | Decane | $C_{10}H_{22}$ | Decene | $C_{10}H_{20}$ | Decyne | $C_{10}H_{18}$ |

**Example 18.6:** What is the molecular formula for the alkane with 15 carbons?

**Solution:** Look at the general formula for alkanes. In this case n = 15. The number of hydrogens is equal to 2n + 2 = 2(15) + 2 = 32. The molecular formula is

$$C_{15}H_{32}$$

Once the number of carbons in an alkane is greater than 3, the alkane can have *isomeric* forms. This is illustrated by the following example.

**Example 18.7:** What are the isomers of $C_5H_{12}$?

**Solution:** Since there are more than 3 carbons, these atoms can be connected in more than one way and still obey all the rules for bonding. Following are the three different ways the carbons and hydrogens can be connected. These three molecules are isomers of each other.

$$CH_3CH_2CH_2CH_2CH_3 \qquad CH_3\underset{\underset{\displaystyle}{|}}{\overset{\overset{\displaystyle CH_3}{|}}{C}H}CH_2CH_3 \qquad CH_3\underset{\underset{\displaystyle CH_3}{|}}{\overset{\overset{\displaystyle CH_3}{|}}{C}}CH_3$$

Because alkanes (and other organic compounds) can exist in so many isomeric forms, a systemic way of naming these compounds has been developed by IUPAC (International Union of Pure and Applied Chemistry).

For any alkane that is not a straight chain of carbons, a special way of naming this compound is used. The branches off the straight chain have specific names. Some of the common ones are shown in Table 18.4. These groups are called **alkyl** groups. They have a stem name that matches the stems in Table 18.3 but have one hydrogen missing. This is the location where the carbon is bonded to another carbon on the chain.

**Table 18.4** Alkyl Groups Used in Naming Organic Compounds

| Group Structure | Name |
|---|---|
| $CH_3-$ | Methyl- |
| $CH_3CH_2-$ | Ethyl- |
| $CH_3CH_2CH_2-$ | Propyl- |
| $CH_3CH_2CH_2CH_2-$ | Butyl- |

The simple rules for nomenclature are summarized below. Each rule is demonstrated by a correct example and sometimes an incorrect example.

**Rule 1: Find the Parent Chain.** This is the longest continuous carbon chain. It can be in any direction.

**Correct** (longest has 8 C)    **Incorrect** (longest has only 6 C)

**Rule 2: Identify any Alkyl Branch Groups.** See Table 18.4

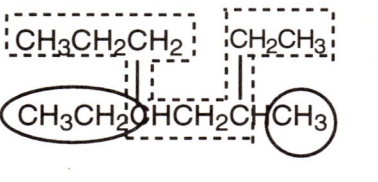

CH₃CH₂CH₂   CH₂CH₃
         |      |
CH₃CH₂CHCH₂CHCH₃

Ethyl group       Methyl group

**Rule 3: Add Location Numbers for Any Substituent Groups.** Begin numbering the longest chain closest to the first branch.

```
8  7  6       2  1                1  2  3        7  8
CH₃CH₂CH₂   CH₂CH₃               CH₃CH₂CH₂    CH₂CH₃
         |      |                          |        |
CH₃CH₂CHCH₂CHCH₃                 CH₃CH₂CHCH₂CHCH₃
      5   4  3                          4   5  6
```

**Correct** (begins numbering so closest branch is C3)

**Incorrect** (begins numbering so closest branch is C4)

**Rule 4: Write the Name of the Complete Alkane.** The root name is the longest chain.

The root for the alkane used in Rule 3 is **octane** because there are eight atoms in the longest chain.

Alkyl substituents will be written before the root name in alphabetical order.

The two alkyl substituents are **ethyl** and **methyl,** which are also written in the correct alphabetical order. The name without location numbers is

**ethylmethyloctane**

Prefix the substituent with the number of the carbon on the straight chain. These numbers are from the location of the alkyl group on the chain and are followed by a dash.

**5-ethyl-3-methyloctane**

Note the correct punctuation and writing of the compound.

If multiple branches of the same group are present, use prefixes *di, tri, tetra,* etc. In this case, separate the numbers by commas. For example, if both of the branches in the compound had been methyl as shown below, the correct name would 3,5-dimethyloctane. Note that a number is still given for each branch alkyl group, but they are grouped with a prefix. The prefixes do **not** count when alphabetizing.

```
      8   7   6       2   1
     CH₃CH₂CH₂       CH₂CH₃
            |           |
          CH₃CHCH₂CHCH₃
            5   4   3
```

**3,5-Dimethyloctane**

The last part of this section deals with reactions of alkanes. These are of two types.

- Combustion (burning of the alkane in oxygen)
- Substitution

Two examples illustrate these two types of reactions.

**Example 18.8:** Write a balanced equation for the combustion of $CH_4$.

**Solution:** Since combustion is the burning of a compound in oxygen to produce carbon dioxide and water, the following balanced equation is fairly easy to write.

$$CH_4 + 2O_2 \rightarrow CO_2 + 2H_2O$$

**Example 18.9:** Write an equation for the reaction of chlorine and hexane (assume conditions of the reaction favor a monochlorinated product).

**Solution:** This is a new type of reaction that you have not encountered before in this course. The important thing to remember is that a chlorine atom is *substituted* for a hydrogen atom in this reaction. Look at the hexane molecule. There are three unique types of hydrogen atoms in this molecule.

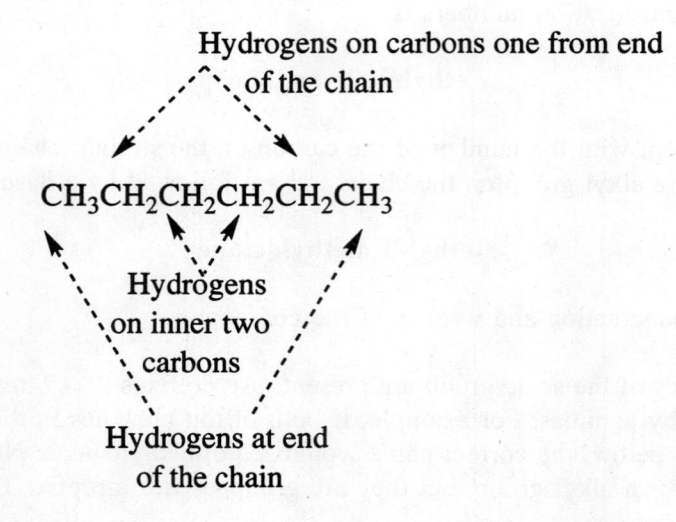

Hydrogens on carbons one from end of the chain

$CH_3CH_2CH_2CH_2CH_2CH_3$

Hydrogens on inner two carbons

Hydrogens at end of the chain

The solution to this problem has three correct equations (for each type of hydrogen), which are written below.

$$CH_3CH_2CH_2CH_2CH_2CH_3 + Cl_2 \rightarrow CH_3CH_2CH_2CH_2CH_2CH_2Cl + HCl$$

$$CH_3CH_2CH_2CH_2CH_2CH_3 + Cl_2 \rightarrow CH_3CH_2CH_2CH_2CHClCH_3 + HCl$$

$$CH_3CH_2CH_2CH_2CH_2CH_3 + Cl_2 \rightarrow CH_3CH_2CH_2CHClCH_2CH_3 + HCl$$

## Practice Problems

**18.14** What is the molecular formula for the alkane with 16 carbons?

**18.15** What are two structural isomers of $C_4H_{10}$?

**18.16** What are the structural formulas for $C_6H_{14}$?

**18.17** Name each isomer in problem 18.16.

**18.18** Name:

$$\begin{array}{c} CH_3CH_2 \\ | \\ CH_3CH_2CHCHCH_2CH_2CH_3 \\ | \\ CH_3 \end{array}$$

**18.19** Name:

$$\begin{array}{c} CH_3 \quad CH_3 \\ | \quad\quad | \\ CH_3CCH_2CHCH_3 \\ | \\ CH_3 \end{array}$$

**18.20** Draw 2,5-dimethylhexane.

**18.21** Draw 5-ethyl-4-propyldecane.

**18.22** Write the balanced equation for the reaction of propane and oxygen.

**18.23** Write a balanced equation for the combustion of hexane.

**18.24** Write a balanced equation for the production of each monochlorinated alkane from the reaction of propane and chlorine.

**18.25** Write a balanced equation for the production of each monochlorinated alkane from the reaction of pentane and chlorine.

## 18.4 Alkenes and Alkynes

These two groups of organic molecules also have only carbon and hydrogen. Their nomenclature (for straight-chain molecules) is summarized in Table 18.3. They each have an identifying suffix. Some of the following attributes are important to understand.

- Alkenes have the general formula $C_nH_{2n}$.
- Alkynes have the general formula $C_nH_{2n-2}$.

## 314  Chapter 18

- Alkenes have at least one double bond.
- Alkynes have at least one triple bond.
- Sources of these compounds are alkanes (cracking).
- They undergo many reactions, including addition.

There are simple rules for naming alkenes and alkynes with branches.

**Rule 1: Find the Parent Chain.** This will be the longest chain that contains both carbons with the double or triple bond.

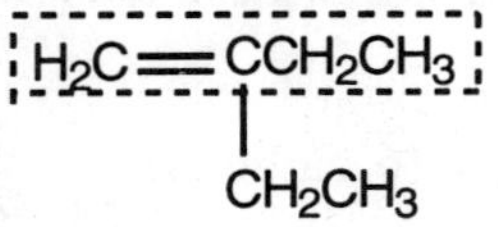

$$H_2C=CCH_2CH_3$$
$$|$$
$$CH_2CH_3$$

**Correct** (longest C chain containing double bond is 4.)

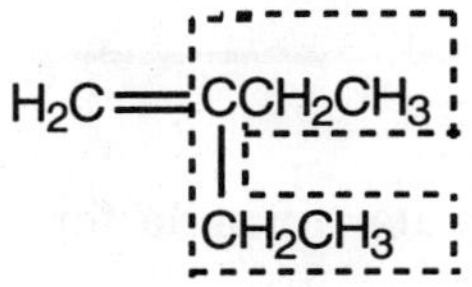

$$H_2C=CCH_2CH_3$$
$$|$$
$$CH_2CH_3$$

**Incorrect** (this chain of 5C does not contain both atoms of the double bond.)

**Rule 2: Identify Branches.** Do this as you did with alkanes.

For the molecule above, there is an *ethyl* branch.

**Rule 3: Add Location Numbers.** Begin numbering as close to the first carbon that contains the multiple bond. The number for the alkene/alkyne will be the first carbon that is in the multiple bond.

For example, a double bond between carbons 2 and 3 in a straight chain of six carbons is **2-hexene**.

In the example used in Rules 1 and 2, the chain would be numbered as follows:

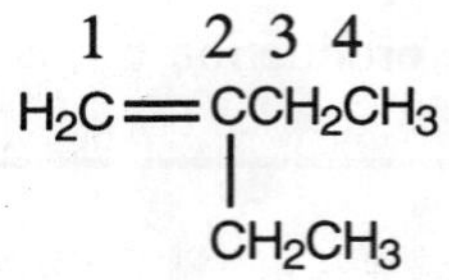

$$\overset{1}{H_2C}=\overset{2\;3\;4}{CCH_2CH_3}$$
$$|$$
$$CH_2CH_3$$

**Rule 4: Name Entire Molecule.** The root name has the number of the multiple bond and the correct number of carbons in the longest chain. The branches are numbered as prefixes, just as they were in alkanes.

The root name for the above molecule is **1-butene**.

The entire name is derived by adding the branch, with its location number as a prefix.

**2-ethyl-1-butene**

Example 18.10: Write the formula for 2-methyl-3-heptene.

Solution: The longest chain must have 7 carbons from the root name *hep-*. This molecule has a double bond (from the suffix *-ene*). The double bond begins at carbon 3, and a methyl group is located on carbon number 2. The molecule is drawn below.

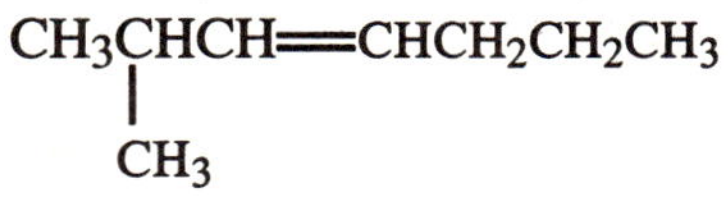

$$\text{CH}_3\text{CHCH}=\text{CHCH}_2\text{CH}_2\text{CH}_3$$
$$|$$
$$\text{CH}_3$$

Example 18.11: What is the name of the following molecule?

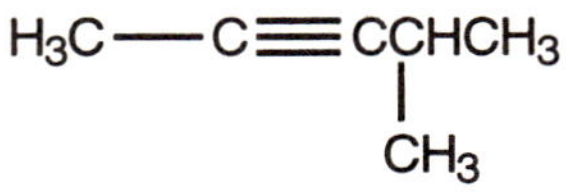

$$\text{H}_3\text{C}-\text{C}\equiv\text{CCHCH}_3$$
$$|$$
$$\text{CH}_3$$

Solution: The triple bond is between carbons 2 and 3. The longest chain of carbons is 5. There is a methyl substituent on carbon 4. Therefore, the name is

**4-methyl-2-pentyne**

Some examples will illustrate the reactions that produce alkenes and alkynes, and the reactions that these molecules undergo.

Example 18.12: Write the reaction for propane being heated to form propene and hydrogen gas.

Solution: This simply involves transferring the word description to a chemical equation and then balancing the equation.

$$\text{CH}_3\text{CH}_2\text{CH}_3 \xrightarrow{\Delta} \text{CH}_2=\text{CHCH}_3 + \text{H}_2$$

Example 18.13: The reaction of 2-butene with water produces $\text{CH}_3\text{CH}_2\text{CHOHCH}_3$. Write the equation for this reaction.

Solution: This is an addition reaction, which means that the hydrogen from water is added to one atom of the double bond and the -OH from the water is added to the other carbon of the double bond. Identify where these pieces of water were added to 2-butene in the following reaction.

$$CH_3CH=CHCH_3 + H_2O \rightarrow CH_3CH_2CHOHCH_3$$

Water is not the only molecule that can add to multiple bonds. Bromine, chlorine, hydrochloric acid, hydrobromic acid, etc. will also undergo addition reactions with alkenes and alkynes.

**Example 18.14:** The reaction of 2-butyne with $Br_2$ produces $CH_3CBr_2CBr_2CH_3$. Write the equation for this reaction.

**Solution:** With an excess of bromine, two molecules will add to the alkyne. You can think of this addition as one that occurs in two steps as shown below.

$$CH_3C\equiv CCH_3 + Br_2 \rightarrow CH_3CBr=CBrCH_3 + Br_2 \rightarrow CH_3CBr_2CBr_2CH_3$$

---

**Practice Problems**

**18.26** What is the name of $CH\equiv CCH_2CH_2CH_3$?

**18.27** What is the name of $CH_3CH_2CH_2CH_2CH=CHCH_3$?

**18.28** Draw the structure of 3-methyl-2-octene.

**18.29** Draw the structure of 4-ethyl-2-nonyne.

**18.30** Calcium carbide ($CaC_2$) plus water reacts to form aqueous calcium hydroxide plus ethyne (acetylene). Write a balanced equation for this reaction.

**18.31** Write a balanced equation for the reaction of HCl and ethyne.

**18.32** Write a balanced equation for the reaction 2-butene and bromine.

**18.33** Write a balanced equation for the reaction of 3-hexene and water.

---

## 18.5 Polyalkene Polymers

Some important aspects of polymers are listed below.

- These are very large molecules.
- They are made up of repeating units, called **monomers**.
- Polymers are formed from two methods.
    Addition (this section)
    Condensation (Section 18.8)

Addition polymers form when the electrons in a double bond connect to another molecule of the same type, many times. These molecules vary in total size but have enormous molar masses.

**Example 18.15:** Show the structure for polypropylene.

**Solution:** The repeating unit in this molecule is propene (notice the connection to propylene, which is common nomenclature, although slightly different from IUPAC). The solution shown below only demonstrates the connection of three units, but imagine that the actual reaction is thousands of units long.

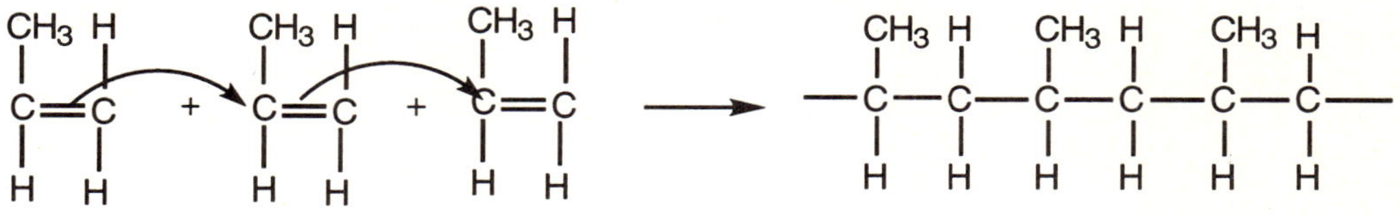

$$\underset{H}{\overset{CH_3}{C}}=\underset{H}{\overset{H}{C}} + \underset{H}{\overset{CH_3}{C}}=\underset{H}{\overset{H}{C}} + \underset{H}{\overset{CH_3}{C}}=\underset{H}{\overset{H}{C}} \longrightarrow -\underset{H}{\overset{CH_3}{C}}-\underset{H}{\overset{H}{C}}-\underset{H}{\overset{CH_3}{C}}-\underset{H}{\overset{H}{C}}-\underset{H}{\overset{CH_3}{C}}-\underset{H}{\overset{H}{C}}-$$

---

### Practice Problems

**18.34** What is the name of $-CH_2-CH_2-CH_2-CH_2-$?

**18.35** Draw the polymer of the monomer of $CH_2=CHOH$.

---

## 18.6 Aromatic Hydrocarbons

The use of the hexagon with the circle inside is the representation for **benzene**. This representation means that there is one carbon with an attached hydrogen at each vertex of the hexagon. The molecular formula is $C_6H_6$. There are alternating double bonds between the carbons in the ring. These electrons are **delocalized** so that the double bonds are always alternating.

Whenever some other atom or group is substituted onto the ring, one hydrogen is replaced. Table 18.5 shows some aromatic compounds, their molecular formulas, and some nomenclature.

**Example 18.16:** What is the molecular formula for *meta*-xylene?

**Solution:** Look for this compound in Table 18.5. *Meta*-xylene, *m*-xylene, is really 1,3-dimethyl benzene, and it has 8 carbons and 10 hydrogens (3 on each methyl group and 4 on the benzene ring).

### Table 18.5 Aromatic Compounds

| Structure | Molecular Formula | IUPAC Name | Common Name |
|---|---|---|---|
|  ⬡ | $C_6H_6$ | Benzene | Benzene |
|  ⬡⬡ | $C_{10}H_8$ | Naphthalene | Naphthalene |

| Structure | Formula | Name | Common Name |
|---|---|---|---|
| C₆H₅– (phenyl) | $C_6H_5-$ | Phenyl group | Phenyl group |
| 1,2-dimethylbenzene | $C_8H_{10}$ | 1,2-Dimethyl benzene | o-Xylene |
| 1,3-dimethylbenzene | $C_8H_{10}$ | 1,3-Dimethyl benzene | m-Xylene |
| 1,4-dimethylbenzene | $C_8H_{10}$ | 1,4-Dimethyl benzene | p-Xylene |
| chlorobenzene | $C_6H_5Cl$ | Chlorobenzene | Phenyl chloride |
| nitrobenzene | $C_6H_5NO_2$ | Nitrobenzene | Nitrobenzene |
| 2,4,6-trinitrotoluene | $C_7H_5N_2O_6$ | 1-methyl-2,4,6-trinitrobenzene | 2,4,6-Trinitrotoluene (TNT) |

> **Practice Problems**
>
> **18.36** What is the molecular formula of 1,3,5-triethylbenzene?
>
> **18.37** Write the reaction for benzene with chlorine and iron(III) chloride catalyst.

## 18.7 Alcohols and Ethers

Both these functional groups can be thought of as deriving from water. If one hydrogen in water is replaced with an alkyl group (R), the compound is an alcohol. If both hydrogens are replaced the compound is an ether. See Figure 18.3.

**Figure 18.3** General Formulas for Alcohols and Ethers

$$H-O-H \qquad R-O-H \qquad R-O-R'$$

water  alcohol  ether

You have probably noticed that alcohols contain an —OH group (hydroxyl group). This group is covalently bonded to carbon and does not behave as it would in bases such as NaOH or KOH. Alcohols are not ionic.

There are several easy rules for naming alcohols.

**1. Rule 1: Find Parent Chain.** (This must contain the carbon with the alcohol group.) In the example shown below, the longest chain is along the horizontal carbons.

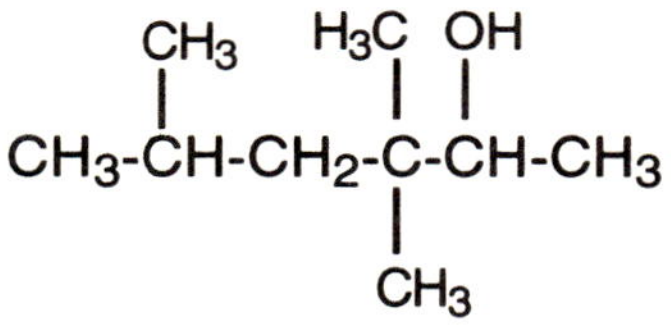

```
        CH3   H3C OH
         |     |   |
CH3-CH-CH2-C-CH-CH3
              |
              CH3
```

**2. Rule 2: Find Any Branches and Identify Them.** (Same method used with hydrocarbon nomenclature.) This molecule has three *methyl* branches.

**3. Rule 3: Add Location Numbers.** (Start numbering with the carbon closest to the —OH group.)

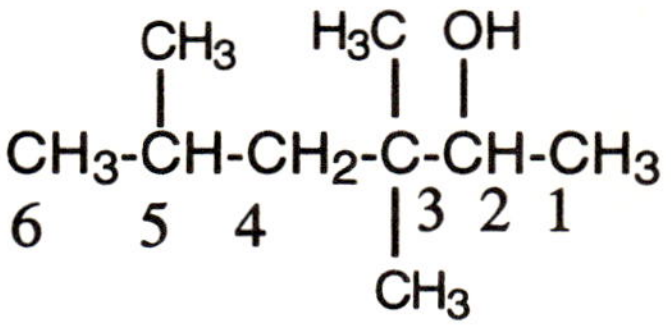

```
        CH3   H3C OH
         |     |   |
CH3-CH-CH2-C-CH-CH3
 6   5   4   |3  2  1
             CH3
```

**4. Write Complete name:** (Use the following steps.)
  a. Identify the root name: In this case the longest carbon chain of 6 carbons would be hexane. When naming an alcohol, the *e* is replaced with the suffix *ol*. This is a

**hexanol**

## 320   Chapter 18

b. Identify the carbon number with the functional group. In this case, the —OH is on carbon number 2, so it is

**2-hexanol**

c. Add the prefixes for any alkyl branches.

**3,3,5-trimethyl-2-hexanol**

Table 18.6 Some Examples of Organic Compounds Containing Oxygen

| No. of C | Alcohol | Aldehyde | Ketone | Ether | Acid |
|---|---|---|---|---|---|
| 1 | $CH_3OH$ <br> Methanol <br> (methyl alcohol) | H–C(=O)–H <br> Methanal <br> (formaldehyde) | None | None | HO–C(=O)–H <br> Methanoic acid <br> (formic acid) |
| 2 | $CH_3CH_2OH$ <br> Ethanol <br> (ethyl alcohol) | H–C(=O)–$CH_3$ <br> Ethanal | None | $CH_3OCH_3$ <br> Methyl ether | HO–C(=O)–$CH_3$ <br> Ethanoic acid <br> (acetic acid) |
| 3 | $CH_3CH_2CH_2OH$ <br> 1-Propanol | H–C(=O)–$CH_3CH_2$ <br> Propanal | $CH_3$–C(=O)–$CH_3$ <br> Propanone <br> (acetone) | $CH_3OC_2H_5$ <br> Methyl ethyl ether | HO–C(=O)–$CH_3CH_2$ <br> Propanoic acid |
|  | $CH_3CHOHCH_3$ <br> 2-Propanol |  |  |  |  |
| 4 | $CH_3(CH_2)_3OH$ <br> 1-Butanol | H–C(=O)–$CH_3CH_2CH_2$ <br> Butanal | $CH_3$–C(=O)–$C_2H_5$ <br> Butanone | $C_2H_5OC_2H_5$ <br> Ethyl ether | HO–C(=O)–$CH_3CH_2CH_2$ <br> Butanoic acid |
|  | $CH_3CHOHC_2H_5$ <br> 2-Butanol |  |  | $CH_3OC_3H_7$ <br> Methyl proply ether |  |

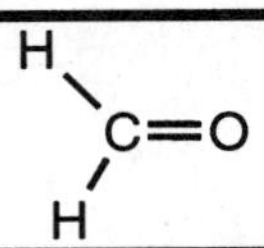

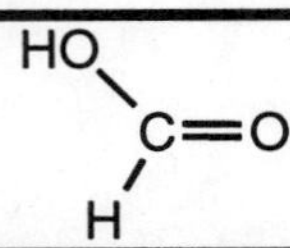

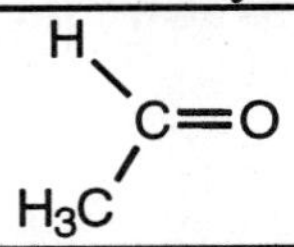

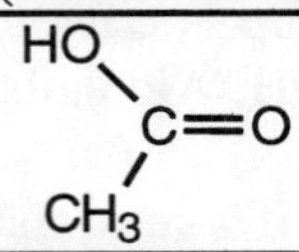

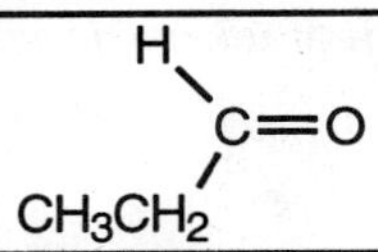

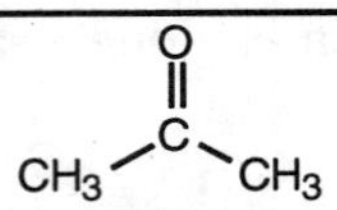

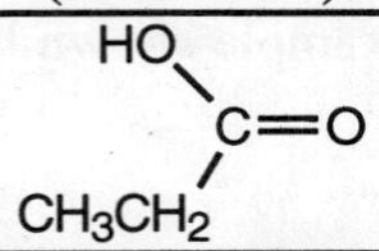

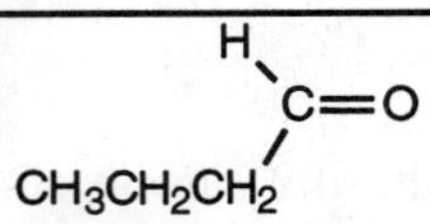

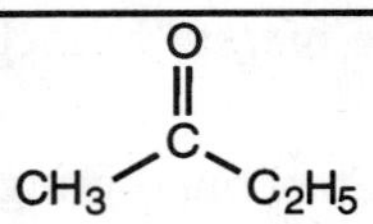

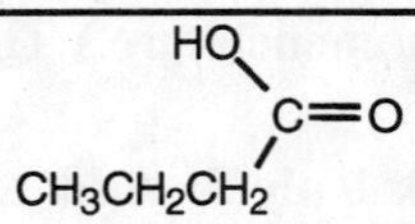

**Example 18.17:** What is the name of $CH_3CH(CH_3)CH_2OH$?

**Solution:** Notice that in this condensed formula the longest chain with the alcohol group is three carbons long. There is a methyl branch at carbon number 2 in the chain. The name is

**2-methyl-1-propanol**

There are three reactions that are representative of the chemistry of alcohols.

- Combustion
- Oxidation to aldehydes or ketones
- Oxidation to carboxylic acids

Examples of these reactions are shown in the equations below (EQ 18.1 to 18.4).

$$2CH_3OH + 3O_2 \rightarrow 2CO_2 + 4H_2O \qquad \text{(EQ 18.1)}$$

$$CH_3OH + (O) \rightarrow CH_2O \qquad \text{(EQ 18.2)}$$
$$\text{formaldehyde}$$

$$CH_3CH_2OH + (O) \rightarrow CH_3COOH \qquad \text{(EQ 18.3)}$$
$$\text{acetic acid}$$

$$CH_3CH(OH)CH_3 + (O) \rightarrow CH_3C(O)CH_3 \qquad \text{(EQ 18.4)}$$
$$\text{acetone}$$

## Practice Problems

**18.38** What is $CH_3CH_2CH_2CH_2OH$?

**18.39** What is $CH_3CHOHCH_2CH_2CH_2CH_3$?

**18.40** Draw 2-methyl-3-octanol.

**18.41** Write an equation for the reaction of 2-butanol burning in oxygen.

**18.42** Write an equation for the reaction of 2-butanol oxidized to a ketone.

**18.43** Write an equation for the reaction of 1-butanol oxidized to a aldehyde.

**18.44** Write an equation for the reaction of 1-butanol oxidized to a carboxylic acid.

## 18.8 Aldehydes and Ketones

These two classes of molecules contain a carbonyl group (C=O). For the aldehydes, the double bond to the oxygen is on the carbon at the end of a chain. For the ketones, it is on an inner carbon.

Aldehydes and ketones are often the product of oxidation of alcohols.

- Aldehydes have the suffix *al* replacing the *e* in the parent alkane chain.

## 322   Chapter 18

- Ketones have the suffix *one* replacing the *e* in the parent alkane chain.
- Aldehydes have the general formula RCHO.
- Aldehydes have the general formula RC(O)R'.

Their nomenclature is illustrated in Table 18.6 (page 320).

**Example 18.18:** What is the structure of 2-methylbutanal?

   **Solution:** Use the previous rules of nomenclature and Table 18.6 to solve this.

$$CH_3CH_2CH(CH_3)-CHO$$

## Practice Problems

**18.45** What is $CH_3CH_2CH_2CH_2CH_2CHO$?

**18.46** What is the name of $CH_3CH_2CH_2C(O)CH_2CH_3$?

**18.47** Draw three ketones with 5 carbons and name them.

**18.48** Draw three aldehydes with 5 carbons and name them.

### 18.9  Carboxylic Acids and Esters

These two classes of molecules are characterized by the carboxylate group. This is a carbon double bonded to one oxygen and also single bonded to another oxygen.

```
        O                          O
        ||                         ||
    R—C—OH                     R—C—OR'

   RCOOH or RCO₂H              RCOOR' or RCO₂R'
   Carboxylic Acid                  Ester
```

Note the carboxylate group in both the generic molecules above. The R can be either a hydrogen or an alkyl group. The R' is an alkyl group.

When naming these two types of compounds, do the following:

1. Find the longest chain containing the C=O, and determine the alkane name for a chain of that length.

2. If the compound is an **acid** (abbreviation used in organic chemistry for a carboxylic acid), name the compound by substituting the suffix *oic* for the *e* in the alkane. If R in the above generic formula is $CH_3CH_2-$, the acid is

**propanoic acid**

3. If the compound is an **ester**, name the compound by substituting the suffix *oate* for the *e* in the alkane. A prefix must be added, by naming the R' alkyl group. If the R in the above generic formula is $CH_3CH_2-$ and R' is $CH_3-$, the ester is

**methyl propanoate**

There are several reactions that are important in this section. They are best learned by looking at the following examples.

**Example 18.19:** Write an equation for the reaction of acetic (ethanoic) acid and water.

**Solution:** This is an acid-base equilibrium reaction. All the carboxylic acids are weak acids, which only partially ionize in water.

$$CH_3COOH + H_2O \rightleftharpoons H_3O^+ + CH_3COO^-$$

**Example 18.20:** Write an equation for the condensation reaction of butanoic acid and methanol.

**Solution:** This is a reaction where two molecules combine to form a larger molecule, and a small molecule (in this case water). The larger molecule is an ester. Its name is methyl butanoate. The methyl group came from the methanol.

$$CH_3CH_2CH_2COOH + CH_3OH \rightarrow H_2O + CH_3CH_2CH_2COOCH_3$$

---

**Practice Problems**

**18.49** Propanoic acid reacts with sodium hydroxide to produce what products?

**18.50** Draw hexanoic acid.

**18.51** Draw proplybutanoate

**18.52** What ester does the reaction of benzoic acid ($C_6H_5COOH$) plus ethanol produce?

---

## 18.10 Amines

The major feature of this class of molecules and the amides (Section 18.11) is the presence of a nitrogen atom. This is fairly important, because the lone pair of electrons on nitrogens bonded to three other groups is basic.

The lone pair can accept a proton, just like ammonia, $NH_3$. Also like ammonia, amines are weak bases and do not behave in the strong manner of ionic hydroxides.

The nomenclature of amines is similar to that of previous organic molecules. Some are summarized in Table 18.7. When only one hydrogen is replaced, the alkyl substituent is named as a prefix to the root name, **amine**.

Table 18.7 Some Amines and Their Accepted Nomenclature

| Formula/Structure | Name | Formula/Structure | Name |
| --- | --- | --- | --- |
| $NH_3$ | Ammonia | $CH_3CH_2CH_2CH_2NH_2$ | Butylamine |
| $CH_3NH_2$ | Methylamine | 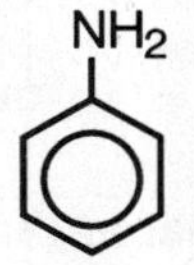 C₆H₅NH₂ | Phenylamine (aniline) |
| $CH_3NH(CH_2CH_3)$ | Ethylmethylamine | 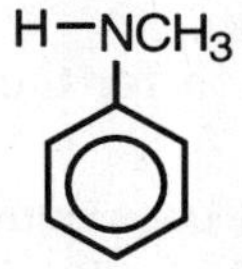 C₆H₅-NHCH₃ | N-Methyl aniline |

**Example 18.21:** What happens when triethylamine reacts with hydrochloric acid?

**Solution:** This is the reaction of a weak base and a strong acid. It will lead to an amine cation and a chloride anion.

$$(CH_3CH_2)_3N + HCl \rightarrow (CH_3CH_2)_3NH^+ + Cl^-$$

---

**Practice Problems**

18.53 Draw diethyl amine.

18.54 What is the molecular formula?

18.55 What is the straight-chain isomer of this compound?

18.56 Write the reaction of butylamine with hydrobromic acid.

---

## 18.11 Amides and Polyamides

There is only one letter that is different in the spelling of amide versus amine. The molecules both have nitrogen, but their properties are somewhat different. Amides are weaker bases than amines. This is a result of their structure. The nitrogen in amides is connected to a carbonyl carbon.

Some typical amides are shown in Table 18.8. These compounds are formed from the condensation reaction of a carboxylic acid (or derivative of an acid, such as an acid chloride) and an amine. An example and several problems will demonstrate this type of reaction.

One interesting reaction covered in your textbook is worth examining. It involves the reaction of a diamine [NH$_2$(CH$_2$)$_6$NH$_2$] with a diacetyl chloride molecule to form nylon. Nylon is a polymer with many repeating amide bonds.

Table 18.8 Some Common Amides

| Structure/Formula | Name | Structure/Formula | Name |
|---|---|---|---|
| HC(O)NH$_2$ | Methanoamide (formamide) | CH$_3$C(O)NH$_2$ | Ethanoamide (acetamide) |
| 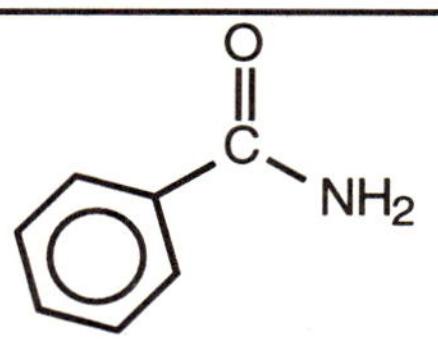 C$_6$H$_5$C(O)NH$_2$ | Benzamide | 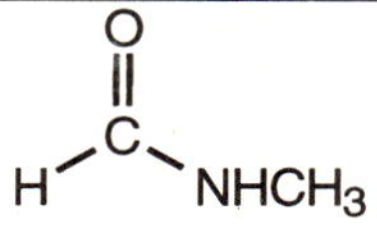 HC(O)NHCH$_3$ | N-Methylformamide |
| CH$_3$CH$_2$CH$_2$C(O)NH$_2$ | Butanoamide | 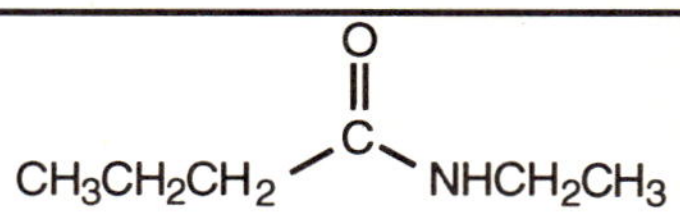 CH$_3$CH$_2$CH$_2$C(O)NHCH$_2$CH$_3$ | N-Ethylbutanoamide |

**Example 18.22:** Write the equation for the reaction of propanoic acid and ammonia to form propanoamide.

**Solution:** Write out the starting molecules. These will combine, while eliminating water to form the corresponding amide (propanoamide).

$$NH_3 + CH_3CH_2COOH \rightarrow CH_3CH_2C(O)NH_2 + H_2O$$

---

**Practice Problems**

**18.57** What is the product of the reaction of acetyl chloride plus ammonia?

**18.58** What is the product the reaction of dimethyl amine and propanoic acid?

**18.59** NH$_2$(CH$_2$)$_6$NH$_2$ + ClC(O)(CH$_2$)$_4$C(O)Cl →

---

## Integration of Multiple Skills

This will integrate principles of stoichiometry with organic reactions.

**Example 18.23:** If 1-pentene (0.785 g) is reacted with an excess of bromine, what is the theoretical yield of product?

**Solution:** To solve this problem, first write a balanced equation, and then apply rules of stoichiometry to find the theoretical yield.

326   Chapter 18

$$CH_2=CHCH_2CH_2CH_3 + Br_2 \rightarrow CH_2BrCHBrCH_2CH_2CH_3$$

$$0.785 \text{ g } C_5H_{10} \times \frac{1 \text{ mol } C_5H_{10}}{70.0 \text{ g } C_5H_{10}} \times \frac{1 \text{ mol } C_5H_{10}Br_2}{1 \text{ mol } C_5H_{10}} \times \frac{229.8 \text{ g } C_5H_{10}Br_2}{1 \text{ mol } C_5H_{10}Br_2} =$$

$$2.58 \text{ g } C_5H_{10}Br_2$$

## Practice Problems

**18.60** If 1-pentene (2.42 g) is reacted with an excess of bromine, what is the theoretical yield of product?

**18.61** If an excess of 1-pentene is reacted with 1.25 mL of bromine (density = 3.102 g/mL), what is the theoretical yield of product?

## Self-Exam

**I. (4 ea) Write T (true) or F (false) in each corresponding blank.**

_____1. All organic compounds contain carbon.

_____2. The simplest hydrocarbons are benzenes.

_____3. An alkyne is an oxygen containing molecule.

_____4. A chlorine atom can be substituted for a hydrogen atom in alkanes.

_____5. Alkanes can be heated to form alkynes and alkenes.

_____6. Hydrocarbons and alcohols undergo combustion to produce water and $CO_2$.

_____7. An alcohol contains a carboxyl group.

_____8. All aromatic compounds have rings.

_____9. Polyamides are produced from addition reactions.

_____10. The aldehyde and acid functional groups are always located on the terminal end of a carbon chain.

**II. (4 ea) Multiple Choice. Place the letter of the BEST answer in the corresponding blank.**

_____1. The molecular formula for 2,2,4-trimethylpentane is
    A. $C_5H_{10}$    B. $C_5H_{12}$    C. $C_8H_{18}$    D. $C_8H_{12}$    E. $C_8H_{16}$

_____2. What is the molecular formula of 3-pentyne?
    A. $C_5H_8$    B. $C_5H_{10}$    C. $C_6H_{10}$    D. $C_5H_{12}$    E. $C_6H_{12}$

___3. Which of the following is **not** an isomer of 2,2-dimethylbutane (shown below)?

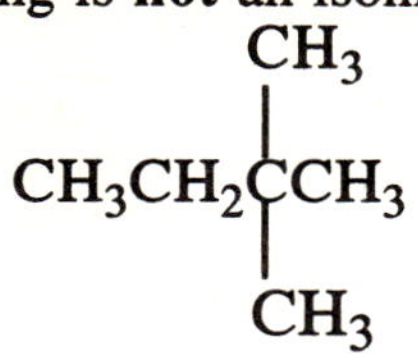

$$CH_3CH_2C(CH_3)_2CH_3$$

A. 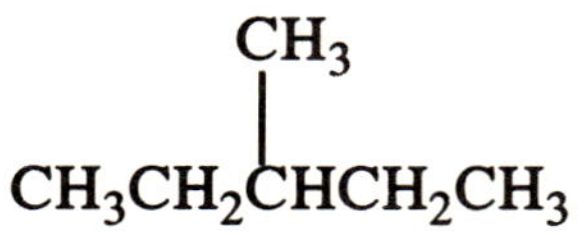 $CH_3CH_2CH(CH_3)CH_2CH_3$

B. 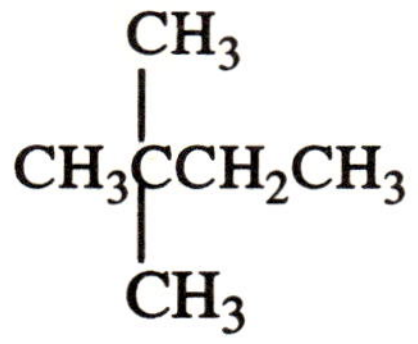 $CH_3C(CH_3)_2CH_2CH_3$

C. 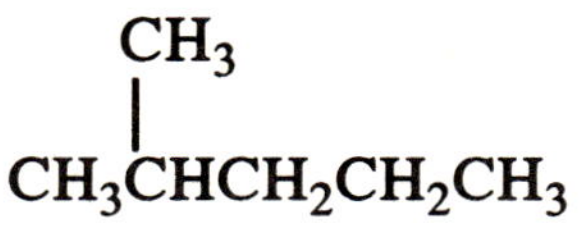 $CH_3CH(CH_3)CH_2CH_2CH_3$

D. 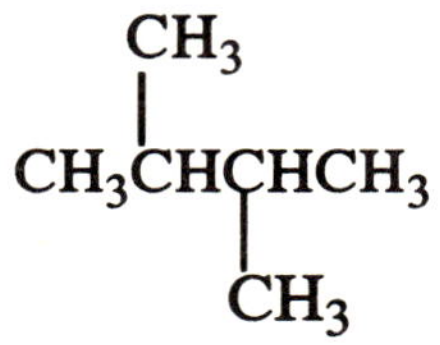 $CH_3CH(CH_3)CH(CH_3)CH_3$

E. $CH_3CH_2CH_2CH_2CH_2CH_3$

___4. Which of the following molecules is 3,4-dimethyl-3-hexene?

A. 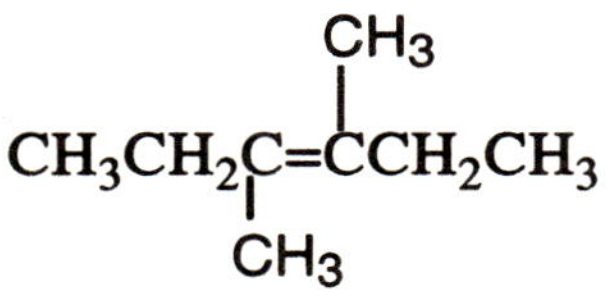 $CH_3CH_2C(CH_3)=C(CH_3)CH_2CH_3$

B. $CH_3CH_2CH=CHCH_2CH_3$

C. 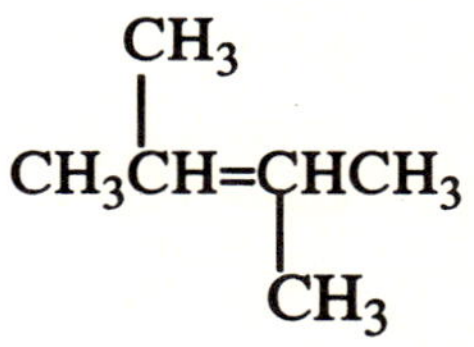 $CH_3CH=C(CH_3)CH(CH_3)CH_3$ — with CH₃ groups

D. 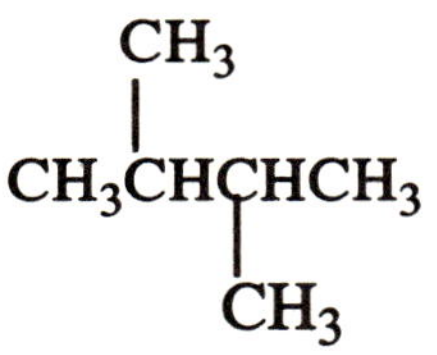 $CH_3CH(CH_3)CH(CH_3)CH_3$

E. 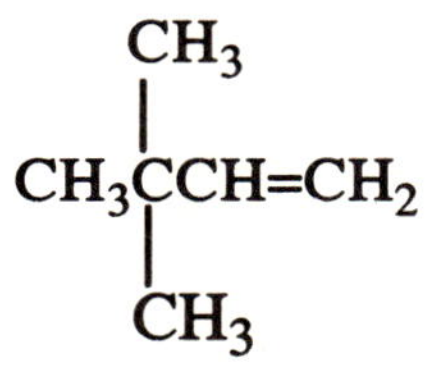 $CH_3C(CH_3)_2CH=CH_2$

___5. How can a mixture of alkanes be separated?
    A. Filtration      B. Distillation      C. Acid-Base reaction
    D. Precipitation reaction      E. Chromatography

___6. Which of the following contains an ethyl group attached to an oxygen atom?
    A. Methanol      B. Ethanal      C. Ethyl acetate
    D. Propylethanoate      E. Acetamide

___7. Which functional group has a nitrogen?
    A. Alkyne      B. Ester      C. Alcohol
    D. Amide      E. Benzene

___8. The reaction of an alcohol plus a carboxylic acid produces water and
    A. An ether      B. An aldehyde      C. A ketone
    D. An amide      E. An ester

## 328  Chapter 18

_____9. How many unique isomers are there for $C_6H_{14}$?
   A. 2     B. 3     C. 4     D. 5     E. 6

_____10. Which of the structures below is an isomer of o-xylene (1,2-dimethylbenzene)?

A. 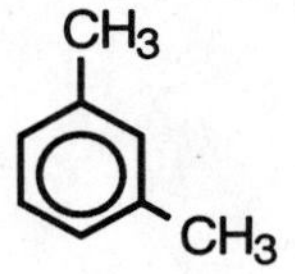 1,2,3-trimethyl... (CH3 at positions with another CH3)    B. 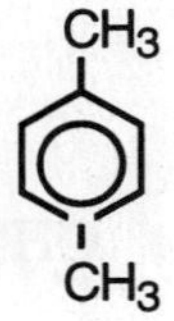 (CH3 para to CH3)    C. 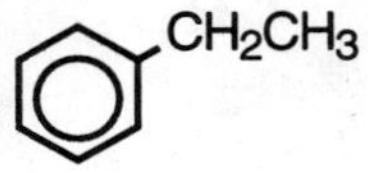 ethylbenzene (CH2CH3)    D. None    E. All

### III. Provide Complete Answers

1. (7 pts) Show all the monochlorinated products for the reaction of butane plus $Cl_2$.

2. (7 pts) Draw the polymer of styrene.

3. (6 pts) Show two oxidation products of ethanol.

## Answers to Practice Problems

18.1   2 bonds, 2 l.p.                  18.2   1 bond, 0 l.p.

18.3   1 bond, 3 l.p.                   18.4   $C_2H_2 + 2Br_2 \rightarrow C_2H_2Br_4$

18.5   $CH_3CH_2OH + 3O_2 \rightarrow 2CO_2 + 3H_2O$

18.6   H—C—H with H above and below (methane)

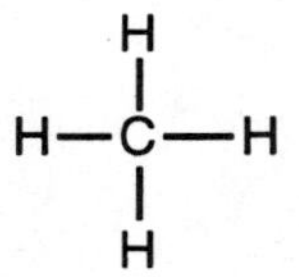

18.7   H₂C=O (formaldehyde Lewis structure)

18.8   :Cl—C(—Cl)(—Cl)—Cl:  (CCl₄ Lewis structure)

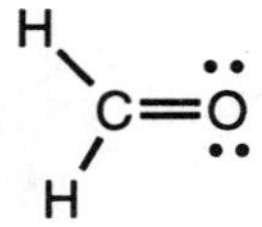

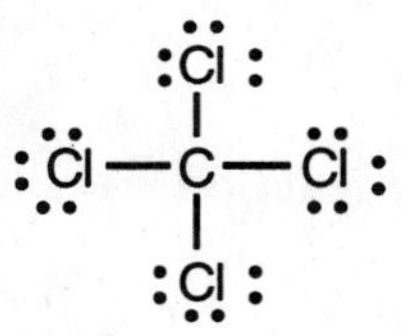

18.9   H—C(H)(H)—O—H (methanol Lewis structure)

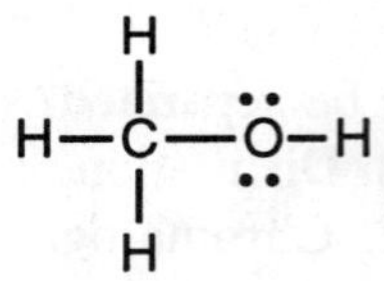

18.10  $CH_3CH_2CH_2Cl$ and $CH_3CHClCH_3$

18.11  $CH_3CH_2CH_2OH$, $CH_3CHOHCH_3$, and $CH_3CH_2OCH_3$

18.12  $C_4H_8BrO$                      18.13   $C_4H_{10}O$

18.14  $C_{16}H_{34}$

18.15  $CH_3CH_2CH_2CH_3$ and $CH_3CH(CH_3)_2$

18.16  a: CH₃CH₂CH₂CH₂CH₂CH₃

b: CH₃CH(CH₃)CH₂CH₂CH₃

c: CH₃CH₂CH(CH₃)CH₂CH₃

d: CH₃CH(CH₃)CH(CH₃)CH₃

e: CH₃C(CH₃)(CH₃)CH₂CH₃

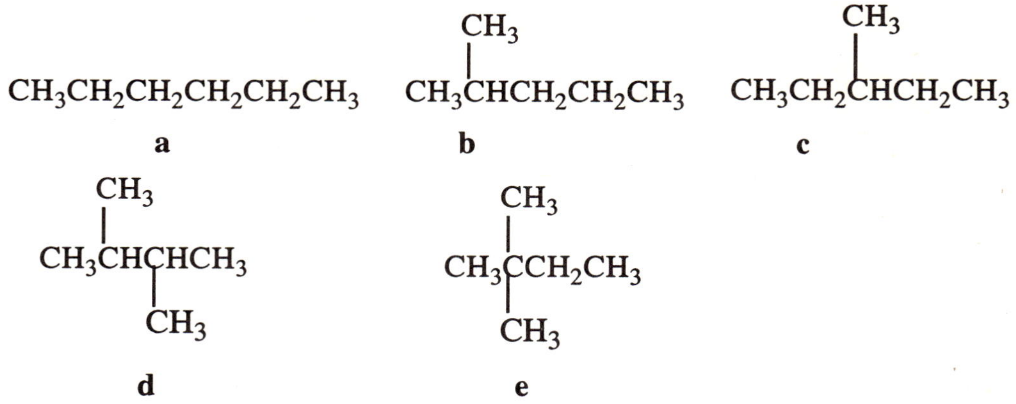

18.17  Hexane (a); 2-methylpentane (b); 3-methylpentane (c); 2,3-dimethylbutane (d); 2,2-dimethylbutane (e)

18.18  3-Ethyl-4-methylheptane     18.19  2,2,4-Trimethylpentane

18.20  CH₃CH(CH₃)CH₂CH₂CH(CH₃)CH₃

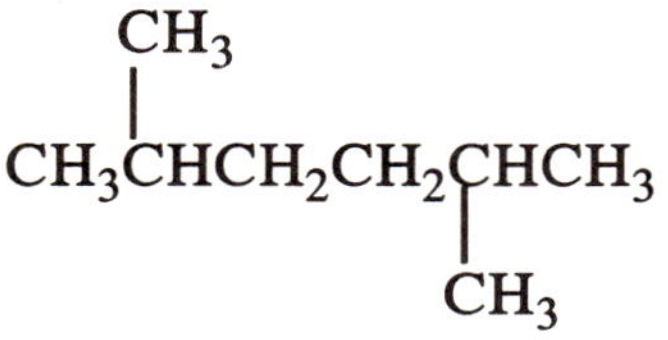

18.21  CH₃CH₂CH₂CH(CH₂CH₃)CH(CH₂CH₂CH₃)CH₂CH₂CH₂CH₃

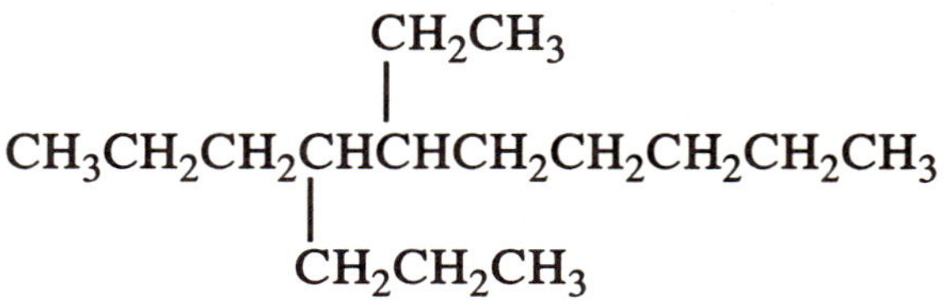

18.22  $C_3H_8 + 5O_2 \rightarrow 3CO_2 + 4H_2O$

18.23  $2C_6H_{14} + 19O_2 \rightarrow 12CO_2 + 14H_2O$

18.24  $C_3H_8 + Cl_2 \rightarrow CH_3CH_2CH_2Cl + HCl$   and   $C_3H_8 + Cl_2 \rightarrow CH_3CHClCH_3 + HCl$

18.25  $C_5H_{12} + Cl_2 \rightarrow CH_3CH_2CH_2CH_2CH_2Cl + HCl$
$C_5H_{12} + Cl_2 \rightarrow CH_3CH_2CH_2CHClCH_3 + HCl$
$C_5H_{12} + Cl_2 \rightarrow CH_3CH_2CHClCH_2CH_3 + HCl$

18.26  1-Pentyne                 18.27  2-Heptene

18.28  CHCH₃=C(CH₃)CH₂CH₂CH₂CH₂CH₃     18.29  H₃C—C≡C-CH(CH₂CH₃)-CH₂-CH₂-CH₂-CH₂-CH₃

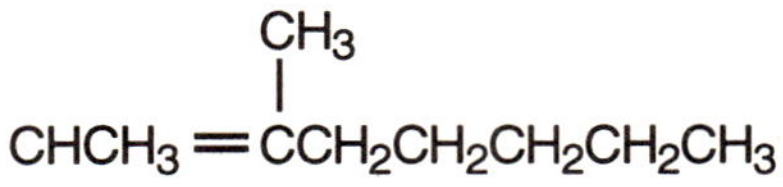

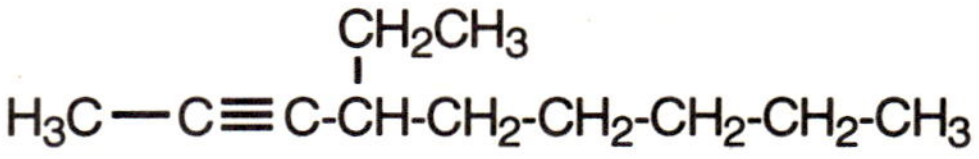

18.30  $CaC_2(s) + 2H_2O(l) \rightarrow CH \equiv CH(g) + Ca(OH)_2(aq)$

18.31  $CH \equiv CH(g) + HCl \rightarrow CH_2=CHCl$

18.32  $CH_3CH=CHCH_3 + Br_2 \rightarrow CH_3CHBrCHBrCH_3$

18.33  $CH_3CH_2CH=CHCH_2CH_3 + H_2O \rightarrow CH_3CH_2CH_2CHOHCH_2CH_3$

**18.34** Polyethylene (polyethene)

**18.35** –CHOH–CH$_2$–CHOH–CH$_2$–CHOH–CH$_2$–CHOH–CH$_2$–

**18.36** C$_{12}$H$_{18}$

**18.37** C$_6$H$_6$ + Cl$_2$ $\xrightarrow{FeCl_3}$ C$_6$H$_5$Cl + HCl

**18.38** 1-Butanol

**18.39** 2-Hexanol

**18.40**
$$CH_3CH(CH_3)CH(OH)CH_2CH_2CH_2CH_3$$

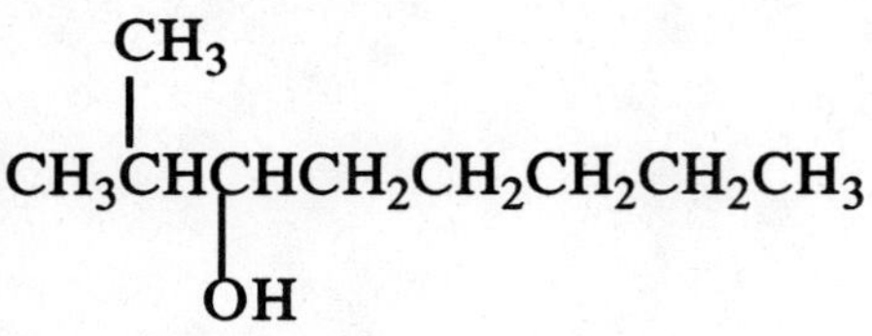

**18.41** CH$_3$CHOHCH$_2$CH$_3$ + 6O$_2$ → 4CO$_2$ + 5H$_2$O

**18.42** CH$_3$CH(OH)CH$_2$CH$_3$ + (O) → CH$_3$C(O)CH$_2$CH$_3$ (butanone)

**18.43** CH$_3$CH$_2$CH$_2$CH$_2$OH + (O) → CH$_3$CH$_2$CH$_2$CHO (butanal)

**18.44** CH$_3$CH$_2$CH$_2$CH$_2$OH + (O) → CH$_3$CH$_2$CH$_2$COOH (butanoic acid)

**18.45** Hexanal

**18.46** 3-Hexanone

**18.47** CH$_3$CH$_2$–C(=O)–CH$_2$CH$_3$  3-pentanone

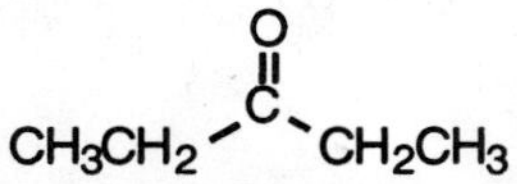

CH$_3$–C(=O)–CH$_2$CH$_2$CH$_3$  2-pentanone

CH$_3$–C(=O)–CH(CH$_3$)CH$_3$  3-methyl-2-butanone

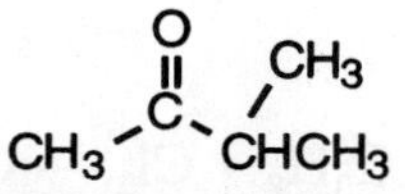

**18.48** CH$_3$CH$_2$CH$_2$CH$_2$–CH=O  pentanal

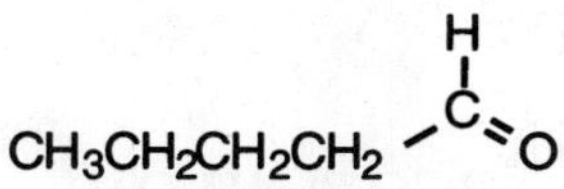

H$_3$CH$_2$CCH(CH$_3$)–CH=O  2-methylbutanal

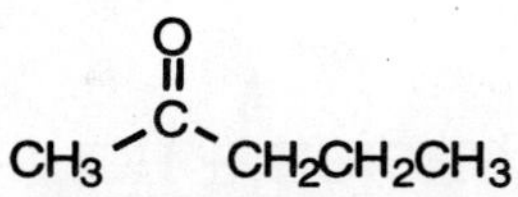

(H$_3$C)$_2$CHCH$_2$–CH=O  3-methylbutanal

**18.49** CH$_3$CH$_2$–C(=O)–O$^-$ Na$^+$ + H$_2$O

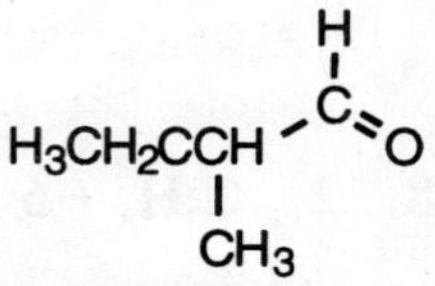

**18.50** CH$_3$CH$_2$CH$_2$CH$_2$CH$_2$–C(=O)–OH

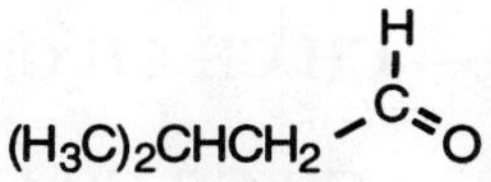

**18.51** CH$_3$CH$_2$CH$_2$–C(=O)–OCH$_2$CH$_2$CH$_3$

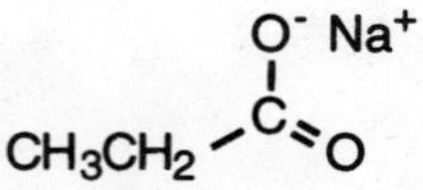

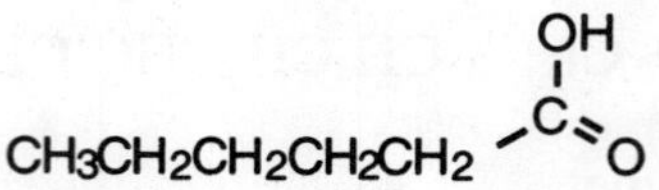

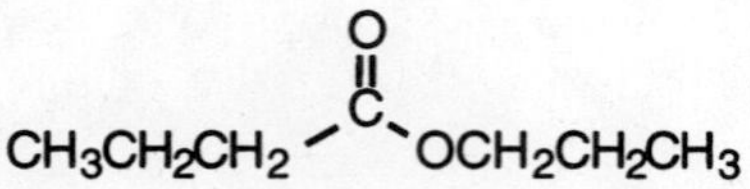

**18.52** C$_6$H$_5$–C(=O)–O–CH$_2$CH$_3$   ethyl benzoate

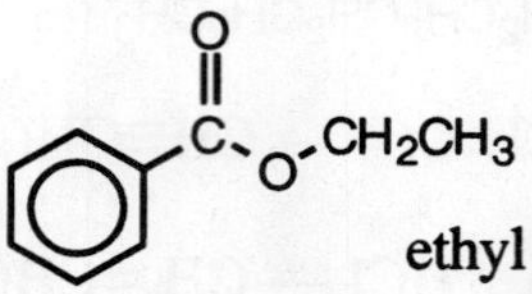

**18.53** CH$_3$CH$_2$N(H)CH$_2$CH$_3$

**18.54** C$_4$H$_{11}$N

**18.55**  CH₃CH₂CH₂CH₂NH₂, butylamine (also C₄H₁₁N)

**18.56**  CH₃CH₂CH₂CH₂NH₂ + HBr → CH₃CH₂CH₂CH₂NH₃⁺ + Br⁻

**18.57**  Acetamide (ethanoamide)       **18.58**   N,N-Dimethylpropanoamide

**18.59**  Nylon-6,6                    **18.60**   7.94 g C₅H₁₀Br₂

**18.61**

$$1.25 \text{ mL Br}_2 \times \frac{3.102 \text{ g Br}_2}{\text{mL Br}_2} \times \frac{1 \text{ mol Br}_2}{159.8 \text{ g Br}_2} \times \frac{1 \text{ mol C}_5\text{H}_{10}\text{Br}_2}{1 \text{ mol Br}_2} \times \frac{229.8 \text{ g C}_5\text{H}_{10}\text{Br}_2}{1 \text{ mol C}_5\text{H}_{10}\text{Br}_2} =$$

$$5.58 \text{ g C}_5\text{H}_{10}\text{Br}_2$$

## Answers to Self-Exam

**I. T/F**
  1. T   2. F   3. F   4. T   5. T   6. T   7. F   8. T   9. F   10. T

**II. MC**
  1. C   2. A   3. B   4. A   5. B   6. C   7. D   8. E   9. D   10. E

**III. Complete Answers**
  1. CH₃CH₂CH₂CH₂Cl and CH₃CHClCH₂CH₃

  2.

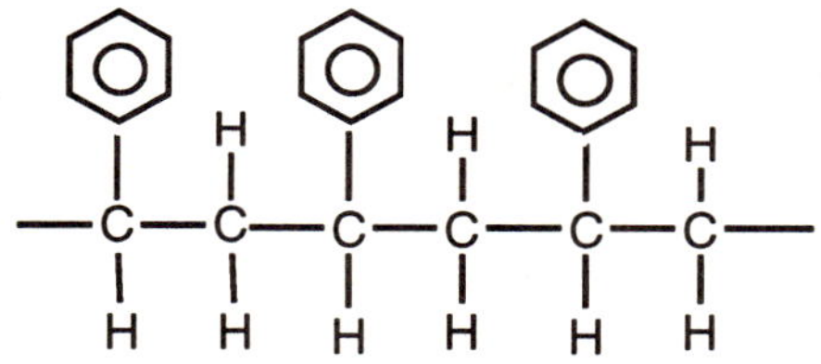

  3. Ethanal and ethanoic acid

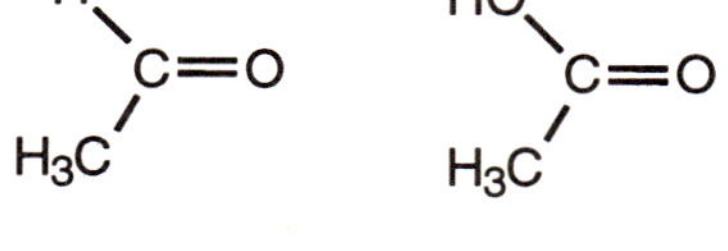

# 19
# Biochemistry

This chapter will look at the chemistry that occurs in biological systems. You will see an overlap of the science of biology and chemistry.

Keys for success in this chapter:

- Learn the new terms.
- Learn the types of molecular structures in the four groups of biomolecules.
- Notice the types of chemical reactions that can occur in biomolecules.

## Summary of Verbal Knowledge

**amino acid:** An organic molecule that contains an amine group and a carboxylic acid group.
**protein:** A biological polymer whose monomer units are amino acids linked by peptide (amide) bonds.
**peptide (amide) bond:** A bond linking the carbon atom and the nitrogen atom in the amide group of a protein molecule.
**primary structure:** The sequence of amino acids in a protein.
**denaturation:** The loss of a protein's three-dimensional shape through the unfolding and uncoiling of the protein as a result of the breaking of the weak forces (such as hydrogen bonding) that hold the protein in its normal three-dimensional shape.
**carbohydrate:** A substance that is either a polyhydroxy aldehyde or polyhydroxy ketone or else a substance that yields such compounds if the carbohydrate hydrolyzes (reacts with water).
**monosaccharide:** A simple sugar. It's a carbohydrate that cannot be broken down by hydrolysis into simpler carbohydrates.
**oligosaccharide:** An oligomer or short polymer of monosaccharides (simple sugars).
**polysaccharide:** A polymer of monosaccharides (simple sugars).
**nucleotide:** A molecule consisting of a sugar, either ribose or 2-deoxyribose, attached to a phosphate group and a nitrogen-containing base.
**deoxyribonucleic acid (DNA):** A polymer of deoxyribonucleotides.
**complementary base pairing:** The pairing, through hydrogen bonding, of certain bases in DNA or RNA.
**ribonucleic acid (RNA):** A polymer of ribonucleotides.
**protein biosynthesis:** The building of protein molecules in a cell.
**lipid:** A biological substance belonging to one of several structurally different classes of substances that dissolves in organic solvents such as chloroform, $CHCl_3$.
**triacylglycerol:** An ester of glycerol (trihydroxy alcohol) and three fatty acids (long-chain carboxylic acids). These are also known as *triglycerides*.
**saponification:** The general term for the base-catalyzed hydrolysis of an ester, especially a fat.
**phospholipid:** A lipid compound that contains a phosphate group.

## Review of Mathematical and Calculator Skills

There are no mathematical skills that you need to review for this chapter. However, you should recall the functional groups from Chapter 18 that will be used in biological molecules.

**Example 19.1:** What does the functional group of an aldehyde look like?

**Solution:** Recall that an aldehyde has a alkyl group attached to a carbonyl (C=O) which also has a hydrogen attached to the carbonyl carbon. Its general formula is

**RCHO**

---

## Practice Problems

**19.1** What is the structure of a carboxylic acid?

**19.2** What is the structure of an amine?

**19.3** What is the structure of an amide?

**19.4** What is the structure of an ester?

---

## Application of Skills and Concepts

### 19.1 Cell Structure

This section contains a summary of cells. This is recapped in the following outline.

I. Prokaryotic cells
   A. Only found in single-celled organisms
   B. Contain a cell membrane.
   C. Operation of the cell is directed by the nuclear region.

II. Eukaryotic cells
   A. Some single-celled organisms
      1. Yeasts
      2. Algae
   B. All multicelled organisms
   C. Generally larger than prokaryotic cells
   D. Contain organelles
      1. Separated from the rest of the cell by a membrane
      2. Perform various functions
   E. Operation of the cell (including organelles) is directed by the nuclear region.

## 19.2 Biological Molecules: An Overview

A summary of the four major classes of biological molecules will be found in Table 19.1. Each class will be examined in detail in the following sections.

**Table 19.1** Classes of Biological Molecules

| Class | Main Elements | Main Functional Groups | Some Polymeric Bonds in These Compounds | Biological Function |
|---|---|---|---|---|
| Proteins | C, H, N, O, S | Amines<br>Carboxylic acid<br>Amides | Amide | Catalysis<br>Transport<br>Structural support<br>Movement |
| Carbohydrates | C, H, O | Alcohol<br>Ketone<br>Aldehyde | Glycoside | Energy source<br>Energy storage |
| Nucleic Acids | C, H, N, O, P | Alcohol<br>Cyclic amines<br>Phosphate | Phosphate | Store genetic information<br>Protein synthesis |
| Lipids | C, H, O, P | Esters<br>Phosphate<br>Choline | Ester | Membrane material<br>Energy storage<br>Dissolve nonpolar molecules |

## 19.3 Amino Acids

The basic structure of an amino acid is shown in Figure 19.1. The name is self-explanatory when you consider that the molecule has a carboxylic acid group and an amine group. The body constructs proteins from the 20 amino acids shown in Table 19.2. The only difference in each unit is the identify of R.

**Figure 19.1** The Basic Amino Acid

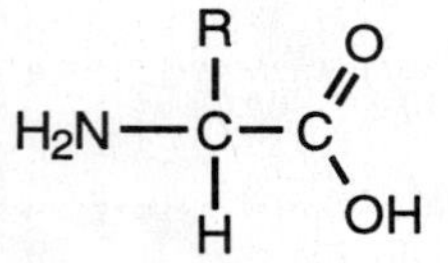

$$H_2N-\underset{\underset{H}{|}}{\overset{\overset{R}{|}}{C}}-\overset{\overset{O}{\|}}{C}-OH \quad = \quad NH_2CRHCO_2H$$

**Example 19.2:** What is the name of $H_2N-CH(CH_3)-CO_2H$?

**Solution:** Identify R as being $CH_3$, and then go to Table 19.2 to find that this amino acid is **alanine.**

Table 19.2 Amino Acids Found in Most Proteins

| Name | R | Abbreviation |
|---|---|---|
| Glycine | —H | Gly |
| Alanine | —CH$_3$ | Ala |
| Valine | —CH(CH$_3$)$_2$ | Val |
| Leucine | —CH$_2$CH(CH$_3$)$_2$ | Leu |
| Isoleucine | —CH(CH$_2$CH$_3$)(CH$_3$) | Ile |
| Proline | —CH$_2$CH$_2$CH$_2$—* | Pro |
| Phenylalanine | —CH$_2$C$_6$H$_5$ | Phe |
| Tryptophan | —CH$_2$–(indole) | Trp |
| Methionine | —CH$_2$CH$_2$SCH$_3$ | Met |
| Serine | —CH$_2$OH | Ser |
| Cysteine | —CH$_2$SH | Cys |
| Threonine | —CH(CH$_3$)(OH) | Thr |
| Aspartic acid | —CH$_2$COO$^-$ | Asp |
| Glutamic acid | —CH$_2$CH$_2$COO$^-$ | Glu |
| Tyrosine | —CH$_2$-p-C$_6$H$_5$OH | Tyr |
| Asparagine | —CH$_2$C(O)NH$_2$ | Asn |
| Glutamine | —CH$_2$CH$_2$C(O)NH$_2$ | Gln |
| Histidine | —CH$_2$–(imidazole) | His |
| Lysine | —CH$_2$CH$_2$CH$_2$CH$_2$NH$_2$ | Lys |
| Arginine | —CH$_2$CH$_2$CH$_2$NHC(NH)(NH$_2$) | Arg |

* Proline is unique in that the R group is attached at the central C and on the N of the amine.

## Practice Problems

**19.5** What is H$_2$N—CH(CH$_2$C$_6$H$_5$)—CO$_2$H?

**19.6** What is H$_2$N—CH(CH$_2$CH$_2$CH$_2$CH$_2$NH$_2$)—CO$_2$H?

**19.7** Write the structural formula for histidine.

**19.8** Write the structural formula for methionine.

## 19.4 Primary Structure of a Protein

The primary structure of a protein is determined by the sequence of amino acids. These are connected by amide bonds.

- Two amino acids are connected by a peptide linkage.
- A peptide linkage is an amide, which is a condensation reaction.
- A peptide is a small polymer of amino acids.
- A protein is a large polymer of amino acids.

**Example 19.3:** Write the structure of a dipeptide of glycine and phenylalanine.

**Solution:** To solve this, draw each structure separately and connect the two by an amide linkage. Primary sequences are listed from left to right, with the amino end of the molecule on the left and the carboxylic acid end on the right.

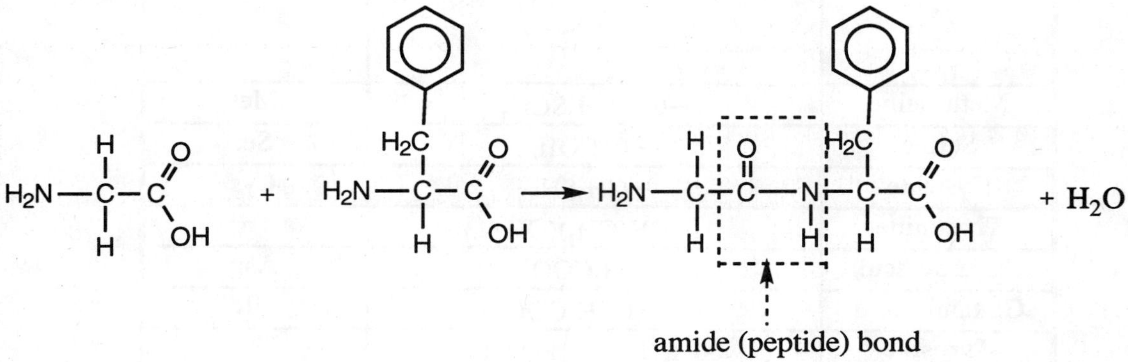

amide (peptide) bond

gly            phe            gly-phe

**Example 19.4:** Write the reaction of two cysteine groups that are oxidized to a disulfide bond.

**Solution:** The two —SH ends are oxidized and hooked together as shown below.

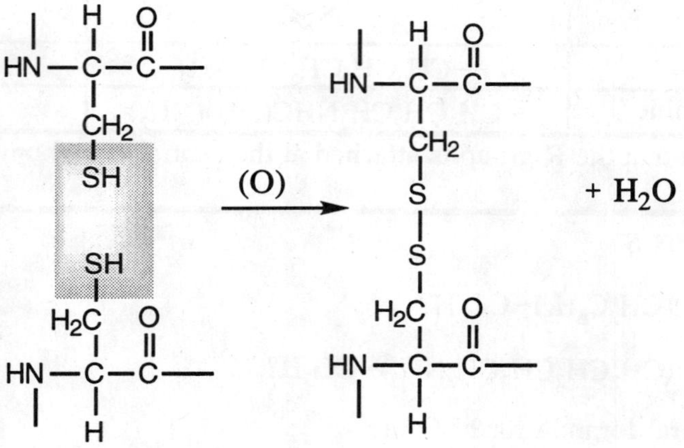

# Biochemistry

> **Practice Problems**
>
> **19.9** Write the structure of the dipeptide ala-asp.
>
> **19.10** Write the structure of the dipeptide asp-ala.
>
> **19.11** Write the structure of the tripeptide his-lys-trp.

## 19.5 Three-Dimensional Structure of a Protein

Once the primary structure of the protein is put together, it will form a three-dimensional structure due to hydrogen bonding between the hydrogen atom on a nitrogen and an oxygen on another amino acid's carboxyl group.

The three dimensional structures will have two general forms, helix and globular. When a protein is denatured (by heat or chemicals), the hydrogen bonds are disrupted and the protein loses its form.

> **Practice Problems**
>
> **19.12** What are two ways to denature a protein?
>
> **19.13** What type of bond holds a protein in a helix form?

## 19.6 Monosaccharides

Below are some characteristics of this class.

- Have general formula of $C_n(H_2O)_n$.
- Cannot be broken down to simpler molecules by hydrolysis.
- Contain carbonyl group (aldehyde in aldoses or ketone in ketoses).
- Contain multiple hydroxyl groups (—OH).
- Have suffix *-ose*.

Some straight chain forms of simple sugars are shown in Figure 19.2. Some ring forms are illustrated in Figure 19.3. Notice that they are numbered from the top of the carbon chain and come in a variety of carbon lengths. Glucose, mannose, galactose, and fructose are all isomers of each other. Confirm that they each have the same number of carbon, hydrogen, and oxygen atoms. Take a few minutes to note their structural differences. Also try to find other isomers.

## Figure 19.2 Some Straight Chain Monosaccharides

### Aldoses

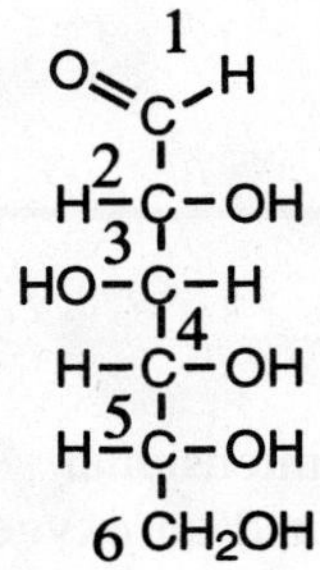

**D-Glucose**

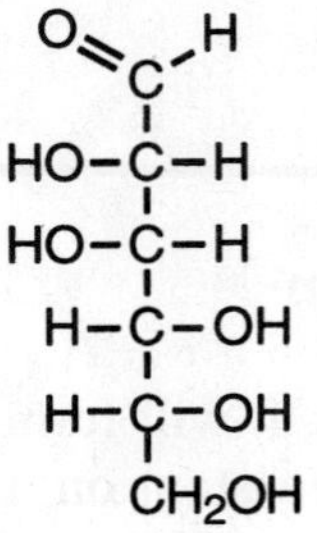

```
      1
   O=C-H
      |
   2  |
   H-C-OH
   3  |
  HO-C-H
   4  |
   H-C-OH
   5  |
   H-C-OH
      |
   6  CH₂OH
```

**D-Mannose**

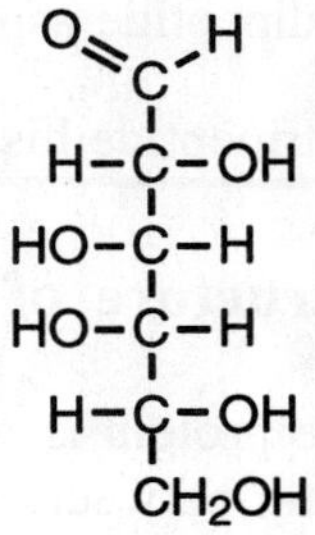

```
   O=C-H
      |
  HO-C-H
      |
  HO-C-H
      |
   H-C-OH
      |
   H-C-OH
      |
     CH₂OH
```

**D-Galactose**

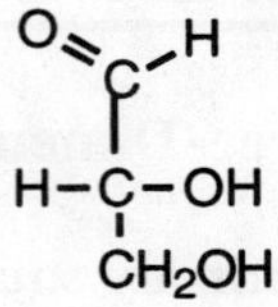

```
   O=C-H
      |
   H-C-OH
      |
  HO-C-H
      |
  HO-C-H
      |
   H-C-OH
      |
     CH₂OH
```

**D-Glyceraldehyde**

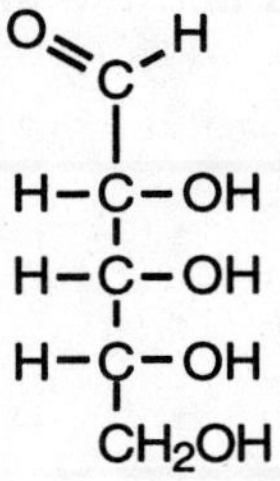

```
   O=C-H
      |
   H-C-OH
      |
     CH₂OH
```

**D-Ribose**

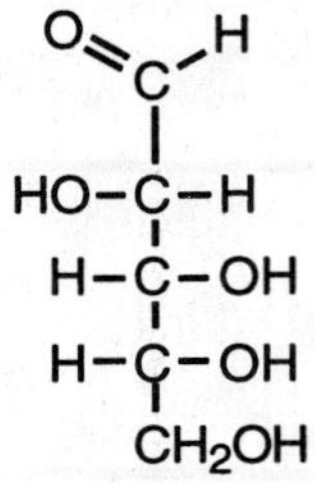

```
   O=C-H
      |
   H-C-OH
      |
   H-C-OH
      |
   H-C-OH
      |
     CH₂OH
```

**D-Arabinose**

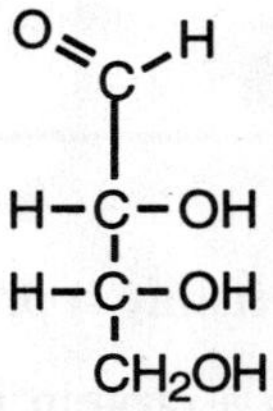

```
   O=C-H
      |
  HO-C-H
      |
   H-C-OH
      |
   H-C-OH
      |
     CH₂OH
```

**D-Erythrose**

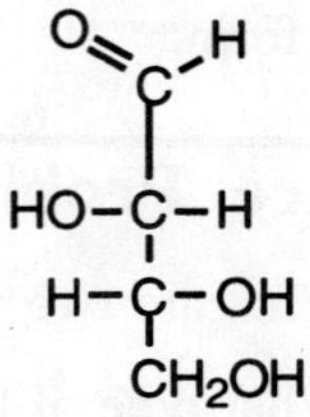

```
   O=C-H
      |
   H-C-OH
      |
   H-C-OH
      |
     CH₂OH
```

**D-Threose**

```
   O=C-H
      |
  HO-C-H
      |
   H-C-OH
      |
     CH₂OH
```

### Ketoses

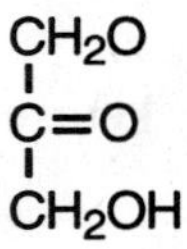

**Dihydroxyacetone**

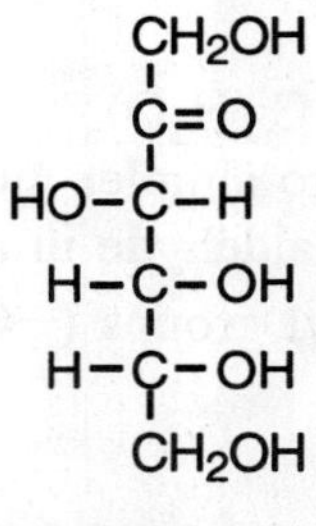

```
   CH₂O
    |
   C=O
    |
   CH₂OH
```

**D-Fructose**

```
   CH₂OH
    |
   C=O
    |
  HO-C-H
    |
   H-C-OH
    |
   H-C-OH
    |
   CH₂OH
```

Figure 19.3 illustrates the ring form of some of the sugars. Try to follow which carbons formed the ring (from the straight chain in Figure 9.2), and confirm that the molecular formulas for the rings are the same as their straight-chain counterparts.

**Figure 19.3** Some Monosaccharides Rings

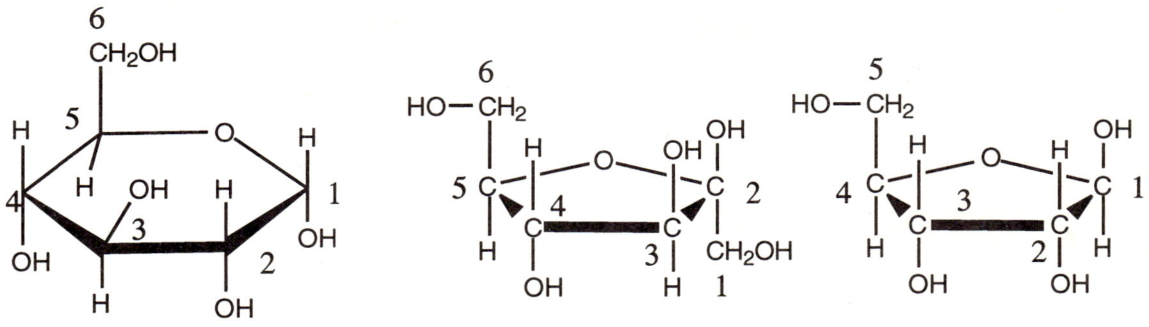

D-Glucose           D-Fructose           D-Ribose

---

### Practice Problems

**19.14** Name a 6-carbon aldose. What is its molecular formula?

**19.15** Name a 6-carbon ketose. What is its molecular formula?

**19.16** What is the formula of ribose in a ring form?

**19.17** What is the formula of deoxyribose in a ring form?

---

## 19.7 Oligosaccharides and Polysaccharides

These are polymers of simple sugars. Each polymer bond is made by a condensation reaction, where a molecule of water is eliminated.

**Example 19.5:** If sucrose is a disaccharide made up of glucose and fructose, write an equation for its formation.

**Solution:** Look at Figure 19.2, and you will note that both fructose and glucose have the molecular formula of $C_6H_{12}O_6$. The reaction is shown below. Note that sucrose has a formula which is the sum of the atoms in glucose and fructose minus water.

$$C_6H_{12}O_6 + C_6H_{12}O_6 \rightarrow C_{12}H_{22}O_{11} + H_2O$$

---

### Practice Problems

**19.18** Write an equation for the hydrolysis of lactose (a dimer made of one glucose molecule and one galactose molecule).

**19.19** Write an equation for the formation of maltose which is made up of two glucose molecules.

**340 Chapter 19**

> **19.20** Write an equation for the formation of sucrose.
>
> **19.21** Write an equation for the hydrolysis of an amylose fragment that has the formula $C_{60}H_{102}O_{51}$ and is reacted with nine molecules of water.

## 19.8 Nucleotides

These are the basic building units that are polymerized to form DNA and RNA. There are three parts to each nucleotide.

- Sugar (ribose in RNA and deoxyribose in DNA)
- Phosphate group
- Nitrogen base

There are several types of bases in DNA and in RNA. These are shown in Figure 19.4.

**Figure 19.4** Bases for Nucleic Acids

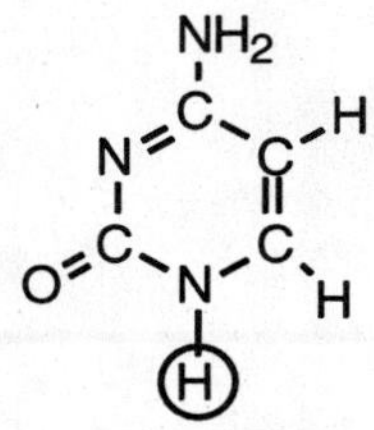

cytosine (C) 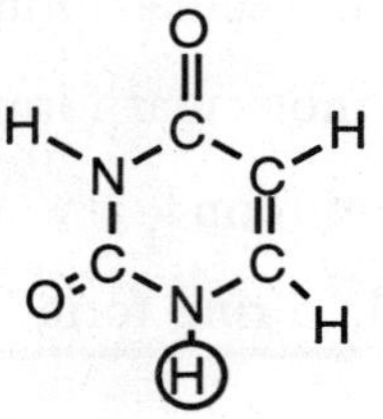    uracil (U) 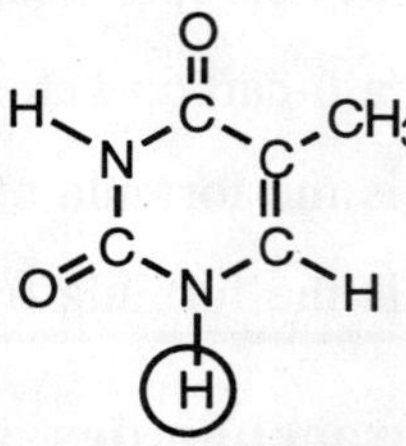    thymine (T)

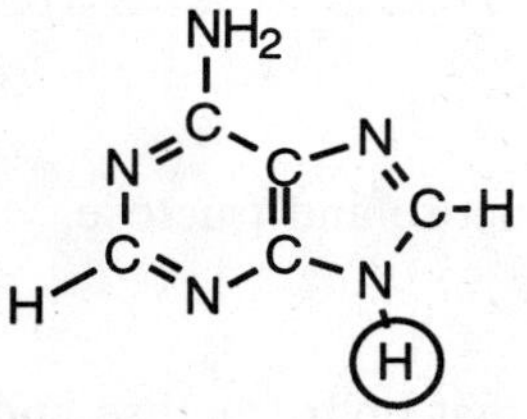

adenine (A) 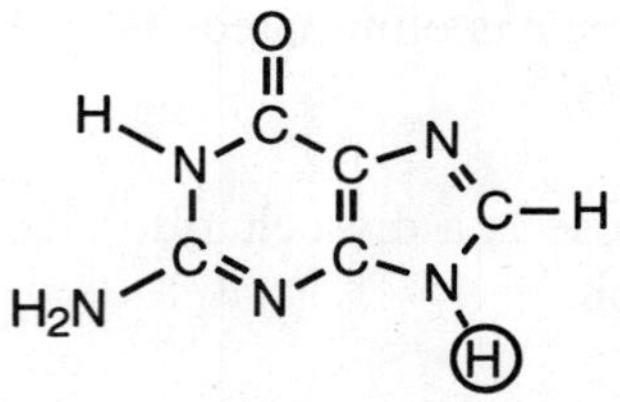    guanine (G)

The circled hydrogens are replaced by a bond to the ribose at the 1' carbon when forming a nucleotide.

Below are the basic DNA and RNA monomers. Also listed are the four different bases that might be attached. When forming a polymer, an oxygen on the phosphate group of one nucleotide bonds to the 3' position of another ribose ring.

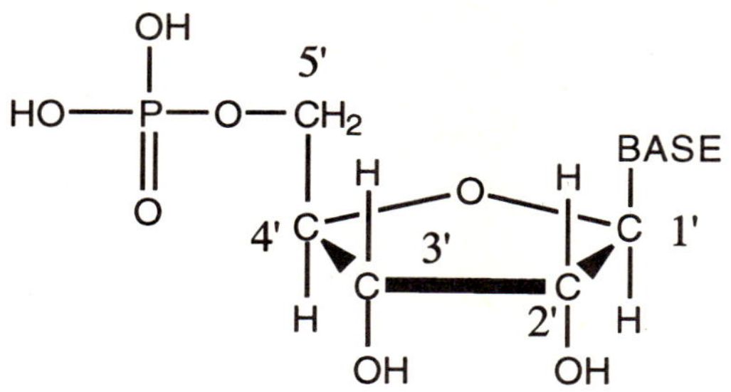

Ribose nucleotide (acidic form)

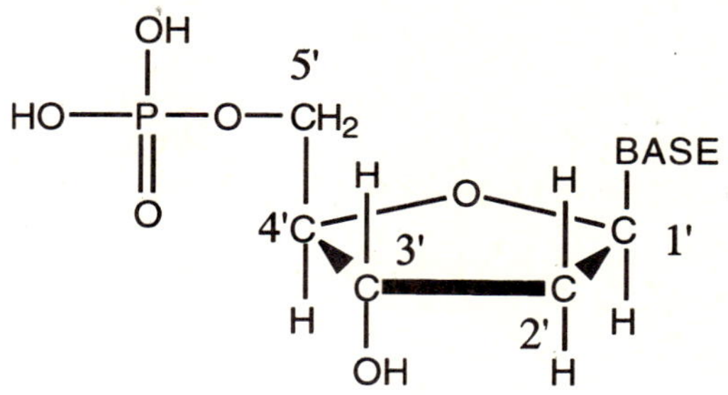

Deoxyribose nucleotide (acidic form)

Base
Cytosine (C)
Adenine (A)
Uracil (U)
Guanine (G)

Base
Cytosine (C)
Adenine (A)
Thymine (T)
Guanine (G)

**Example 19.6:** What is the molecular formula of adenosine-5'-monophophate in the acidic form (i.e., the nucleotide containing an adenine base)?

**Solution:** To solve this, look at the molecule that would result when adenine is attached at the 1' carbon of the ribose. Remember that the hydrogen on the base is replaced by the bond to the ribose carbon. The molecular formula is

$$C_{10}H_{14}N_5O_7P$$

---

**Practice Problems**

**19.22** Which (RNA or DNA or both) would contain the nucleotide thymine?

**19.23** Which (RNA or DNA or both) would contain the nucleotide uracil?

**19.24** Which (RNA or DNA or both) would contain the nucleotide guanine?

**19.25** What is the molecular formula of deoxythymidine-5'-monophophate in the acidic form (i.e., the nucleotide containing a thymine base and deoxyribose)?

---

### 19.9 Deoxyribonucleic Acid (DNA) and the Double Helix

This section looks at the primary structures (order of nucleotides) and secondary structures of DNA.

In each, the sequence will be represented by one of four letters (for each nucleotide unit possible).

The DNA is a double helix, where the two strands are held together by hydrogen bonding. The pairs always match up in a predictable way:

T hydrogen bonds to A
C hydrogen bonds to G

**Example 19.7:** If a segment of one strand of DNA has the sequence T-T-A-C, what is the sequence on the complimentary strand?

**Solution:** Use the information above. The complimentary strand must be

A-A-T-G

## Practice Problems

**19.26** What is the complimentary sequence of A-T-G?

**19.27** What is the complimentary sequence of C-C-G-T?

**19.28** What is the complimentary sequence of C-A-T-C?

### 19.10 Ribonucleic Acid (RNA) and Protein Biosynthesis

The RNA macromolecule is a single stranded molecule. It has three types.

- **Messenger RNA (mRNA):** This copies information from DNA. The bases from mRNA match up to DNA in a predictable fashion. Uracil (U) pairs to thymine (T) and thymine to uracil; guanine (G) pairs to cytosine (C) and cytosine to guanine. The mRNA copies a coded message from the DNA for protein production.
- **Ribosomal RNA (rRNA):** This makes up the ribosomes, where proteins are made.
- **Transfer RNA (tRNA):** Transfers the amino acid to the protein being made, based on the code (3 sequence) that it is programmed to add next.

There are three nucleotide sequences from the mRNA that direct the order of amino acid synthesis. Many of the amino acids have more than one code for their incorporation into the protein being made. See Table 19.3.

**Example 19.8:** What is a possible sequence for mRNA that would provide the following sequence of amino acids (omit begin and end codons)?
ala-gly-phe

**Solution:** To solve this, look up the codes in Table 19.3. One possibility is listed below.

GCUGGUUUU

**Table 19.3** Genetic Codes for Amino Acids

| Amino Acid | Code | Amino Acid | Code |
|---|---|---|---|
| Begin codon | AUG | His | CAU, CAC |
| Phe | UUU, UUC | Asn | AAU, AAC |
| Leu | UUA, UUG, CUU, CUC, CUA, CUG | Lys | AAA, AAG |
| Ile | AUU, AUC, UAU | Asp | GAU. GAC |
| Met | AUG | Glu | GAA, GAG |
| Val | GUU, GUC, GUA, GUG | Cys | UGU, UGC |
| Ser | UCU, UCC, UCA, UCG, AGU, AGC | Trp | UGG |
| Pro | CCU, CCC, CCA, CCG | Arg | CGU, CGC, CGA, CGG, AGA, AGG |
| Thr | ACU, ACC, ACA, ACG | Gln | CAA, CAG |
| Ala | GCU, GCC, GCA, GCG | Gly | GGU, GGC, GGA, GGG |
| Tyr | UAU, UAC | End codon | UAA, UAG, UGA |

## Practice Problems

**19.29** What are two other sequences for these three amino acids?

**19.30** What is a code for val-pro-lys?

**19.31** What is another code for val-pro-lys?

## 19.11 Triacyloglycerols

These are often referred to as **triglycerides**. They are really **triesters**. In Chapter 18 you learned that esters are formed from a carboxylic acid and an alcohol. Glycerol has three alcohol branches, and each one can form an ester group.

When these branches are attached to fatty acids (these are acids with a long chain of carbons), they form triacyloglycerol fats. This reaction is shown below. R is a long (6C or more) chain.

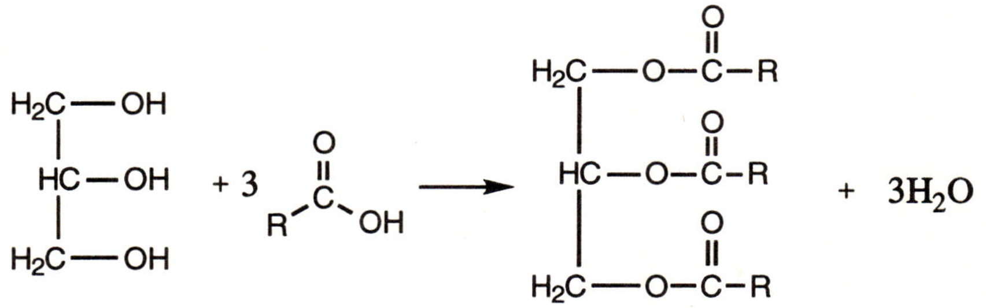

$$\begin{array}{c} H_2C-OH \\ | \\ HC-OH \\ | \\ H_2C-OH \end{array} + 3\ R-\overset{O}{\underset{}{C}}-OH \longrightarrow \begin{array}{c} H_2C-O-\overset{O}{\underset{}{C}}-R \\ | \\ HC-O-\overset{O}{\underset{}{C}}-R \\ | \\ H_2C-O-\overset{O}{\underset{}{C}}-R \end{array} + 3H_2O$$

A triacyloglycerol can be used to make soap. This is really the base catalyzed reverse of the reaction above. Since this reaction occurs in a basic condition, the acid is an ion. This ionic end

**344    Chapter 19**

of the fatty acid is soluble in water. The long chain carbon part of the molecule will be soluble in oils and grease. This makes this molecule a good cleaning agent.

$$\begin{array}{c} H_2C-O-\overset{O}{\overset{\|}{C}}-R \\ HC-O-\overset{O}{\overset{\|}{C}}-R \\ H_2C-O-\overset{O}{\overset{\|}{C}}-R \end{array} + 3NaOH \longrightarrow \begin{array}{c} H_2C-OH \\ HC-OH \\ H_2C-OH \end{array} + 3\ R-\overset{O}{\overset{\|}{C}}-O^-Na^+$$

---

**Practice Problems**

**19.32** Show the structure of the product when glycerol is reacted with $CH_3(CH_2)_{18}COOH$.

**19.33** Show the product of saponification of the product in problem 19.32.

---

## 19.12 Phospholipids

This section looks at one other group of lipids. When a phosphate group is attached as an ester to one of the carbons of the glycerol, a compound with some of the same properties as soap results.

## Integration of Multiple Skills

There are many topics which could be included here, and a few are selected.

**Example 19.9:** What functional groups are in triglycerides (triacyloglycerols)?

> **Solution:** Examine theses structures and you will notice the **ester** functional group.

**Example 19.10:** How many unique pentapeptides can be formed from five different amino acids?

> **Solution:** The number of permutations for putting different amino acids together is equal to the number factorial. This means that for 5 amino acids there are
>
> $$1 \times 2 \times 2 \times 4 \times 5 = 125 \text{ different combinations}$$

---

**Practice Problems**

**19.34** What functional groups are in glucose?

**19.35** What functional groups are in peptides?

**19.36** What functional groups are in nucleic acids?

**19.37** Which of the amino acids (Table 19.2) are good bases? (*Hint*: Bases are good proton acceptors)

**19.38** How many unique hexapeptides can be made from six different amino acids?

## Self-Exam

**I. (4 ea) Write T (true) or F (false) in each corresponding blank.**

_____1. All multicelled organisms have eukaryotic cells.

_____2. Proteins are also called *carbohydrates*.

_____3. Nucleic acids are subunits for DNA and RNA.

_____4. The synthesis of proteins begins with transcription onto mRNA.

_____5. Amide bonds are found in proteins.

_____6. Monosaccharides with 5 carbons have 12 hydrogens and 6 oxygens.

_____7. Phospholipids have a polar and a nonpolar end like soaps.

_____8. Saponification of a triglyceride produces triglycerol and long chain carboxylate ions.

_____9. Primary protein structures are determined by hydrogen bonding.

_____10. A polysaccharide is a polymer of amino acids.

**II. (4 ea) Multiple Choice. Place the letter of the BEST answer in the corresponding blank.**

_____1. A ketose has which functional group(s)?
  A. Ketone     B. Aldehyde     C. Alcohol     D. Both A and C     E. All

_____2. What features of an amino acid gives it a unique identity?
  A. The identity of the R group
  B. The length of the chain between the amino end and the carboxylic acid end
  C. The number of alkyl groups on the nitrogen
  D. The number of alkyl groups on the carboxylic acid
  E. All are necessary to identify the amino acid.

_____3. Which is an isomer of glucose?
  A. Ribose     B. Fructose     C. Glyceraldehyde     D. Threose     E. Sucrose

_____4. Which of these biomolecules are polymers?
  A. Cellulose     B. Protein     C. RNA     D. DNA     E. All

_____5. What is the name of the cell particle where protein is made?
  A. Mitochondria     B. Cell membrane     C. Nucleus
  D. Cytoplasm        E. Ribosomes

_____6. What molecule is responsible for bringing the next amino acid to be attached in a protein sequence?
  A. DNA     B. mRNA     C. tRNA     D. rRNA     E. None

____ 7. Which of the following bases is not present in DNA?
   A. Thymine   B. Adenine   C. Uracil   D. Guanine   E. Cytosine

____ 8. Which is the largest polymeric biomolecule?
   A. DNA   B. mRNA   C. tRNA   D. rRNA   E. Fructose

____ 9. How many unique tetrapeptides can be made from four different amino acids?
   A. 32   B. 24   C. 16   D. 12   E. 8

____ 10. If a DNA sequence is G-A-T-A, what is the base sequence on the complimentary strand?
   A. G-A-T-A   B. C-T-A-T   C. C-T-U-T   D. A-T-A-G   E. T-C-G-C

### III. Provide Complete Answers

1. (6 pts) Show a ketose and an aldose each having 3 carbons.

2. (7pts) Provide an example of a triglyceride containing 45 carbons.

3. (7 pts) Show the structure of a tripeptide made of ala-phe-gly, given the following condensed formulas:

$$ala: H_2N-CH(CH_3)-COOH$$
$$phe: H_2N-CH(CH_2C_6H_5)-COOH$$
$$gly: H_2N-CH_2-COOH$$

## Answers to Practice Problems

19.1  RCOOH

19.2  $RNH_2$ or $RN(R')H$ or $RN(R')(R'')$

19.3  $RC(O)NH_2$ where $C(O)$ is a carbonyl

19.4  RCOOR'

19.5  phenylalanine

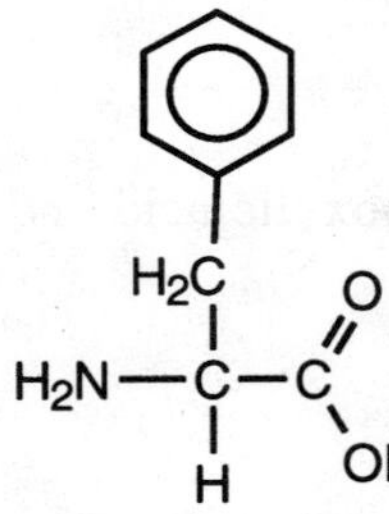

19.6  lysine

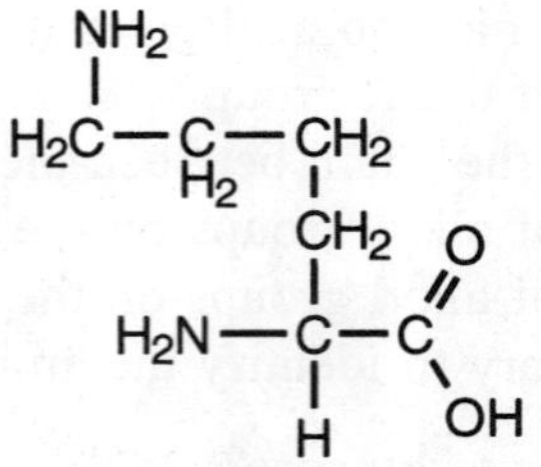

19.7

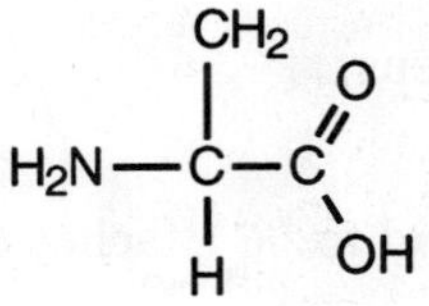

19.8

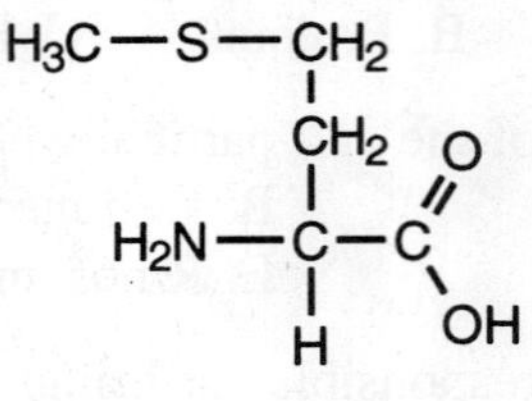

# Biochemistry

**19.9** [structure: H₂N–CH(CH₃)–C(=O)–NH–CH(CH₂–CO₂H)–COOH]

**19.10** [structure: H₂N–CH(CH₂–CO₂H)–C(=O)–NH–CH(CH₃)–COOH]

**19.11** [tripeptide structure with histidine, lysine, and tryptophan side chains]

| | | | |
|---|---|---|---|
| **19.12** | Heat or chemical treatment | **19.13** | Hydrogen bonding |
| **19.14** | Glucose, $C_6H_{12}O_6$ | **19.15** | Fructose, $C_6H_{12}O_6$ |
| **19.16** | $C_5H_{10}O_5$ | **19.17** | $C_5H_{10}O_4$ |

**19.18** $C_{12}H_{22}O_{11} + H_2O \rightarrow$ [glucose structure] + [glucose structure]

**19.19** $C_6H_{12}O_6 + C_6H_{12}O_6 \rightarrow$ [disaccharide structure] $+ H_2O$

**19.20** $C_6H_{12}O_6 + C_6H_{12}O_6 \rightarrow$ [disaccharide structure] $+ H_2O$

**19.21** $C_{60}H_{102}O_{51} + 9H_2O \rightarrow 10 C_6H_{12}O_6$

**348 Chapter 19**

19.22 DNA
19.23 RNA
19.24 Both
19.25 $C_{10}H_{15}N_2O_7P$
19.26 T-A-C
19.27 G-G-C-A
19.28 G-T-A-G
19.29 GCCGGCUUC or GCAGGGUUU
19.30 GUUCCUAAA
19.31 GUACCCAAG

19.32

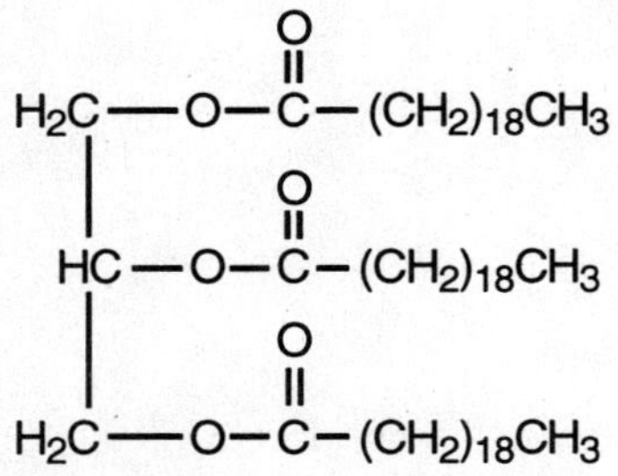

$$H_2C-O-\overset{\overset{O}{\|}}{C}-(CH_2)_{18}CH_3$$
$$HC-O-\overset{\overset{O}{\|}}{C}-(CH_2)_{18}CH_3$$
$$H_2C-O-\overset{\overset{O}{\|}}{C}-(CH_2)_{18}CH_3$$

19.33 $2\ CH_3(CH_2)_{18}COO^-\ Na^+$
19.34 Aldehyde, alcohol
19.35 Carboxylic acids, amines
19.36 Phosphate esters, cyclic amines hydroxides
19.37 Look for those with nitrogen: lys, arg, his, asn, gln
19.38 720

## Answers to Self-Exam

**I. T/F**
1. T  2. F  3. T  4. T  5. T  6. F  7. T  8. T  9. F  10. F

**II. MC**
1. D  2. A  3. B  4. E  5. E  6. D  7. C  8. A  9. B  10. C

**III. Complete Answers**

1.

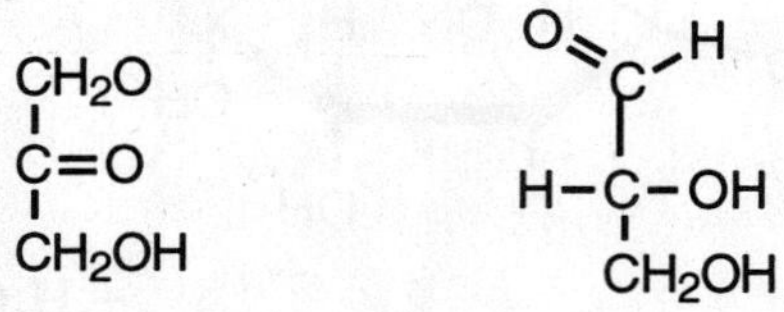

$$\begin{array}{c} CH_2O \\ | \\ C=O \\ | \\ CH_2OH \end{array} \qquad \begin{array}{c} O{=}C{-}H \\ | \\ H-C-OH \\ | \\ CH_2OH \end{array}$$

2. The molecular formula is $C_{45}H_{86}O_6$

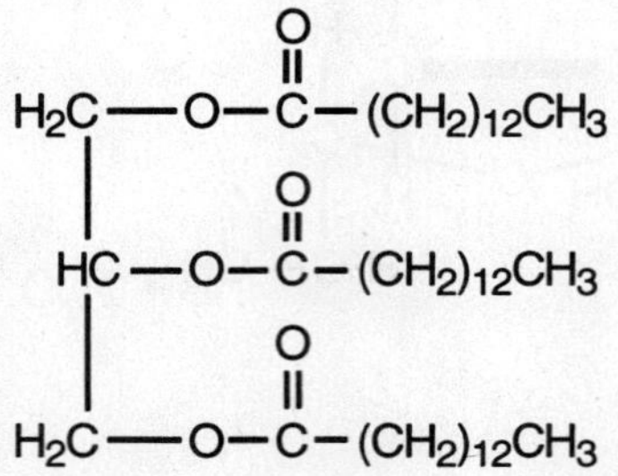

$$H_2C-O-\overset{\overset{O}{\|}}{C}-(CH_2)_{12}CH_3$$
$$HC-O-\overset{\overset{O}{\|}}{C}-(CH_2)_{12}CH_3$$
$$H_2C-O-\overset{\overset{O}{\|}}{C}-(CH_2)_{12}CH_3$$

3.

$$H_2N-\underset{H}{\overset{H_2C}{C}}-\overset{O}{\overset{\|}{C}}-\underset{H}{N}-\underset{H}{\overset{CH_2-\phi}{C}}-\overset{O}{\overset{\|}{C}}-\underset{H}{N}-\underset{H}{\overset{H}{C}}-\overset{O}{\overset{\|}{C}}-OH$$

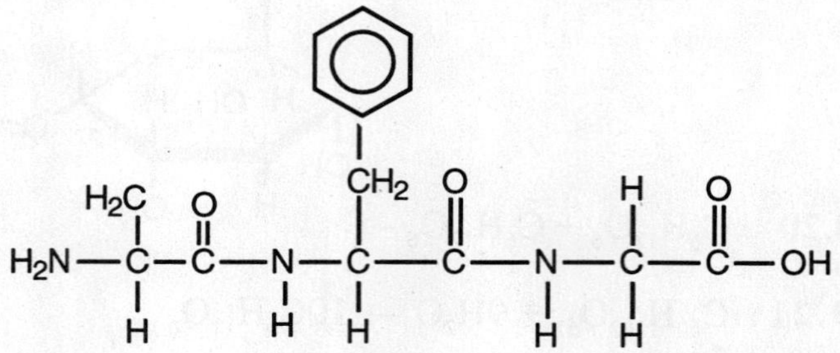